EAST

AFRICAN MAMMALS

Jonathan Kingdon

EAST AFRICAN MAMMALS

An Atlas of Evolution in Africa

Volume II Part A (Insectivores and Bats)

The University of Chicago Press

The University of Chicago Press, Chicago 60637
Academic Press, London

93 92 91 90 89 88 87 86 85 84   54321

Library of Congress Cataloging in Publication Data

Kingdon, Jonathan
    East African Mammals
    Reprint. Originally Published: London, New York:
Academic Press 1971–1974.
    Bibliography, P.
    Includes Indexes.
    Contents: V. 1 (No title)—V. 2, Pt.A. Insectivores and Bats—
V. 2, Pt.B. Hares and Rodents.
    1. Mammals—Africa, East—Evolution. I. Title.
QL731.E27K56   1984   599.09676   83-24174
ISBN 0-226-43718-3 (V. I)
        0-226-43719-1 (V. IIA)
        0-226-43720-5 (V. IIB)

Printed in Great Britain by
BAS Printers Limited, Over Wallop, Hampshire

Preface

The aim of this book is to picture a wonderful variety of animals and at the same time to provide a long overdue inventory and atlas of the mammals of East Africa.

Both the variety of mammals and their distribution are manifestations of the evolutionary process and so it is evolution that is the central theme of this book.

It is in search of further information about the process that I have essayed into the behaviour, ecology and anatomy of species. It cannot be said that the inclusion of these topics will bring the volume any nearer to being comprehensive, but they may perhaps serve to increase awareness of the magnitude and magnificence of evolution.

The book is also intended to provide a broad background for the student of East African mammals, with information on local names, breeding, measurements, food and so on. As the animals have economic, medical and veterinary importance to the East African countries and a scientific value for the world at large, I have also included some data on these aspects.

Whether one is interested in their conservation, their exploitation or their control, a practical approach towards mammals in East Africa must be based on biological knowledge and I hope this work may be found useful by all those with an interest in this fauna.

The prime stimulus for the drawings, however, has been the contemplation of physical beauty in mammals; this is a reward in itself. Drawing is the discipline in which I am trained, and it has been a chosen form of note-taking and a useful adjunct in the study of mammals. The making of a drawing is not only a matter of technique for there is a constructive effort to "figure" the animal; looking at drawings can also be an active retracing of this figuring process and it is in this that I hope others will share the pleasure of looking at animals.

East Africa is not a natural geographic or faunal region, so the fauna discussed here really belongs to a very much wider area, and in many ways is broadly representative of the fauna of Africa as a whole.

Sub-saharan Africa is occupied by the fauna of two biotic extremes, moist forests and dry open savannas (see Maps 1 and 2), and East Africa has been an ancient theatre for the excursions of these habitats and their fauna. If maps of the forest and savanna faunal zones of contemporary Africa are superimposed, the result is a broad overlapping area in central and East Africa. At the present time forest mammals are confined in this region to numerous small islands of forest, but there have undoubtedly been several periods when the forest was very much more extensive, and other periods when arid conditions were widespread. Mammal populations have therefore been subjected to isolation and gradual but extensive climatic change. On a large continent with relatively few physical barriers to the movement of animal populations, climatic fluctuations leading to the isolation of populations over millions of years have been an important determinant in the evolution of species.

The "overlap" area in central and East Africa (see the third map, p. vi)

v

contains many endemic species, and these forms have received particular attention in this work.

The patterns of mammalian evolution seen today have been and are being continuously modified by man. Although greatly accelerated today, human interference is nothing new, and hominid fossils testify to the continuous presence of men and pre-men in this area over millions of years.

Our own emergence and survival as a species was within a rich community of mammals such as is found in East Africa today. The interaction of man and wild mammals dominated human culture for millennia, yet today the close and ancient connections with animals have long ceased to be a part of human culture. It is urgent that we gain some insight into this world that is so much a part of our inheritance and so much older than our civilization which is destroying it.

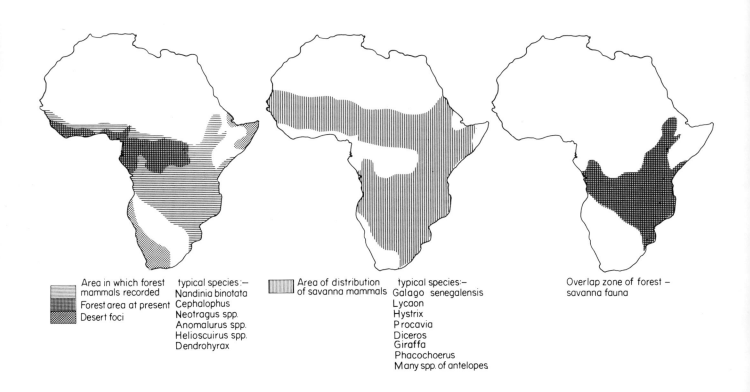

Area in which forest mammals recorded
Forest area at present
Desert foci

typical species:—
Nandinia binotata
Cephalophus
Neotragus spp.
Anomalurus spp.
Helioscuirus spp.
Dendrohyrax

Area of distribution of savanna mammals

typical species:—
Galago senegalensis
Lycaon
Hystrix
Procavia
Diceros
Giraffa
Phacochoerus
Many spp. of antelopes

Overlap zone of forest – savanna fauna

Acknowledgements

In putting together a volume dealing with hundreds of bats, mice and shrews I have had to travel over a very wide area. The Wellcome Trust have kept me on the road in more senses than one and I am grateful to the Trustees and Secretary of the Trust for their assistance. I have also enjoyed the help and hospitality of many kind people and have been joined by enthusiastic friends in the hunt for "small game".

In assembling data and writing up I am deeply indebted both to personal friends and family and also to correspondents who have answered queries, sent material and sometimes given me the benefit of their experience in comments on the text.

I should especially mention the collaboration of A. Archer, E. Balson, P. Clarke, R. Glen, D. L. Harrison, H. Heim de Balsac, J. E. Hill, R. Hughes, the late C. J. P. Ionides, A. McKay, D. Pye, G. Rathbun, A. Root, A. Start, T. Synnot, S. Tomkins, D. Vesey-FitzGerald, J. White and R. White, all of whom have given very generously of their time and talents.

I am also most grateful for the help and information given me by W. F. Ansell, J. Clevedon Brown, M. Coe, G. B. Corbet, M. Delany, F. Dieterlen, G. Durrell, L. Goodwin, G. Harrington, R. W. Hayman, D. Hopcraft, K. Howell, J. Kielland, E. Kulzer, P. Leedal, C. Leys, J. Meester, X. Misonne, P. Morin, F. Mutere, S. Price, U. Rahm, A. Rees, I. Ross, A. Suzuki, A. Walker, R. Wheatear, G. B. White, D. Kock, D. Ebbels and M. McInnery.

I am also very grateful to the Staff of Uganda National Parks, the Uganda Game Department, many Makerere University students and the members of the Wildlife Clubs of Kenya for answering my questionnaires so very informatively.

The following are also gratefully remembered: J. de Vere Allen, W. Banage, P. Boston, K. Brown, R. Carr-Hartley, G. S. Child, M. Doornbos, M. Duncan, B. Foster, F. B. I. Kayanja, P. Kingdon, H. Lamprey, R. Leakey, A. McRae, A. Mence, R. Milburn, J. Mungai, O. Odegaard, R. L. Peterson, F. Petter, M. Prentice, K. B. Robson, D. R. Rosevear, F. Reynaert, J. Sale, H. Tripp, J. Waite, D. Webb, A. Williams and many others not mentioned.

With patience and speed Roger Hargreaves has corrected the proofs. I beg forgiveness for a lack of consistency, using the names Congo and Zaire as interchangeable and for the spelling of some local names and languages, faults that are my own responsibility.

Once again I am indebted to Tag el sir Ahmed for his design work, including the layout of awkward keys.

Academic Press continue to be friendly collaborators in a joint effort to produce a technically difficult book.

Contents

LIST OF KEYS

Introduction

In the introduction to a volume devoted to small and inconspicuous mammals I could perhaps expect to address an audience of specialists, for the study of small mammals has tended to be an unusual and esoteric occupation. Today there are signs that this is changing. Interest in behaviour and ecology is being expressed at many levels of popular culture. With the development of mass media and the rapid growth of environmental consciousness the intellectual and aesthetic possibilities of Natural History have begun to touch people's imagination. Once again people are becoming conscious of sharing their environment with a teeming multitude of other forms of life and this consciousness has found a place on the broad platform of contemporary culture.

Gilbert White, Henry Thoreau and Jean Henri Fabre always had their following of naturalists but Ecology and Behaviour have become new frontiers for our civilization through the writings of Julian Huxley, Conrad Lorenz, Rachel Carson, Desmond Morris and the film and television features of Hugo van Lawick, Jacques Cousteau and Alan Root (to name but a few). As the minutiae of natural communities and details of behaviour become appreciated so the lives of small mammals begin to gain significance and interest.

For the great majority of people it has been childhood glimpses into the lives of small and familiar animals that quickened interest in the natural world. It is therefore easier to use small mammals as the ground for an affirmation of Natural History, not as an esoteric science, nor as an expensive hobby but as an aspect of natural philosophy to be enjoyed on a similar intellectual and aesthetic level to art, archaeology, architecture and other cultural pursuits. I hope that this sort of conception will be found contained herein in spite of a long and piece-meal text.

This volume is in two parts. The first is devoted to Insectivores and Bats, the second to Hares and Rodents. The fragmented form reflects the nature of the subject-matter—a variety of animals belonging to 25 families and some 280 species.

The literature on these small mammals of Africa is extensive and is scattered through many journals and books in several languages. A great part of the recorded biology of the species and subspecies only figures in the literature as a scientific name with an attached description. Where I have no personal experience of the rarer animals there has been little alternative but to add once more an obscure name to a long inventory of names. However, I hope that the discussion and the illustrations of those other species which I have been able to observe alive, will be a reminder that every species is a unique form of life and a worthwhile subject for study.

The pursuit of small mammals has up to the present been dictated mainly by human self-interest, by economy and by convenience, so that the most extensive literature exists on those species that are vectors of disease, destroyers of crops, or readily studied near the large towns. Contemporary communications have now put all species easily within the reach of determined naturalists, and the difficulty of working or living in remote areas is no longer a serious obstacle to the study of a particular species.

There is also, at the present time, a heightened awareness of the finite nature of the world's resources and the huge diversity and abundance of animal life that exists here in East Africa can no longer be seen as the ubiquitous product of a fecund Nature. The natural communities, of which the small mammals are a part, are seriously circumscribed by climate and geography. I have already stressed in Volume I that these communities are increasingly limited by the use to which man puts the land. Land use by indigenous animals and plants is a subject of no less fundamental interest or relevance than agriculture or bovine husbandry and the important role played in all habitats by small mammals is striking. Evolutionary radiation has led to the filling of every conceivable niche by specially adapted forms; furthermore, every single species represents a response to the opportunities offered by an ever more complex environment. This ever-increasing complexity of natural communities reaches its climax in the Tropics and it is this complexity that has contributed to the great number of species. Only by studying the labyrinthine interaction of the animals and plants within these communities can we begin to define the "niche" occupied by any one particular organism. In this respect the small mammals offer innumerable opportunities to scientists and naturalists in East Africa.

It will be obvious that the possibilities of the subject-matter greatly exceed what can be said or illustrated in this volume but I hope that a challenge emerges. I am obliged to elaborate further on the challenge that these natural communities pose, since the value of a species is often diminished by being tabulated and listed in an inventory, and I will have been guilty of a serious offence if I do not stress the absolute value of each species in its own right. In the evolutionary perspective—which is attempted in this work—no animal is more or less interesting than another, because each one represents the realization of a unique possibility of existence. An important role for this work has been to advertise the intrinsic "value" of a species for its own sake. I hope, therefore, that this aim is not obscured by uneven treatment.

One conclusion to be drawn from this study of mammals is that evolution in the tropical zone has led to a huge proliferation of niches and a maximum radiation of species. Seen in the longer perspectives of evolution, this zone also represents the main reservoir of life and innumerable forms that are now adapted to temperate regions can be shown to have a tropical derivation.

Any sensitive ecologist watching the onslaught on all types of ecosystems that is taking place in Africa today must wonder at the quality of the communities that will ultimately replace those that originally evolved there. Some scattered data on indigenous African ecosystems have accumulated and important research is being conducted (notably in the National Parks), but it will be apparent from the following profiles how very little is really known. Perhaps the deficiencies will tempt workers to contribute to our understanding of this neglected field, for the small mammals are very important, though inconspicuous members of every African biotope.

Inadequacies in our ecological knowledge can be matched in every aspect of mammalian biology but, ignorance apart, this work has also had to be kept within reasonable bounds. In order to range over as wide a field as possible I

2

have allowed many individual profiles to become discussions of topics rather than making them into well-rounded portraits. In this I have been guided by the particular curiosities each species has aroused in me. This has inevitably meant the neglect of aspects that may be of special importance to other workers. Apart from my own bias, the circumstances in which the animals have been studied must also colour my judgement. Where I have depended on literature, the second-hand nature of the profile will be evident. Sometimes animals of known economic or medical significance may have received more attention, both because more information is available and also because the need for data is widely felt.

A fragmentary character will also be apparent in the profiles. This is partly because the affinities of many families of rodents, bats and insectivores are quite unknown and it is possible to treat them only as self-contained entities referable to the other families of their order by the tenuous resemblances common to the order. Within smaller groupings, however, there is great interest in the relationship between species and their relative evolutionary levels. Competition and replacement between species at different evolutionary levels can be demonstrated for some rodent species, and the role of highland areas and of southern Africa as refuges for declining species is particularly obvious. Distribution patterns are discussed here and they illustrate in detail some of the more general remarks made in Volume I on zoogeography both within the continent and between Africa and Eurasia.

A new factor has emerged in the course of this study which has far-reaching implications for local zoogeography. Unpublished reports by B.P. Shell Oil Exploration Parties have shown that the Rufiji Valley, which is the major tectonic break in the Tanganyika coast, is probably associated with trough faulting during the Pleistocene. The very low topography of the coast and the flood plain of the Rufiji area suggest that Pleistocene sea levels could well have penetrated very far inland, perhaps as far as Stiegler's gorge (Paul Temple, personal communication). This would have constituted a major barrier. Furthermore, Lake Malawi and the mountains to the North-east of this lake would have effectively sealed off what may be called the "Mozambique zone" from the rest of eastern Africa. Today the Rufiji River is the southern boundary for the range of numerous mammal species and the area south of the river is somewhat impoverished faunally. However, *Tatera inclusa* is one species restricted to this area and *Funisciurus flavivittis* is rare outside it. A particularly interesting feature of this area is that it is a very extensive hybrid zone for two distinct species of squirrel and two forms of elephant shrews. There is a discussion of this in the profile of *Rhynchocyon* and those of the *Funisciurus* species *palliatus* and *cepapi*. This area is one of the least known parts of Africa and further study of the fauna and flora south of the Rufiji would be of the greatest interest.

The discovery of this hybrid zone has entailed some revision of the nomenclature of the animals concerned, and in the squirrels generally I have been unable to accept the current taxonomic position. For the rest I have been largely guided by the authors of the Smithsonian Preliminary Identification Manual. In the complexities of Soricid nomenclature I have been led by Professor Heim de Balsac. I have avoided cluttering the nomenclature with authorities and dates. A complete list of mammals from East Africa with their

3

scientific names, authorities and dates will be appended to Vol. III. In some of the profiles, particularly those of animals showing adaptations, I have discussed form and function at some length. For instance, the profile of the hero shrew (p. 85) suggests some of the possibilities in this field of study.

The structural peculiarities of small mammal species invite very much more intensive interest than they have received to date, but I hope this volume will call attention to the wealth of extraordinary forms that exist here and excite an interest in correlating form and function with the many facets of an individual species' particular way of life.

Finally I hope this volume will be an agreeable introduction to the fascinating but largely hidden world of our small mammals.

Insectivores

INSECTIVORA

The order Insectivora has been described as a "scrap-basket" for small mammals that are not referable to other orders but which are of a generally primitive character.

Several authors have proposed that some of the African groups from this "scrap basket" be re-classified. For instance, Broom (1916) suggested the golden moles become a distinct order, Chrysochloroidea.

The elephant shrews, Macroscelididae, have been the subject of a controversy for half a century, as they have close anatomical resemblances to tree shrews, Tupaiidae. Evans (1942) was of the opinion that the elephant shrews and the tree shrews should not be separated and he thought that there was as much justification for regarding the elephant shrews as primates as there was for the tree shrews. More recently, Butler (1956) and McDowell (1958) have come to similar conclusions and the former author proposed a super-order, Archonta, to include the primates, tree shrews, elephant shrews and the Oriental flying lemur.

Patterson (1965) argues for placing the family Macroscelididae in its own order Macroscelidea, while Corbet (1966) places them in the order Menotyphla.

These arguments ask for a formal recognition that the special diversity of these living insectivore groups is the result of a differentiation which may be as ancient or even more ancient than the divergence of some other distinctive orders. However the convenience of the "scrap-basket" has led most authors to retain it in use.

In spite of their highly modified anatomy, golden moles retain some resemblances to the Tenrecidae. Likewise, while the elephant shrews are not sharply divided from the tree shrews, which in turn have some sort of relationship with the primates, they do, in spite of their peculiarities, display undoubtedly primitive characteristics. Numerous fossils from the Cretaceous to the Oligocene also suggest that the groups traditionally included in the Insectivora share the common distinction of being the most direct descendants of the earliest primitive placentals (Romer, 1945).

Butler (1956) recognized three major divisions within the insectivores: the Soricomorphs, including shrews, tenrecs and golden moles; the Erinacomorphs or hedgehog group; and the Menotyphla, containing the elephant shrews and the tree shrews. The diagram opposite illustrates possible affinities between the various insectivore groups.

In the Insectivora represented in East Africa there are striking specializations which have undoubtedly contributed greatly to their survival. Structural modifications for subterranean life in golden moles, the prickly cape of the hedgehog and the aquatic adaptations of the potamogale are discussed presently in the profiles. The structures of their heads alone are interesting examples of the relationship of form to function and correlations can be made between the skull form, the orientation and function of muscles, the size of the eye and various developments of the nose.

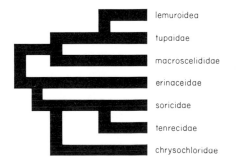

lemuroidea

tupaidae

macroscelididae

erinaceidae

soricidae

tenrecidae

chrysochloridae

In all insectivores a dominant sense is smell, but the nose itself has had its function extended to serve as a wedge in golden moles, a probe in elephant shrews, a sensitive bewhiskered hydrofoil in potamogale and an anchor for pulling its "cape" forward in the hedgehog.

Erinaceus albiventris, Rhynchocyon petersi, Potamogale velox and *Chrysochloris stuhlmanni* will serve as representatives of four major groups of specialized insectivores. The hedgehog, *Erinaceus*, is probably closest in form to the ancient proto-insectivores and is very generalized in some respects. Nonetheless, it is peculiar in having a stout muscle attached to the nose. Its function is to pull forward the circular muscle that supports the cape of spines so that the head is protected when the animal rolls up. The nasal area is widest above the molars and this area provides both attachment for muscles operating the short mobile snout and also buttressing for the principal chewing teeth, which are very broad. A strong development of the chewing apparatus is exhibited in the width of the zygomatic arch and the size of the coronoid and angular processes together with their associated muscles, the masseter and the temporal. As the eye is relatively small the origin of the temporal muscle is spread over the whole cranium.

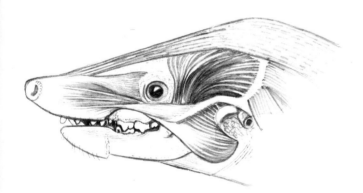

Erinaceus.

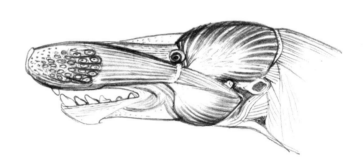

Potamogale velox.

The most striking feature of the head of potamogale is the enormous development of the muzzle, which is flattened so that it projects on either side of the mouth as much as it does in front. The bulk of the muzzle is made up of hair follicles in which the very numerous and stiff vibrissae are embedded; the muscular base for this mass of whiskers is appropriately robust. These muscles arise above the eye and in front of it and from the zygomatic process; some become tendinous at their extremities. Each follicle is served by large nerve fibres deriving from a branch of the fifth nerve, implying that the sensory function of the vibrissae is highly developed. Instead, the ear and auditory bullae are not conspicuous in size and the eye is exceptionally small. The masticatory apparatus is well developed and the small size of the eye allows the temporal muscle to extend until its forward edge actually lies under the eyeball.

Most Soricomorphs have a large vertical coronoid process. It is possible that a vertical action by the temporal muscle could, in phylogeny, render the

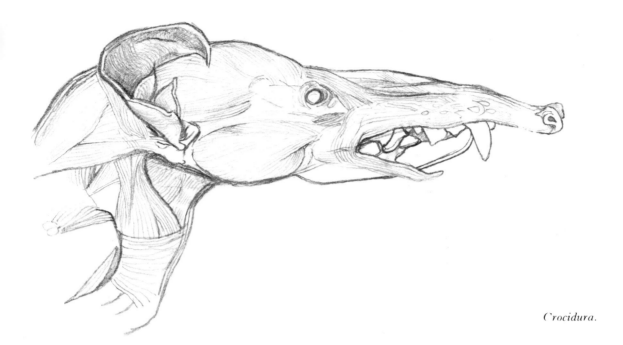

Crocidura.

equivalent vertical action of the masseter obsolete. At any rate, the zygomatic arch has been lost in the Soricomorphs (excepting golden moles) and as a consequence the masseter runs rather horizontally from the zygomatic process to the angular process.

The head of the golden mole, *Chrysochloris*, is striking for its conical form and for its lack of eyes and external ears. The breadth of the back of the skull and well-defined occipital ridge is correlated with very strongly developed neck muscles. The conical wedge of the head derives its motive force as a lever from stout, short muscles connecting the back of the head with the neck and shoulders. One of the principal muscles, the rhomboideus, has a tendinous connection to the occipital ridge or crest and runs directly to the scapula (see drawing). The bony cylinder of the premaxilla provides a firm base for the nose and the muscles and tendons operating the hard gristly snout and the upper lip are attached to the sides and front of the zygomatic arch. Early fusion of the bones or anchylosis ensures a unified skull form capable of withstanding the physical stresses imposed upon it. The head has

Chrysochloris.

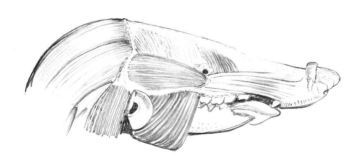

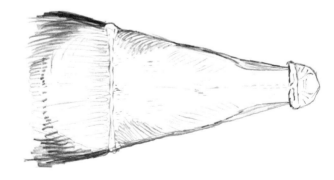

a relatively flat upper surface but the profile falls away sharply from the over-hanging and projecting snout, so that the mouth opens out of a relatively smooth lower surface. It can be seen from this that the architecture of the golden mole's head has been modified by its mechanical use as a wedge.

Elephant shrews have a most distinctive head and skull with very large eyes and an exceptionally long and mobile proboscis. As the nose is constantly being thrust into soft soil or crevices, and as the nostrils are at the tip of the proboscis, there is a need both for flexibility and also for a stout tube that cannot be sealed easily by pressure while the animal is probing about in the leaf litter. In *Rhynchocyon* this tube is lined by thirty cartilaginous rings similar to those found in the larynx. In this species the area enclosing the turbinals and the nasal sensory apparatus is about a third of the skull and is accommodated in a triangular area extending forward from between the eyes. The muscles controlling the proboscis run along the sides of this cone and are anchored along the lower and forward edges of the large round orbit. The prominence of the orbit area is influenced by the need for adequate purchase for the well-developed muscles and tendons of the proboscis. The effect is to emphasize the overall breadth of this part of the head with its large nasal area and broad palate. This development seems to have led to a slight displacement of the eyes backward in *Rhynchocyon*. In correlation with the enlargement of the eye and the brain in elephant shrews, the temporal muscle originates at the back of the cranium and the coronoid process is shallow and steeply sloped. The large eyes and brain together with the finger-like sensitivity of the nose make the elephant shrews contrast with the other insectivores.

The shrews also have one specialist in the hero shrew, *Scutisorex,* which has a uniquely complex backbone.

As a whole, insectivores are of little economic or medical importance but they are of very great scientific interest and many have been shown to be useful subjects for laboratory study.

Rhynchocyon.

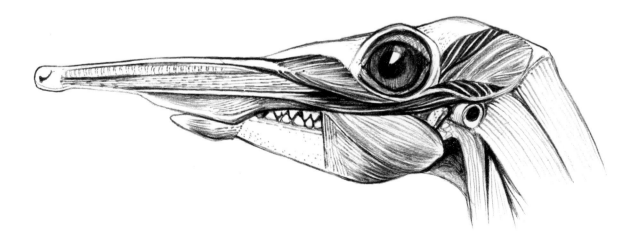

Tenrecs and Potamogales

Tenrecidae and Potamogalinae

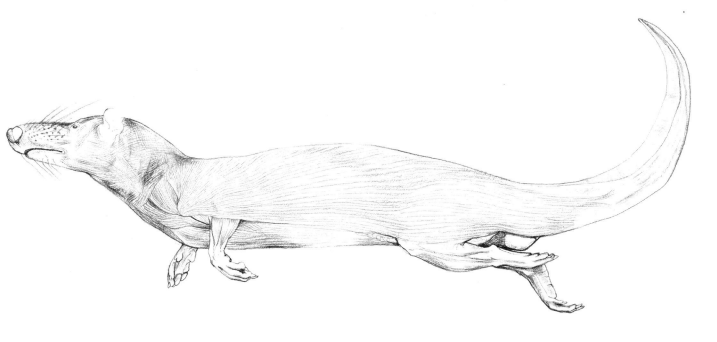

Living members of this family were formerly thought to be restricted to the island of Madagascar, where they have radiated, like the lemurs, into a variety of forms. The existence of living Tenrecidae in Africa was admitted in 1954 (Heim de Balsac, 1954) with the discovery of a West African insectivore *Micropotamogale lamottei* which is very similar to the Madagascar *Oryzoryctes* and *Microgale*.

In 1955 *Micropotamogale ruwenzorii* was described. This animal, although similar to the West African *M. lamottei*, appeared to be adapted to a more aquatic life. These adaptations, however, were of a different nature to those of *P. velox*, a species known since 1860. Because of its modifications, *Potamogale* had always been assigned to a separate family, although Dobson (1882) recognized the close relationship of *Potamogale* with the tenrecs.

Insectivores have sometimes been divided into two groups on the basis of tooth patterns: zalambdodonts and dilambdodonts. Within the Potamogalinae both patterns occur so that this division of the insectivores has been broken down.

Verheyen (1961), Guth, Heim de Balsac and Lamotte (1960) examined some of the differences between *P. velox, Micropotamogale lamottei* and *M. ruwenzorii*. Apart from differences in the structure of the teeth, feet, tail and myology (see drawing above), the wall of the stomach in *P. velox* is unusually thick, while in *Micropotamogale* most of the stomach wall is thin. Also the gall bladders and livers of these two animals differ and the penis of *P. velox* is very long. A shorter sacrum in *P. velox* suggests that the demand for increased mobility in its aquatic life has led to a secondary freeing of the last vertebra of the sacrum. The Potamogalinae are the only insectivores without clavicles and in mammalian phylogeny this is generally to be regarded as an advanced condition.

The tenrecs are an ancient group of mammals that have survived on Madagascar through an early isolation from the competition of more advanced mammals. Two fossil tenrecs, *Protenrec tricuspis* and *Erythrozootes chamerpes* are known from the Miocene deposits in Kenya and Uganda (Butler and Hopwood, 1957).

> "The former species agrees with Potamogale in many respects, but most of the resemblances are shared with at least some of the Malagasy tenrecids and are probably primitive characters of the family . . . Thus *Protenrec* appears to be near the primitive stock from which both the Potamogalinae and the Malagasy tenrecids have been derived . . . The Tenrecidae appear to be a family of African origin, derived from an unknown group of primitive Lipotyphla and not clearly related to any other family of living insectivores, except possibly the Chryso-chloridae. Until more is known about the early history of the Lipotyphla it is not possible to be more precise than this." (Butler, 1969.)

Potamogales therefore represent the last relics in Africa of a very ancient group of mammals. It is interesting that the largest and the most specialized animal is the more successful of the three forms. The exploitation of a very peculiar niche seems to have saved the *Potamogale* from direct competition from more advanced mammals such as the mongooses. Nonetheless, where its habitat suffers any appreciable modification—whether long-term or seasonal —the potamogale's sensory and locomotory limitations make it a non-starter in competition with a versatile animal like the marsh mongoose, *Atilax*.

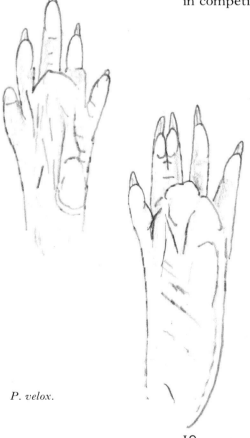
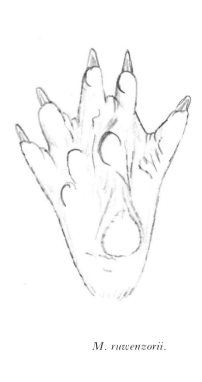
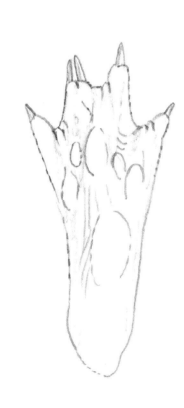

P. velox.

M. ruwenzorii.

Ruwenzori Potamogale, Dwarf Otter Shrew (Micropotamogale ruwenzorii)

Ruwenzori Potamogale, Dwarf Otter Shrew (Micropotamogale ruwenzorii)		
Family	Tenrecidae	
Order	Insectivora	
Local names		
Songi (Lukonjo)		

Measurements
head and body
123—200 mm (ave. 188 mm)
tail
100—150 mm (ave. 135 mm)
weight
135 g

This species has some resemblance to a large shrew. The long tail is sturdily built and very slightly flattened vertically, but is otherwise not unlike that of some rats and shrews. The most distinctive feature is the muzzle, which is very flat and broad, with numerous long whiskers and a curious flat pad or nose shield. The feet are webbed and the hindfeet are immediately recognizable as the second and third toes are bound together or syndactylous. The fur is very dark and has a texture similar to that of the otter.

11

This species was first discovered in 1953, when one was caught in a local basket fish-trap in the river Talya on the western slopes of the Ruwenzori Mountains. Since that date the animal has been caught in a number of other rivers in the Kivu and Ruwenzori region. Most of these were caught by Dr U. Rahm, and the following account is derived entirely from his publications (1960a, b, 1961, 1966). I have subsequently found evidence of these animals' presence in East Africa, in the Dungilia River near Kyarumba on the southern spur of the Ruwenzoris at about 1,220 m.

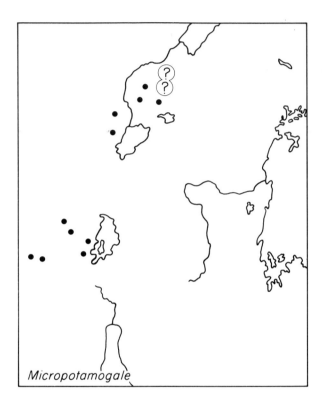

Although its distribution may eventually turn out to be more extensive than is known at present, it is, without doubt, relatively restricted. Nonetheless, neither altitude nor vegetation appears to be of very great importance in its distribution, as it has been found in small rivers and streams running through montane forest, savanna, cultivation and lowland forest. The water temperature of the rivers in which it has been found ranges between 12° C and 21° C.

They are known to dig tunnels and make a sleeping chamber and captives arrange hay and grass into a sort of tunnel with a nesting hollow at the end.

Their favourite foods are worms, *Oligochaetes*, and insect larvae, Hydrophilidae, Hemipterae and Ephemeridae. They also eat crabs, *Potamonautes*, and fish, *Clarias* and *Barbus* species, tadpoles and small frogs. Of fish they showed a preference for cat fish, *Clarias*. Among crabs they avoided individuals in which the carapace is more than 5 cm wide, presumably because of the dangerous pincers. Crabs are tackled from behind with a very fast snap of the jaws across the back of the body, if the crab succeeds in pinching the potamogale, it keeps quite still until the crab lets go. When the crab is caught it is taken onto dry land and overturned, when the soft parts only are eaten. Fish too are brought onto land and are killed by biting them vigorously. The bony head of the cat fish is generally discarded. Worms and insects are scratched out of the soil and from among grass roots. Small prey caught underwater is eaten in a matter of seconds while still in the water, accompanied by a rapid shaking of the head. After emerging from its hollow, between 7.30 p.m. and 8.30 p.m., the animal alternatively rests and hunts throughout the night, in the course of which it may eat up to 80 g of food.

When making a sudden spurt in the water the dwarf otter shrew kicks out simultaneously with all its limbs. When paddling slowly it uses alternate left and right strokes of both the front and back legs. On the surface, it floats with the upper part of the back and much of the head above water. Dives are of very short duration and it comes out of the water frequently. Much time is taken up in cleaning the fur and scratching with the hindlegs. The tunnel to the nest may also perform an important role as a sponge for absorbing the moisture on the fur.

Nothing is known of the social or sexual life of these animals, but Rahm noted that dung and urine deposits were regularly replaced after the potamogale quarters had been cleaned and also on being moved to new quarters.

Although *Micropotamogale* and *Potamogale* are geographically sympatric it is not known if they live together in the same streams. The former seems to prefer small rivers, while the latter is more frequently found in large rivers and even lakes. Nearly all the *Micropotamogale* collected to date have been caught in locally made fish basket traps made from the fronds of *Phoenix reclinata* palms.

One female collected by Rahm in late September was pregnant and contained two embryos, which had conspicuous and well-developed whiskers.

Potamogale tail viewed from the side has 4 distinct ranks of muscle running along. The height of the tail is augmented by a skin connective tissue phlange along its upper margin at the point where the fur gives way to glossy short hair.

Giant Otter Shrew
(Potamogale velox)

Family   Tenrecidae
Order   Insectivora
Local names Kiniabu (Lukiga)

**Measurements
head and body**
290—350 mm
tail
245—290 mm
weight
1 kg (approx.)

Giant Otter Shrew
(Potamogale velox)

This otter shrew is larger than *Micropotamogale* and more specialized for aquatic life. Its most obvious feature is the tail, which is long, thick and laterally flattened to make a vertical blade ending in a point. The long muzzle is very broad and flat, with its bulk made up of hair follicles supporting multitudes of very long, stiff vibrissae; the naked rhinarium on the nose has flaps which conceal the nostrils. The short front limbs have unwebbed fingers and are not used in swimming. The hind feet have the outer side of the sole peculiarly flattened so that in the natural swimming position they lie flat against the tail and give the body a more streamlined shape. Another aquatic feature is the otter-like fur which is dark brown on the back, with a metallic lustre, and white on the underside.

The animal is propelled entirely by lateral movements of the back and tail and the most conspicuous modifications of structure concern the mode of propulsion; the pelvic region and the tail show a greater degree of specialization than those of *Micropotamogale ruwenzorii*.

The caudal vertebrae have well-developed neural spines, metapophyses, transverse processes and also 23 chevron bones which, besides providing a canal for the subcaudal artery, furnish support and attachment for the caudal muscles and contribute to the depth of the tail in cross-section (see overleaf). Therefore, these huge muscles lie on either side of the vertebrae and consist of muscular rather than tendinous tissues right to the end of the tail. Near the body, the upper surface of these muscles is overlaid by a remarkable extension backwards of a part of the *gluteus*; indeed, the base of the tail is of the same thickness as the body and the anus is the only external sign indicating where the body ends and the tail begins.

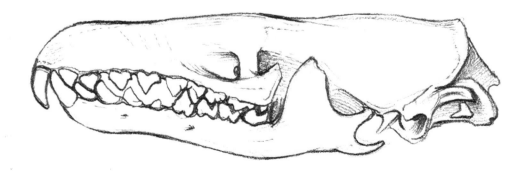

15

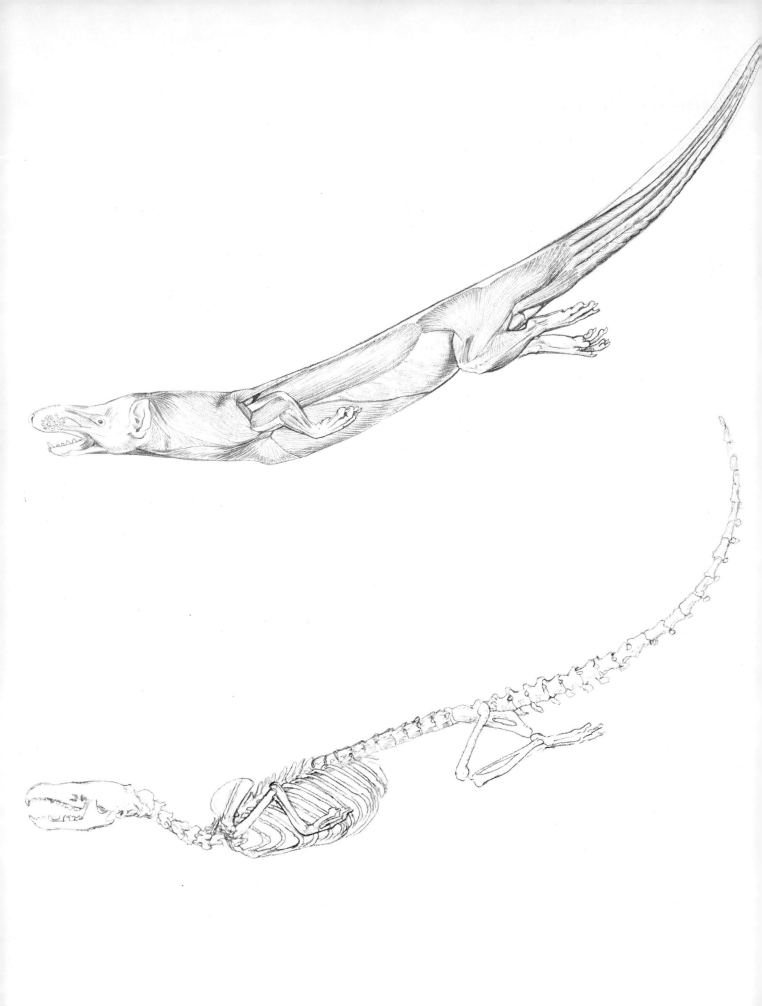

The pelvic region provides the anchorage for most of the tail muscles, some of which are in effect extensions of lumbar muscle groups, notably the *multifidus caudi* and the lateral *externsor caudi*. As the centre of gravity rests in the thorax, the action of the tail is continuous with the sinuous movements of the lumbar and pelvic region.

The horizontal flattening of the muzzle and the spectacular growth of whiskers, while having primarily a sensory function, may also serve as a hydrofoil; in a tubular-shaped body the whiskered muzzle is the principal horizontal structure capable of counteracting rolling and yawing movements resulting from the tail's lateral strokes. During rapid changes of direction the forelegs may come into play, but they are unlikely to act effectively as hydrofoils, and the peculiar flattening of the head probably evolved as a stabilizing device to compensate for the effect of the tail's rapid movements through the water. This is a development that concerns the soft parts of the muzzle only, and there is no tendency towards flattening in the skull.

The otter shrew ranges from the Cross River in Nigeria across the Congo basin to western Uganda. Its northern limits are the Oubangi and Uelle Rivers and it does not occur south of about 14° S. In Uganda it is known from Kalinzu and the Dura and Kashasha Rivers.

The short range of the potamogale's senses, very poor sight and clumsy gait on land have limited it to regions in which the depth of the rivers and the abundance of food supply are not subject to critical fluctuations. The past climatic instability of East Africa is probably the principal reason for its very limited range in this part of the continent.

In West Africa the otter shrew has been found in rivers of various sizes, ranging from fast flowing mountain streams to sluggish coastal rivers and swamps. Like *Micropotamogale* it makes its burrow in a river bank, often with the entrance below the water level.

According to tests made on captives, the potamogale's favourite food consists of fresh-water crabs, but examination of their dung suggests that fish too are commonly eaten. Frogs and water molluscs are also part of their diet, while Rahm (1966) found that the smaller species evinced no interest in molluscs.

The enormous whiskers play an important part in hunting, which is guided primarily by scent and touch. They have been observed to turn crabs over, killing them with a swift bite, and then to tear out and eat the body and the larger claws. Food captured in the water is eaten on land and they may eat 20—25 crabs in a night.

Movement on land is rather clumsy, whereas the smaller otter shrew is quite nimble; this limitation also stresses the degree to which *P. velox* has adapted to a thoroughly aquatic existence.

Durrell (1953) describes finding one at night on a sand bank and mistaking it for a crocodile; it churned up a small wake when pursued and, after being caught by the tail, it climbed up its own body and bit its captor severely. When caught it hissed like a snake and it also hissed and grunted while chasing crabs in its tank. It fouled the water very rapidly and both Durrell's captives died suddenly without any sign of injury or distress.

The study of these animals in the field appears to be fraught with difficulties and has not been attempted to date. They are found solitary or in pairs but nothing is known of their social or sexual behaviour. Ansell (1960) reports foetuses from the southern parts of its range in October.

Golden Moles

Chrysochloridae

The golden moles have some dental resemblances with the tenrecs but they appear to have adapted to digging at a very early date and have evolved many characteristics not found in other insectivores. Their mode of digging and the structures associated with this way of life are fundamentally different from those found in the moles, Talpidae.

Golden moles are blind and have a wedge-shaped skull, the bones of which fuse at an early age; they have large bullae and an unusual hearing apparatus. Their very robust shoulders and forelimbs—with extraordinary double-phalanged digits—are accommodated by a deep hollowing out of the thorax, which is exceptionally long—19 vertebrae—in contrast to the lumbar region which is short and has only 4 vertebrae. The neck also is very short and strong muscular connections ensure that the head and shoulders become a compact unit with maximum rigidity. The clavicles are stout and long; all these adaptations are associated with digging. Dobson (1882) described the anatomy of the golden mole in detail.

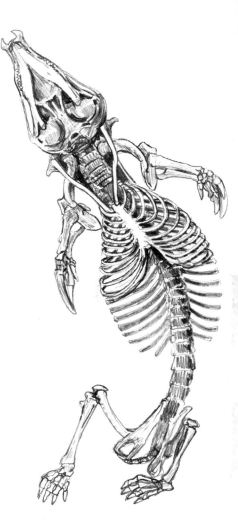

The family is restricted to Africa, principally to the southern part of the continent, where several genera have adapted to a wide range of habitats, ranging from desert to forest. Species of the genus *Amblysomus* are found in South Africa and another in Somalia: an interesting example of widely separated distribution.

The survival and radiation of numerous forms in southern Africa presents a curious contrast to their limitation to a few montane or forest localities in tropical Africa. Another insectivore, *Myosorex*, has a similar distribution pattern. It would appear that colder conditions on the tropical mountains and in the southern latitudes foster micro-habitats that favour these ancient mammals. However, the Chrysochloridae distribution pattern is not simply the result of survival in relic zones. The fossil, *Prochrysochloris miocaenicus*, from Songhor shows that adaptation to the exacting requirements of the mole niche was already very advanced in the Miocene (see Butler, 1969). Climatic changes since that time may have changed the status of the family, but if there has been any retreat or decline, this must have been due to ecological limitations rather than adaptive inferiority. Any subterranean insectivore is likely to be excluded from tropical habitats that are subject to drastic changes in the condition of the soil and in the abundance of cryptic fauna, whether the changes are seasonal or due to longterm climatic fluctuations. That this is so is suggested by the absence of any direct mole-like competitor in Africa. The family's relative success in the cooler parts of the continent and its virtual absence in other regions seems to be related to physiological, anatomical and ecological limitations associated with excessive specialization.

There is a most remarkable degree of convergence between the golden moles and a marsupial from Australia, *Notoryctes*, which has iridescent silky fur, a nose pad and similar claws to those of the Chrysochloridae. There is also a very close resemblance of skull form between *Chrysochloris* and an

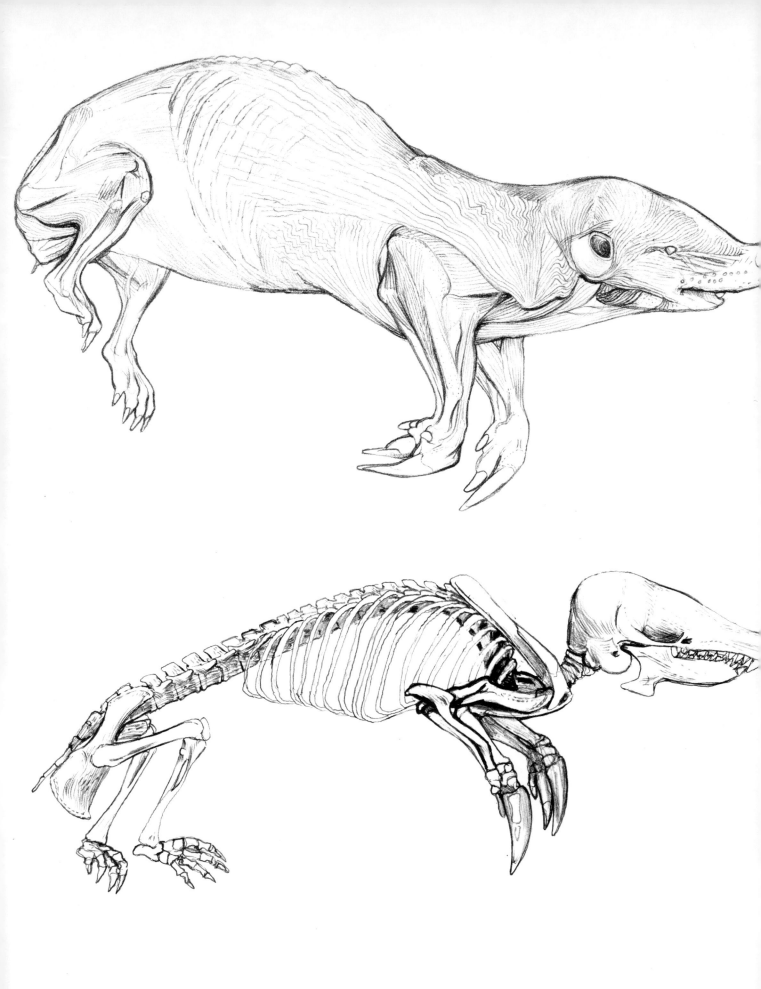

archaic edentate, *Epoicotherium unicum*, which is known from the American Oligocene. This last example of convergence is a reminder of the rigid discipline that the environment exerts on animal form when its primary adaptation is to function as a tool, in this case a digging wedge.

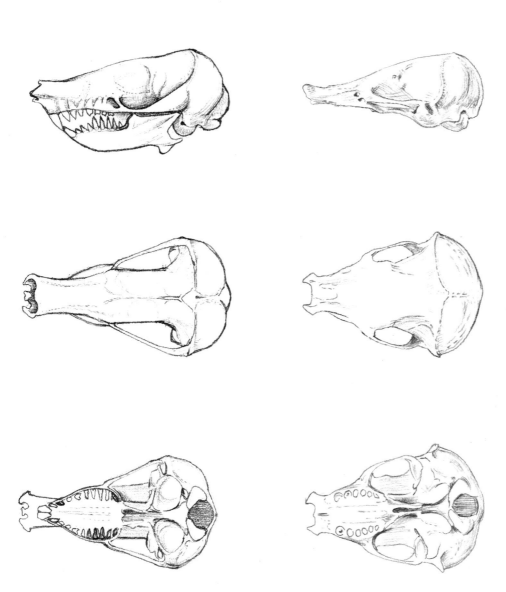

Chrysochloris stuhlmanni.   *Epoicotherium unicum.*

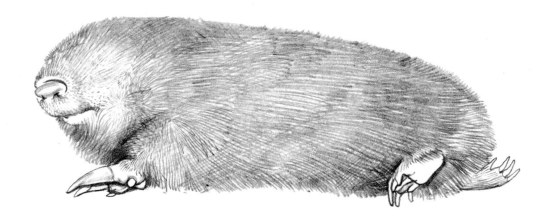

Golden Mole
(Chrysochloris stuhlmanni)

Family Chrysochloridae
Order Insectivora
Local names
Fuko la pesa (Kiswahili), Ifulula
(Kihehe), Fukuzi (Lukiga), Kingiri
(Lukonjo), Kienge (Kikami),
Lisukadope (Kikinga)

Measurements
head and body
105—120 mm
tail
4—5 mm
weight
26—53 g

Golden Mole
(Chrysochloris stuhlmanni)

Races

Chrysochloris stuhlmanni stuhlmanni   Ruwenzori, western Uganda,
Southern Highlands, Tanzania

Chrysochloris stuhlmanni fosteri   Mt Elgon, Cherengani Mountains

Chrysochloris stuhlmanni tropicalis   Ulugurus Mountains, eastern
Tanzania

This species of golden mole is restricted to East Africa and eastern Zaire (Congo). The subspecies *C. s. fosteri* and *C. s. tropicalis* are found on isolated massifs well to the east of the range of *C. s. stuhlmanni*. The extensive range of the latter race is surprising, as mammals with a similar discontinuous distribution generally show racial differences between populations in the Southern Highlands and those found in western Uganda.

The species is most numerous and successful in montane areas but also occurs at lower altitudes in moist habitats. Wet periods in the past may have allowed some sort of ecological connection between the Ruwenzori Mountains and the Southern Highlands, but the golden mole may also have a rather conservative genetic make-up.

Golden moles are known on Ruwenzori at an altitude of 3,500 m, they are

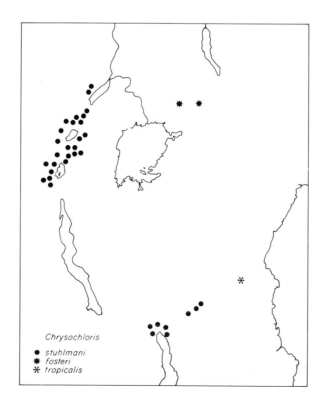

Chrysochloris

● stuhlmani
✳ fosteri
✲ tropicalis

found in the alpine grassland, in tree heath and bamboo zones and as far down as medium altitude forest—800 m in the Congo. They are not conspicuous animals and it is possible that the range of the species may be found to be more extensive than is known at present. The soil they burrow in needs to be soft and moist. Golden moles will invade recent lava flows in volcanic areas, as long as there is enough humus to provide adequate shelter and food.

C. stuhlmanni feeds principally on earthworms, beetles, grubs, slugs and snails. Captives will feed on a variety of insects, including moths but rejecting aposomatic species. Insects are snapped up, chewed vigorously and swallowed; earthworms tend to be swallowed rapidly from one end.

The golden mole is reported to seek food on the surface after rain, but this may be due to flooding of the burrows. When on the surface it rootles over the soil and captives soon learn the limits of their confinement and live almost entirely on the surface even when they have enough soil in which to burrow. Bateman (1959) found that a golden mole, *Amblysomus hottentotus*, ate about 45 g of earthworms each day; this species would respond to the presence of worms with movements of the head and twitching of the snout, as if to get a sense of direction, after which a positive advance towards the worm would follow. Only worms were sensed in this manner; the mole seemed to be unaffected by the smell of insects and indifferent to sound; with food other than worms it relied on actual contact.

Research in European pastures has shown that there may be two tons of earthworms in one acre of land. Some montane habitats in East Africa may have a similarly abundant underground fauna. This rich food supply may well explain the extraordinarily numerous mole tunnels to be found in parts of the Ruwenzori Mountains. The rarity of golden moles in other parts of the range is probably related to the density of cryptic fauna available and to its year-long abundance. However the extent of digging may not be a true indication of the number of moles present; extensive tunnelling may either be the work of a single hungry pair or may represent a larger group. Until the species has been studied in more detail it is difficult to know what density of population is present in areas where tunnelling is conspicuous. One function of the tunnel systems might be that of trapping cryptic fauna and a mole travelling along already dug tunnels may be picking up food; thus galleries would only be extended if the existing network were not trapping enough earthworms or other invertebrates.

Above the ground, the golden mole's gait is a quadrupedal shuffle, but it can scurry along quite fast when frightened. When digging, the head butts continuously up and down; this action is supported by that of the foreclaws digging under and behind the head. In soft soils and vegetable debris, a tunnel is formed by movements of the leathery snout and by the upwards butting of the top of the skull's bony barrel. As the tunnel generally lies just beneath the surface, the soil gives way under this butting and a raised ridge often enables the human observer to follow the course of the mole's subterranean passages without difficulty. At deeper levels the main effect of this butting is to compress the soil and in harder ground the action of the claws becomes more critical to digging. The claws are then brought forward into the crevice which is forced by the wedge-shaped tip of the nose so that head and claws act together on the same spot; a very powerful opening-up action is

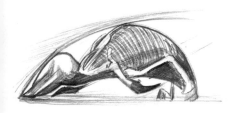

then exerted by the head pushing up and the claws tearing down and back into the earth (see drawing below).

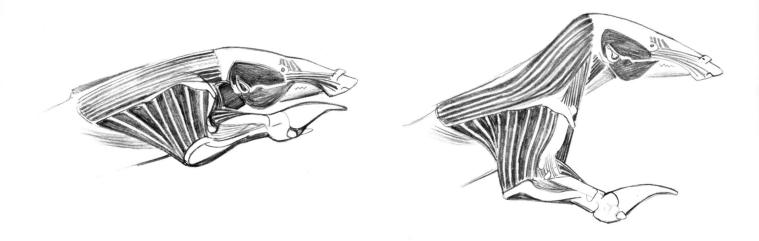

An illustration of the strength of this action is Bateman's report of a 60 g captive golden mole being able to nudge up a 9 kg railway iron covering its container. The forelegs have a short, strong action, which pushes the broken soil back under the body. The hindlimbs, which are much slighter in build, have a long, weaker action, which piles up the soil and then treads and compresses it.

Golden moles are generally silent, but have been reported to make a noise similar to that of a puff-adder. Scent probably plays an important part both in orientation and communication; they have scent glands that emit a strong musky smell. They have been found to be active during the day and night; food supply, temperature and soil conditions are probably the decisive factors determining activity.

An adult male and female with a half grown young were dug out of a burrow in Ruwenzori, but it is not known what social structure or how many moles occur in a single complex of burrows. Rahm and Christiaensen (1963) published a diagram of tunnel networks excavated in the eastern Congo. This shows a meandering central gallery with very numerous offshoots; most tunnels are confined to soils growing grasses and shrubs. Tunnels venturing into open terrain and up steep banks are less numerous and the pattern suggests a distinct preference for the area with grass and shrubs. Schaerffenberg (1942) and Godet (1951) have demonstrated that soil aeration is an important factor in the ecological distribution of European moles and it is possible that this could hold also for *Chrysochloris*. Although I have found mole tunnels in open sward in Ruwenzori, these were relatively rare, and it is possible that in such situations golden moles may be more vulnerable to predators.

I found one golden mole in the pellets of grass owls, *Tyto capensis*. These birds conducted most of their hunting over a high-altitude swamp and in an area of Kigezi where golden moles are rather rare. *Tyto* is able to hunt by sound as well as by sight, which probably gives this bird a special advantage as a predator of the golden mole. An investigation of these birds' pellets from

an area where the moles are common would be interesting.

There is a curious inverse relationship in the relative abundance of mole rats, *Tachyoryctes*, and golden moles in East Africa; on Ruwenzori, where golden moles are very numerous, the mole rats are almost entirely absent, whereas on Mt Elgon, where golden moles are rare, *Tachyoryctes* is very numerous. The two are not mutually exclusive, however, and I have found both species together in a fallow field in West Ankole. The phenomenon may be quite accidental, but it is possible that the very aggressive *Tachyoryctes* leaves little room for *Chrysochloris* when it occurs in dense populations.

Golden moles are often hoed up by cultivators. In Kigezi the skins used to be valued as charms. After human disturbance or excavation, moles may leave the vicinity and they are probably capable of travelling some distance overland before starting a fresh gallery.

Rahm (1966) found a leaf-lined nest of a golden mole at the end of a tunnel under vegetable debris. Loveridge (1937) found a female in a nest made of masses of dry leaves under the root of a large tree. In Ruwenzori I found a nest about a foot deep under a rock; it was lined with one or two leaves and the area surrounding this nest was honeycombed with passages. This contrasts with the ordinary tunnels, which seem in effect to "sample" the leaf litter and surface humus.

The gestation of *Chrysochloris* and other details of its reproduction are unknown. The closely related *Chrysochloris asiatica* from the Cape gives birth in the nest. The young are 47 mm at birth and suckle for two or three months, after which they are almost fully grown. The permanent teeth cut through at about the same time. Breeding patterns are not known. Small, probably subadult animals, have been collected in July from western Uganda and in March from the Congo. Loveridge (1933) found no females in breeding condition in January and February in Rungwe district.

Hedgehogs

Erinaceidae

The best known members of this family are the spiny hedgehogs but it also includes several genera of furry insectivores, Gymnures, which are restricted to South-east Asia.

Erinacids are known from the Eocene to the Recent in Eurasia, from the Miocene to the Recent in Africa and from the Eocene to the early Pliocene in North America—where they no longer exist; the family, therefore, was formerly more widespread and contained a greater variety of types. Romer (1945) lists nearly forty extinct forms within the super-family Erinaceoidea. Six species belonging to four genera have been found in East African Miocene deposits (Butler, 1969). An Upper Cretaceous fossil—about 80 million years old—from Mongolia, *Zalambdalestes*, is a small primitive insectivore with features suggesting that it is ancestral to the Erinaceidae, if not also to other insectivores (Romer, 1945).

The most successful and widespread modern survivors of this very ancient group are the hedgehogs and there can be little doubt that the evolution of a highly effective spiny armour has been the main reason for their survival. Their existence is a striking example of the role of predators in natural selection for, beneath their spiny capes, they have a generalized body plan and their diet is only remarkable for its variety rather than its peculiarity. Their

27

relatives, the Gymnures, are less widely distributed and, in their case, survival seems to have been assisted by the development of obnoxious scents. The smaller genera are rather shrew-like while *Echinosorex*, the largest living insectivore, has acquired black and white aposomatic colouring like skunks and zorillas.

Total world range of Hedgehog genus *Erinaceus*.

The most striking habit of the hedgehogs is their ability to curl up. Certain minor modifications in the skeleton can be correlated with this habit—short, blunt processes on the vertebrae, a rather small thorax and a wide pelvis—but this capacity depends almost entirely upon the development of the skin muscles and in no other mammal are these muscles so developed. There is a circular muscle that runs along the sides and across the rump and neck and which coincides exactly with the extent of the spines that are embedded in it. This muscle called the *orbicularis panniculi* is peculiar to the hedgehogs and has no recognizable homologue in related groups.

When the hedgehog rolls up the *orbicularis panniculi* contracts, forming a bag into which body, head and legs are withdrawn. Meanwhile the spines, which were formerly set obliquely in the skin, become erect. The front margin of this muscle is pulled forward over the head by the *fronto-cuticulares*, while the hindmost margin is attached by the *coccygeo-cuticulares* to the sides of the tail and these muscles also assist the *orbicularis* to pull over the rump; smaller muscles pull down the sides over the shoulders and ears. Finally a simple contraction of the *orbicularis* closes the bag involving both the head and the limbs. When it relaxes the *orbicularis panniculi* is returned to its original position by a variety of very small muscles lying beneath it. Dobson (1882) described the musculature of the hedgehog in detail. The drawing opposite illustrates this muscle in action.

The Hedgehog
(Erinaceus albiventris)

Family   Erinaceidae
Order   Insectivora

Local names
Kalunguyeye (Kiswahili),
Kanungumaria (Ngulu),
Chanungumiya (Kirabai),
Linyungi (Kimatengo), Nunguri
(Chagga), Kilungumigia (Kiramba),
Karianyongu (Kimeru),
Kithangaita (Kikamba),
Karangatiti (Kiembu),
Gathangaraita (Kikuyu),
Ngakakubonwa (Kinyamwezi), Ngekewa (Kingindo,
Kimbunga), Engenelwa (Lugisu), Ngenge
(Kingakyusa), Kenye (Knyatura), Likili (Kipangwa),
Ndirira (Runyankole), Umali (Iruq), Enjolit
(Kimasai), Njoliss (Samburu), Sejese (Kigogo),
Kinyalegese (Kizanaki), Kirasasa (Kirwa), Kisesegede
(Kitaita), Kinyesuke (Kirangi), Kamasi (Kisandawe),
Kabinubinu (Luganda), Kaminamini (Luhaya),
Ngapupu (Itesot, Karamojong), Olapa (Lutoro),
Apinaka (Lugbara), Awutula (Madi), Mabuku
(Ukerewe), Buy u (Alur), Sawayantet (Kisebei),
Okodo (Lwo)

Measurements
head and body
170—230 mm
tail
20—50 mm
weight
270—700 g

The Hedgehog
(Erinaceus albiventris)

The East African or four-toed hedgehog differs from the European species, *E. europaeus*, in being smaller and in having the third upper incisor single instead of double rooted. It also differs from this species and from other African forms in lacking a hallux.

Erinaceus albiventris ranges rather intermittently throughout the savanna and semi-arid zones of the northern half of Africa from Senegal to Somalia and Tanzania. A darker, five-toed species, *Erinaceus frontalis*, replaces it in southern Africa, and another slightly larger species, *Erinaceus algirus*, is distributed along the African shores of the Mediterranean. There are five-toed hedgehogs in Somalia, *Erinaceus sclateri*, which appear to merge with *E. albiventris* in North Kenya; the distinguishing characteristic of four toes may therefore break down in North-east Africa.

Hedgehogs have a vast total range but a very localized distribution and

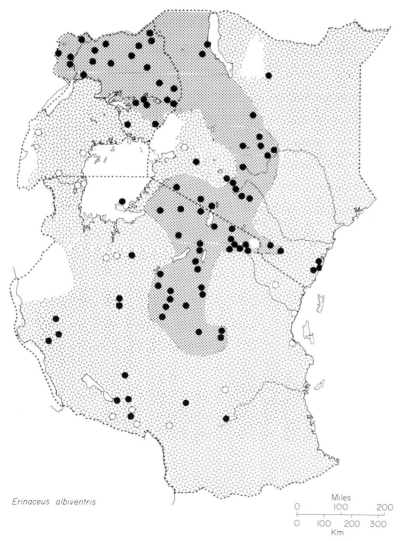

Erinaceus albiventris

Miles
0   100   200

0   100 200 300
Km

●Confirmed records; ○ reported sightings; heavy stippled area: common; light stipple: probably present but rare; blank: apparently absent.

although they live under a variety of climates they have specific temperature preferences. Herter (1963) devised a temperature gradient test and found that the East African hedgehog preferred a temperature of 36·78 C, *E. europaeus* from Berlin 33·28 C and the desert-dwelling *Paraechinus aethiopicus* 40·42 C.

All hedgehogs seem to need dry shelters and are not found in waterlogged marshy country or forest. The East African hedgehog is commonest on sandy well-drained soils in areas where it can sleep in termitaries, rock crevices, buildings or under tangles of brushwood or dry leaf litter. As hedgehogs have restricted home ranges, the habitat must provide an abundance of ground-dwelling insects and other invertebrates. Such requirements are evidently well-met in Nairobi suburban gardens and in many heavily grazed range-lands, where there is an abundance of trampled herbage and dung to support termites and other insects; the leaf litter of thickets and coastal palm groves also seem to be good habitats for hedgehogs. The distribution of hedgehogs in this part of Africa is problematic for there are extensive areas where it is present but very rare and other apparently suitable regions where it is totally unknown. Do predators or competitors play decisive roles? Is it food or the need for a permanent and undisturbed dry shelter which might possibly be related to the habit of aestivating? The family's tolerance of a very wide range of latitudes, altitudes, climates and ecological zones makes it difficult to see gross limitations imposed by the habitat. The most promising solution to the problem of interpreting the hedgehog's scattered distribution probably lies in a very detailed study of its micro-habitat and of its year-round physical needs.

The favourite foods are insects, earthworms, snails and slugs, but a wide range of other animal and even vegetable foods are eaten; eggs and ground-nesting birds, small mammals, frogs, reptiles, crabs, fruit, fungi, roots and groundnuts are reported to have been eaten. This omnivorous diet is reflected in the hedgehogs' molar teeth, which are blunter and broader than those of most other insectivores and in the relatively long gut—about $6\frac{1}{2}$ times the length of the body. A hedgehog will eat a third of its body weight in one night.

It has been shown that these animals hunt mainly by scent: they ignore prey hidden behind a glass cover but can quickly smell food buried 3 or 4 cm in the ground. Hearing may also play a part in their search for food. Immobile foods such as slugs and vegetables are toyed with before eating, but active prey is snapped at and eaten rapidly with much noisy smacking of the jaws. Herter's European hedgehogs learned to pull 30-cm rods through the bars of their cage in order to get at cockchafers attached to the outer end. Lizards and mice are occasionally caught and these are worried and shaken to death. Snakes are approached carefully and then bitten severely; the hedgehogs' spines are all the while directed forward over the nose to expose a minimal area of the body and the snakes generally strike directly onto the spines. The hedgehog attacks repeatedly until it has broken the spinal column or disembowelled the snake; an egg-eating snake, *Dasyplettis scaber*, was killed in this way in Tanzania. Although hedgehogs are not immune to snake venom, they have about 40 times as much resistance as a guinea pig (Bourlière, 1955) and they also seem to be very resistent to insect toxins and to various chemical poisons including cyanide, chloroform, arsenic, opium and corrosive sublimate.

Although their sense of sight is not very good, Herter tested hedgehogs' colour vision and found that they are not colour blind. At night they will ignore a torch shone on them. Individual hedgehogs vary, but noises and smells in the animal's immediate vicinity generally provoke a response such as snorting, sniffing or curling up.

Both the European and the African hedgehog have been seen to react in an extraordinary way to unfamiliar objects with a strong scent. After much sniffing, the object is licked or chewed, whereupon the hedgehog salivates copiously and then plasters the froth over the spines on its sides and back. This saliva-marking behaviour has similarities to the "anointing" ceremony of the mongoose, *Mungos mungo*; for this carnivore the sudden appearance of a foreign smell in the familiar surrounding of the home range provokes alarm followed by a feverish self-anointing by rubbing in the new smell. The act seems to release tension in the mongoose for the new smell is apparently ignored very soon after this ritual, whereas if the smell is inaccessible the excitement is maintained (see Volume III). Scents listed by Herter as provoking salivation in hedgehogs are valerian, human sweat, toadskin, decaying protein, newsprint, perfume, soap, glue, cigarette ends, smoke, varnish and also certain foods such as earthworms. Significant salivation is also released by the appearance of a strange hedgehog. The ritual might therefore release tension aroused by small objects or animals in the home territory and serve to reconcile the animal to the inevitable shocks of change and intrusion in its miniature realm. However, Herter suggests that the action is to cover up the natural scent of the hedgehog to make its discovery more difficult. Harris and Duff (1970), on the other hand, suggest that the saliva may act as an irritant to reinforce the defensive uses of the spines. Poduschka and Firbas (1965) examining the organ of Jacobson in the hedgehog suggest that it operates as an additional sense organ, capable of perceiving various strong stimuli and they think that this organ is involved in the phenomenon of "flehmen" as well as in that of self-anointment.

The legs of the hedgehog are short and its gait is a quaint toddle, in which the hindfoot steps into the impression left by the forepaw, leaving a characteristic spoor. They often chatter when fighting or courting, two activities that involve much snorting and butting with the crown; occasionally they make a soft growl. When in pain they scream loudly.

Hedgehogs are generally active through the greater part of the night and may very occasionally be seen about at dusk or dawn. They sleep curled up or, if quite confident and relaxed, lying out full length.

Numerous observers in East Africa have commented on the virtual disappearance of hedgehogs during the height of the dry season. In Nairobi, they are not seen during the dry months but are very conspicuous between April and June and then again during October and November, at which time they are frequently found run over on the roads. In southern Africa, *E. frontalis* is reported to hibernate during the southern winter—May to September —and to become active again with the onset of summer in October (Smithers, 1966).

When desert hedgehogs from North Africa were taken to Berlin, they

went into hibernation when the temperature fell to between 19—23° C and Herter suggests that, under particular circumstances, all hedgehog species can hibernate. He found that cold and lack of food were the main stimuli inducing some degree of torpor. He also found that, by injecting European hedgehogs with insulin, they could be induced to hibernate during midsummer. Temperature is clearly not an important factor controlling seasonal activity in East African hedgehogs; the most likely trigger for reduced activity is an internal biochemical change associated with a decline in the availability and quality of food. In common with other animals that aestivate, the East African hedgehog acquires thick fat deposits on which it can probably live while inactive. An investigation of this phenomenon in tropical hedgehogs living under natural conditions would be well worth while.

Hedgehogs are generally solitary and both females with young and mated pairs stay together only for a brief period before the young are driven off or the pair lose interest in one another. Very occasionally larger groups have been seen, but the reason for these aggregations is unknown. Herter kept hedgehogs in groups under the artificial conditions of captivity and reported that a pecking order appeared. However, the most striking aspects of the behaviour of wild hedgehogs are their solitary habits and their conservative attachment to a very small home range. On the island of Skikeroog, Herter found European hedgehog pairs patrolling an area with a radius of between

200 and 300 metres surrounding their nest. My own observations over three years of a solitary male *E. albiventris*, living free in a suburban garden suggested a somewhat smaller total range.

It is possible that competition and predation influence the distribution of the hedgehog. Other large insectivores, such as elephant shrews, small carnivores or certain species of birds might compete for resources and thus keep hedgehogs out of some habitats. This factor might possibly influence their success in suburbia, where they are one of the few wild mammals tolerated in gardens.

Large birds of prey and a variety of carnivores are known to prey on hedgehogs. Uttendorfer found their spines in 134 pellets of central European birds of prey—this number, however, is not a significant amount out of a total of 102,000 pellets. In East Africa, Brown (1968) has found that hedgehogs are a favourite prey of Verreaux' eagle owl, *Bubo lacteus*. The bird's talons are capable of killing the animal in spite of its spines; hedgehogs' remains are very common in this bird's pellets and the owl eats everything but the cape. Pieces of spiny skin have also been found impaled on branches, perhaps carried there by some scavenging shrike. They are often killed by dogs.

35

Hedgehogs are parasitized by fleas, ticks and mites, and they are also infected with intestinal and pulmonary worms. It has been suggested (McLaughlin and Henderson, 1947) that the European hedgehog might provide an endemic reservoir for foot and mouth disease and might be implicated in the outbreak of epidemics. Hedgehogs are now used as laboratory animals in virus research (Edwards, 1957). In East Africa, many superstitions surround this animal, its skin or spines are often used as fertility charms. In Karamoja and Teso a bumper harvest is thought to be obtained by placing the skin of the hedgehog on the seed before sowing and for maximum cotton yields its skin should be burned in the cotton field (Watson, 1951).

In Europe, hedgehogs may have two litters in the course of a summer, but breeding seasons have not been noted in East Africa. According to Smithers (1966) in Zambia and Rhodesia the young are born in November, about two months after the animals emerge from their aestivation. The gestation lasts 30—40 days and from two to ten young may be born; the average number is five. The young are born in a nest and Watson found a brood of three in Uganda beneath a pile of sawn timber beside a busy carpenter's bench. Young have also been found in disused rodents' nests.

The courtship of hedgehogs is often an extended ritual; a male will walk round and round a female in oestrus for hours, puffing and extending his snout towards her. J. S. Boys Smith (1967) reports watching a solitary European hedgehog which became a compulsive circler every night for a month, leaving its quiet corner of the garden to spend many hours trotting round and round under the light of street lamps and close to the noise of the traffic. This extraordinary behaviour must have been induced by some strong stimulus, presumably olfactory, which triggered off the first action in a behavioural chain designed to culminate in copulation. The female's response to the male's attention is usually to butt him vigorously and snort angrily and she will continue to reject him, perhaps for days. However, the male's persistence triumphs eventually, the female relaxes her spines and stretches back her hindlegs until her genitalia are exposed. The male's very long penis reduces the danger of his being hurt. In common with many other small mammal species, the ancillary glands to the male's reproductive system secrete a gum-like plug which seals the vagina after copulation. The young are born blind and almost naked with a few soft spines. At birth the mother licks the young, eats the afterbirth and, picking up the young in her mouth one by one, places them on her belly to suckle. Within two or three days dark spines begin to grow among the pale prickles. At this stage the great development of their muscular mantle can be appreciated, but the young are not able to roll up until they are a fortnight old, by which time the back is covered in dark spines. The white infant spines are not shed till after a month and the eyes are reported to open between 8 and 18 days after birth. In their struggle for milk, week-old hedgehogs will resort to butting with erected head spines. The blunt foetal shape of the infant changes rapidly and by the age of one month they resemble small scale adults and accompany the mother on her foraging trips; they eat an increasing amount of solid foods and are weaned by 40 days of age. The family then breaks up and the young animal wanders off and will itself be able to breed by the following year.

Elephant Shrews

Macroscelididae

The elephant shrews are a very distinctive group of animals exclusive to the African continent. They resemble the Oriental tree shrews more closely than any other living group, but a totally different way of life is reflected in their form. Their limbs and tails are not unlike those of some long-legged rodents, e.g. *Malacomys* and *Deomys*. The head is distinguished by a long proboscis and large eyes with well-developed nictitating membranes. Aspects of the functional anatomy of an elephant shrew's head were illustrated and discussed earlier.

Long legs enable elephant shrews to jump well but they are quadrupedal and only appear to hop when running at great speed. As the action is very fast it is difficult to follow the locomotory pattern without the aid of a high-speed camera. The forelegs are proportionately longest in *Petrodomus* and shortest in *Rhynchocyon*. The weight-bearing role of the forelegs is betrayed by the well-developed vertebral spines at the shoulder, bipedal hopping animals have diminutive spines and short front legs.

The main division within the elephant shrews is between the Rhyncho-cyoninae and the Macroscelidinae. The former have eight lumbar vertebrae to the latter's seven, a longer more developed tail, long sharp upper canines and a larger braincase with a peculiarly broad, flat skull. *Rhynchocyon* lose their upper incisors with age, the lateral digits of the forefeet have been reduced and they have no carpal pad on the hindfeet (see drawing overleaf). Glandular areas are found on all species, but the placing of the glands varies; on the under surface of the tail, on the chest as in *Elephantulus fuscipes*, or on the rump.

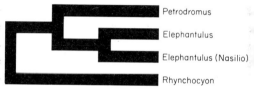

Petrodromus
Elephantulus
Elephantulus (Nasilio)
Rhynchocyon

Fossils that are undoubtedly Macroscelid are limited to African deposits, but *Anagale*, an Oligocene insectivore from the Gobi Desert in Asia, possessed all the characteristics shared by Tupaiidae and Macroscelididae (Simpson, 1931) although probably closer to the tree shrews. A mandible of *Metoldobotes*, from the Oligocene of Fayum, is not dissimilar from that of *Petrodomus* or *Rhynchocyon*. The genus *Rhynchocyon* was already recognizable in the Miocene of Kenya. This species, *Rhynchocyon clarki* was smaller than living forms. Various Pleistocene fossils resembling *Elephantulus* and *Macroscelides* are known from South Africa, but the fossils do not greatly assist in the classification of the living species. Corbet and Hanks (1968) used minimal differences between living forms to display phenetic relationships in the Macroscelidinae. The margin diagram is derived from their paper.

The most successful genus is *Elephantulus* with nine species which are distributed from Somalia to the Cape and with a very isolated species in North-west Africa. This form, *E. rozeti*, is similar to the southern species and its isolation is not thought to be very ancient.

Van der Horst (1946) suggested that the ancient Egyptian god Set was a deification of the elephant shrew and that the Nile Valley might therefore have linked the Mediterranean populations of *Elephantulus* with those of eastern Africa. The okapi has also been identified as Set (Wendt, 1956) but

37

Orycteropus is actually the most likely contender for this title. On the other hand, the rock paintings of the Sahara mountains are a human record of an "East African" fauna extending far to the north and west of its present limits. The Sahara uplands connect Morocco (home of *Elephantulus rozeti*) with the southern Sudan (northwestern limit of *E. fuscipes* and *E. rufescens*) and this connection accords more closely with the distribution pattern of the genus, which is almost exclusively that of an animal of relatively elevated altitudes, absent from all the more extensive area of lowland (see map on p. 61). It is not known what factor in the life history of *Elephantulus* limits its overall distribution but it seems likely to be associated with altitude in some way. An altitudinal factor would help explain its absence from West Africa, and would debar the Nile Valley as a natural dispersal route. Within the genus, some species of *Elephantulus* have a very sharply defined ecological distribution. Further discussion of these patterns and other aspects of the zoogeography of elephant shrews are discussed in the profiles of the genera and species.

Van der Horst (1944) examined the ovaries of various species of Macroscelidinae and found differences within the subfamily that cut across the present classification. *Petrodomus tetradactylus*, *Elephantulus rupestris* and *Elephantulus intufi* are conservative in the number of eggs released at ovulation, while other South African species of *Elephantulus* and *Macroscelides* release great numbers of ova. Other differences of egg chemistry and relative size in the *corpora lutea* and Graafian follicles have also been observed.

The smaller elephant shrews have been kept successfully in laboratories and are easily handled as experimental animals. A type of non-human malaria has been discovered in *Elephantulus rufescens* and this species has subsequently been used for medical research.

Opposite: *Rhynchocyon*.
Below: Skeleton of *Elephantulus*.

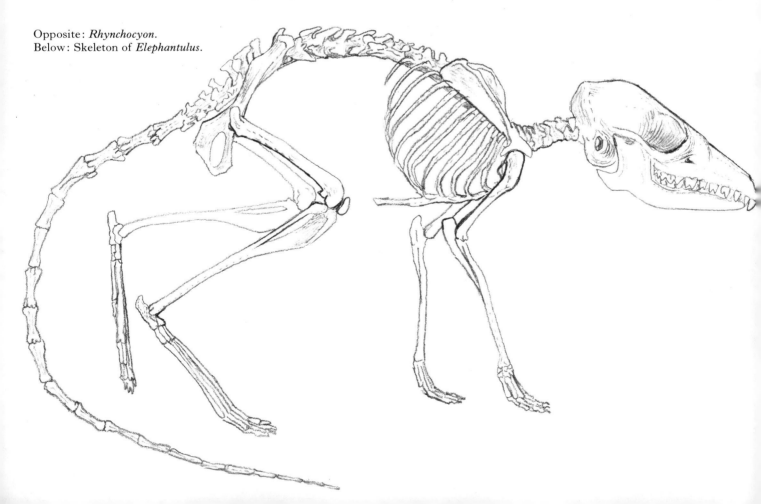

Giant Elephant
Shrew
(Rhynchocyon cirnei)

Family   Macroscelididae
Order   Insectivora
Local names
Njule (Kiswahili), Fugu (Giriama),
Zagari (Kikami), Koni (Kisambaa),
Ngombo (Kinguja), Kasonde
(Kinyakyusa), Songi (Kihehe),
Kabegwe (Kuamba)

Measurements
head and body
235—315 mm
tail
190—263 mm
weight
408—440 g

Giant Elephant Shrew
(Rhynchocyon cirnei)

Races

Rhynchocyon cirnei stuhlmanni	Uganda
Rhynchocyon cirnei reichardi*	Western Tanzania
Rhynchocyon cirnei petersi†	Eastern Tanzania and South Kenya coast
Rhynchocyon cirnei chrysopygus	North Kenya coast

Rhynchocyon is a highly distinctive animal larger and heavier than the other elephant shrews; its long-legged quadrupedal form can be somewhat reminiscent of a miniature antelope, while its hunched back and probing snout sometimes give it a faintly pig-like silhouette. The head is a long tapering cone, a shape typical of ant and termite-eating animals; its long tail is very rat-like. However these allusions do not give much of an idea of the character of this gentle, elegant creature which is little known partly because of its extreme delicacy in captivity.

There is very little difference in size or morphology over a very wide area. There are, however, four distinctly coloured forms and an examination of the various colours and patterns of *Rhynchocyon* raises several interesting problems.

Patterns can be employed as part of an intra-specific signal system, they can be directed towards other species (generally predators or competitors), they can be a physiological response to climatic or ecological conditions. It is also possible that pattern can have some sort of phylogenetic implication and where there is no selective pressure against it, the retention of pattern as a vestige might be possible; for instance, in addition to their cryptic role, the spots of young lions or the stripes of wild piglets suggest a recapitulation of some ancestral pattern (see Portmann, 1952). All these factors might play a part with *Rhynchocyon*. Spots or stripes occur on the young of all forms and there are smudges of lighter or darker colour even on forms where it only distracts from another pattern type, "spoiling" the contrast of yellow rump and dark back in *chrysopygus* and diluting the contrast of a pale tail on the very dark *stuhlmanni*. The persistence of this type of pattern in relatively isolated populations not only underlines a common origin for all forms but also suggests a common ancestral pattern of bold white spots and black stripes down the back (exemplified most perfectly among living forms in *reichardi*).

As to the role of predation in relation to pattern, there is an interesting point to be made here; *R. c. reichardi* occupies the driest range of habitats in which the species can be found and for part of the year it lives in long grass

---

* Dark forms from montane forest in Rungwe district are often classified as *R. c. hendersoni* (see Swynnerton and Hayman, 1951). Typical *R. c. reichardi* figured opposite.

† Animals from Zanzibar and Mafia islands have been described as *adersi*. A variety of hybrids between the black and red *petersi* and the spotted types have been described as *macrurus*.

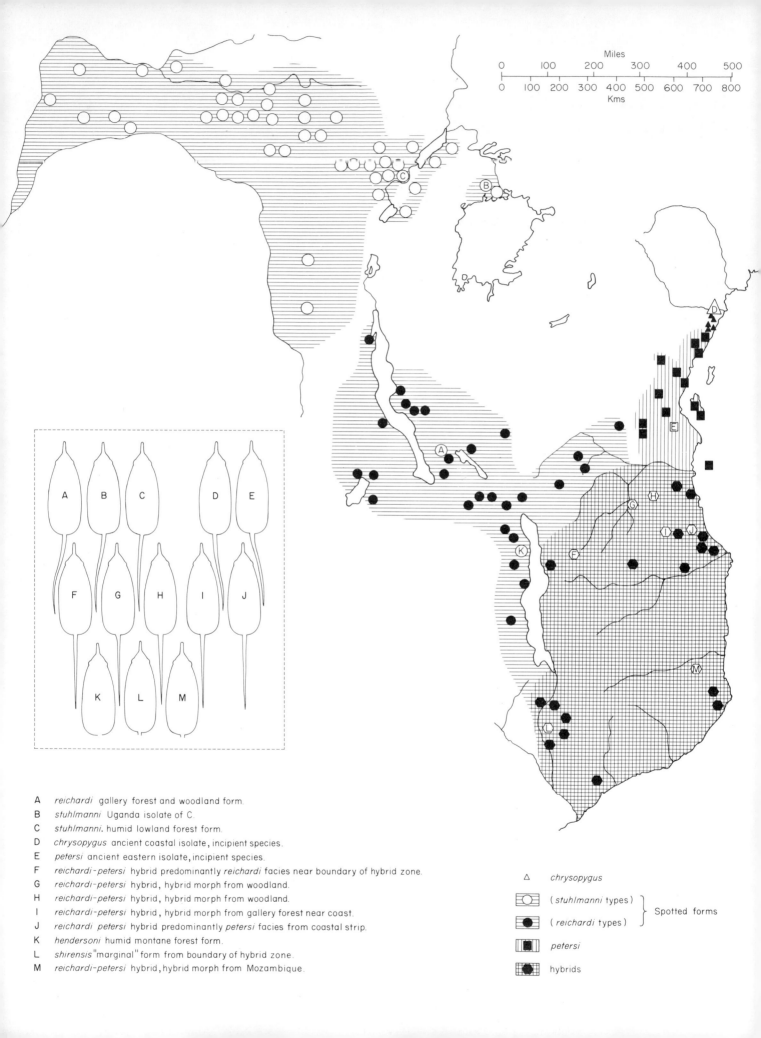

A   *reichardi* gallery forest and woodland form.

B   *stuhlmanni* Uganda isolate of C.

C   *stuhlmanni*. humid lowland forest form.

D   *chrysopygus* ancient coastal isolate, incipient species.

E   *petersi* ancient eastern isolate, incipient species.

F   *reichardi-petersi* hybrid predominantly *reichardi* facies near boundary of hybrid zone.

G   *reichardi-petersi* hybrid, hybrid morph from woodland.

H   *reichardi-petersi* hybrid, hybrid morph from woodland.

I   *reichardi-petersi* hybrid, hybrid morph from gallery forest near coast.

J   *reichardi petersi* hybrid predominantly *petersi* facies from coastal strip.

K   *hendersoni* humid montane forest form.

L   *shirensis* "marginal" form from boundary of hybrid zone.

M   *reichardi-petersi* hybrid, hybrid morph from Mozambique.

savanna and woodland. There is a striking resemblance between the dappled back of *reichardi* and that of the grass mouse *Lemniscomys*. Could predator selection by hawks be more important for populations living in more open habitats? The hazards of attack would presumably be diminished in thicket or forest, but even here the dark colouring of *stuhlmanni* and *petersi* are appropriate to the poorly lit forest floor. However this darkening of colour also seems to express obedience to Gloger's rule, for especially dark forms appear independently in several areas of high humidity. This is illustrated in the colour plate (previous page), where the similarities between *stuhlmanni* and *hendersoni* can be appreciated. Very dark animals occur in the high rainfall

R. c. stuhlmanni.

zone at the north end of Lake Malawi and at Livingstonia (*R. c. hendersoni*). Immediately outside this mountainous area the paler *reichardi* occurs, with only altitude to separate the two forms. Between the lakes Malawi, Rukwa and Tanganyika there are other examples of ecologically contrasting races or species. There seems to be little doubt that the differences are genotypic and not simply phenotypic adaptations. These species seem to have differentiated into ecologically separate forms in spite of the absence of a physical barrier between populations living within a very small area. If such ecological sub-speciation is possible within such a small area it is scarcely surprising that larger differences can appear in larger and geographically separate populations elsewhere.

   This point becomes important when considering the separation of two neighbouring forms of *Rhynchocyon* on the East African coast. The golden-rumped *chrysopygus* is one of a number of isolated mammal populations from

44

the North Kenya coast (see Vol. I, p. 75) but it clearly derives from a typical spotted and striped form, for the newly born young have a fully developed pattern of spots right down the back (see drawing). The paedomorphic development of a light rump in *chrysopygus* with the darker pattern elements progressively suppressed can be contrasted with the development in *petersi*, where the darkest elements dominate and the whole rump is pure black in adults and only the faintest trace of a pattern can be seen in some young animals.

In Volume I (p. 75) I pointed out that different populations of closely related mammals can be found within distinctive zones of the East African coast. Furthermore the peculiarity of these two *Rhynchocyon* populations has some similarity with the situation of *Colobus badius*, in which a Tana River population (resembling a Uganda form) contrasts with a very different Zanzibar colobus. Today the elephant shrew populations are not as severely isolated from one another as the colobus, but there can be little doubt that they must owe their peculiarities to a decisive break in gene-flow in the past. Perhaps the ancestral *chrysopygus* focus centred on the forested banks of a river (or, if sea levels have changed, it might have survived on islands). Forests on the ancient mountain blocks of Usambara and Uluguru probably sheltered the ancestral *petersi*. As with *Colobus badius* there is the possibility that these populations have had separate connections with populations further west. *R. c. petersi* has the least in common with all other forms of *Rhynchocyon*, this could signify that it is genetically less conservative or it may have suffered the longest period of isolation.

It has been suggested that a cline links the spotted shrews with *petersi* (Corbet and Hanks, 1968; Corbet, 1970). At first sight there does appear to be a gradation in the shrews from southern Tanzania but these linking forms are the result of hybridization and do not represent a cline. Animals from the coastal strip south of the Rufigi are predominantly of *petersi* type, while those from the uplands of Songea and Ulanga are predominantly *reichardi*. C. J. P. Ionides observed and collected a large series of morphs of peculiar colour from gallery forests and bush in the Ruponda, Tunduru and Newala districts. He found various combinations of *mixed* characteristics in shrews from the same area and remarked (personal communication) that scarcely any two animals are identical in pattern; a genetic instability typical of hybrids. It is interesting to note that even the hybrids appear to have adapted to their local environment in that those collected in the driest habitats tend to be paler. Other animals exhibiting mixed characters have been collected in the "Mozambique zone" (see p. 42). The nominate race *cirnei* occurs in the extreme south of *Rhynchocyon's* range; this and the southern Malawi *shirensis* probably represent two more or less stabilized populations from the margins of the hybrid zone. (The colour plate illustrates the dorsal pattern on some of these shrews and indicates their provenance.)

The very extensive "hybrid zone" between *petersi* and *reichardi* cannot be matched to the north where the ranges of *petersi* and *chrysopygus* approach each other. The two populations only meet over a very narrow corridor of suitable habitat, with *petersi* recorded from Rabai near Mombasa and *chrysopygus* from the forested hills some 20 km further north. It is impossible to judge the nature of their interaction on present evidence. It is possible that

R. c. chrysopygus at birth (complete cirnei pattern).

Above: running sequence from film. After Rathbun, 1973.

the golden rump-patch of *chrysopygus* is linked with sexual behaviour so that an ethological barrier has been erected between the two forms.

I prefer to regard both *petersi* and *chrysopygus* as incipient species deriving from parent stock similar to the existing *reichardi-stuhlmanni* complex. While gene-flow has been re-established between *petersi* and *reichardi*, *chrysopygus* has had less opportunity to escape its coastal cul-de-sac, but it has retained more pattern elements from the parent stock than *petersi* has.

Colour and pattern apart, *Rhynchocyon cirnei* appears to be a very stable and uniform species. Represented in the Miocene of Kenya by a rather smaller species, *Rhynchocyon clarki*, the genus must be an ancient and conservative type of elephant shrew.

The restriction of *Rhynchocyon* to the highly fragmented forests of southeastern Africa and to the Central Forest Refuge area in eastern and northern Congo (Zaire) is an unusual distribution but it has some similarity with that of the mitis monkey, *Cercopithecus mitis*. Its absence south of the Congo River is odd when one remembers that *Petrodromus* occurs there but is absent to

Below and opposite: *R. c. petersi*.

the north. The two elephant shrews therefore appear to be mutually exclusive in this part of their range although sympatric in other areas. The explanation may lie in the climatic history of the Congo basin which is thought to have been desiccated in the south during certain periods of the late Pleistocene (see Chapter IV, Vol. I). Indeed, climatic events of the past may have much to do with the present distribution pattern of *Rhynchocyon* which is becoming increasingly fragmented as its habitat becomes cultivated. Large rivers are probably important barriers to dispersal but the total absence of *Rhynchocyon* (or indeed any elephant shrews) from West Africa is inexplicable on present knowledge.

Rhynchocyon has a wide altitude range: from sea level to 2,300 m but its ecological plasticity is more apparent than real for the species is entirely dependent on shaded leaf-litter. Some races also represent ecologically discrete populations, *hendersoni* is a montane forest form, *stuhlmanni* is restricted to lowland forest and *chrysopygus* is found in thick vegetation in a very small coastal area. *R. c. reichardi* can be found in tall grassland and open woodland but such extensions of individual ranges are largely limited to the temperate period of the rainy season while insects and cover are abundant. I have observed savanna-ranging *reichardi* regularly retreat into riverine vegetation when cover is burnt off by dry season grass fires.

Rhynchocyon finds its food by rootling and sniffing with its probe-like nose and by scuffing with its rather fragile-looking clawed toes. Once the prey has been exposed, the long facial vibrissae appear to come into play and food may be identified as much by touch as it is by scent. The flat undercut lower jaw is hidden beneath a very broad rostrum and this arrangement is probably related to it being a surface feeder, for instance insects on twigs even a few centimetres above ground level are generally ignored.

The principal foods are ants and termites although beetles, worms and other cryptic fauna have been recorded, and even small lizards have been taken by captives. The relatively large size of *Rhynchocyon* and its specialized method of hunting minute insects are limitations that suggest the animal can only survive under adequate cover and in areas where the leaf litter is rich in insects or other invertebrate fauna throughout the year. Outside the forest these conditions are rather restricted but where there is thick vegetation along water courses giant elephant shrews can survive in savanna and woodland. While feeding they explore and turn over the litter with the nose but use the forefeet when they appear to have sensed something, scuffing the surface with several rapid scratches and immediately returning the nose and mouth to the spot, darting the very long tongue in and out to sweep up the insects together with much earth. The animals often leave evidence of their hunting on the forest floor in the form of little saucer or arrow-shaped clearings in the debris.

An important field study of *chrysopygus* has been conducted by G. Rathbun at the Gedi Historical Monument and in the Sokoke-Arabuko forest. By colour-ringing animals Rathbun has accumulated much information on their behaviour and ecology. Rings have allowed individuals to be recognized in the field, and daily records to be kept over a long period of time. Preliminary results indicate that the average range of an animal in this habitat is in the region of one and a half to two hectares, it also appears that they are terri-

48

Sketches of foetal *R. c. chrysopygus*.

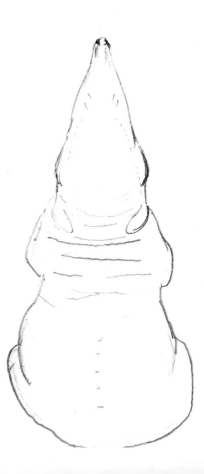

torial. This is one of the first intensive studies to be made on the behaviour of a small mammal in East Africa, the final results will be available in a thesis presented to the University of Nairobi.

Elephant shrews seem to be active at any time of the day and they are occasionally seen at night. Brown (1964) thought they were most active during the midday hours, particularly between 2 and 3 p.m.

Their gait is not unlike that of a miniature antelope, particularly when they run fast. They appear to be given to sudden bouts of violent activity, running in a wide circle back to their companion or to the spot they started from and then continuing to trot about exploring the forest floor. Rathbun (personal communication) has twice observed them adopt a peculiar gait, which he likens to an antelope's "stotting":

"They leave the ground with all four feet at once and land on all fours at once. My impression is the gait has a high amplitude and low frequency—tall, short leaps."

If this behaviour is associated with mild alarm could it have the function of breaking the scent trail and also keeping the source of alarm in sight? They sometimes follow paths in the forest, for I was told by an acquaintance, who was perhaps unwittingly sitting on a "path", that an elephant shrew came running along and found itself in his lap. On steep slopes *Rhynchocyon* always follows the contours.

49

Foetal *R. c. chrysopygus* fully formed but
hairless. Note proportions.

R. c. chrysopygus recently born.

Rhynchocyon ranges into grassland during the wet season (observed in the Southern Highlands and in the southern region in Tanzania), at this time it is opportunistic and follows the runs and paths made by other animals, but it seems to establish its own regular paths in a greatly contracted home range during the dry season. The range of *Rhynchocyon* is therefore likely to vary with the seasons, with availability of food and with the type of habitat.

They are often found as pairs and these associations are usually a male and a female, or a female with a young or a subadult animal. Rathbun's observations confirm this as the commonest social structure, although Copley (1950) and Brown (1964) have recorded small parties of four to six *chrysopygus* foraging together on the forest floor. An ear missing in a specimen caught by Loveridge (1942) and tears noted in the ears of other specimens might be regarded as evidence of fighting, but this has not actually been observed.

Rathbun has concluded that each pair defends a stable joint territory. Spending about a quarter of their time together the male invariably trails the female (often led only by her scent). He will chase only males, while the female pursues her own sex. Intruders are approached at a low crouch, ears forward, nose twitching, tail rapping; a fast chase may ensue. Of many chases observed by Rathbun "only once was the trespasser not quite fast enough and then there was a blur of flying leaves and tumbling animals."

Rhynchocyon have a gland at the base of the tail, and they use it to mark twigs, stones etc., Rathbun has frequently observed both sexes marking as they walked about. Sometimes the animals smell strongly of musk (a smell which reminds me of the palm-civet, *Nandinia*). Sniffing at their immediate surroundings seems to be continuous; olfactory and tactile stimuli are probably exclusive means of finding food. The sense of smell does not seem to be important in scenting danger, instead the animals appear to be very sensitive to movements; sound also seems to engage their attention very easily at medium to close range and a foraging animal can walk up to the boots of an observer who is perfectly silent and still.

Like some other elephant shrews they rap their tails in alarm and pairs or small parties keep in touch with little squeaks. The teeth are frequently clicked together and a loud squeal or shriek is made when the animal is hurt. Rathbun noticed one female to be particularly vocal, making a soft, low chatter whenever she was captured.

The excreta are black oval pellets usually containing earth or humus which seems to be ingested freely in the course of feeding.

Rhynchocyon is sometimes found to be quite heavily infested with parasites; tapeworms, fleas and mites have been recorded. Ionides (personal communication) records the following ticks on animals from southern Tanzania: *Ixodes pilosus*, *Rhipicephalus appendiculatus* and *Haemaphysalis parmata*. Since a malarial cycle has been observed that involves *Elephantulus*, it may be of interest to note that I have seen *stuhlmanni* in Uganda being bothered by a mosquito which is very numerous on the forest floor and which also bites man, but never much above the ankle. Similar persecution from mosquitoes, particularly during the rains, has been noticed by Rathbun at Gedi.

Leaf nests are used throughout the year and Loveridge found one provided with an entrance between the aerial root of two saplings. More typically the nest "is composed of a shallow depression in the soil, which the elephant shrew digs with his forefeet, and a carefully constructed lining of leaves gathered from the immediate area. To line its nest the elephant shrew spreads its forelegs stiffly forward and then lurching backward drags and accumulates a pile of leaves under its belly until it reaches the nest site where it arranges its haul. A finished nest appears as an indistinct mound on the forest floor, and unless one is familiar with the terrain it is very difficult to detect. At dusk a lone elephant shrew cautiously approaches its nest, pauses to sniff the air, smells the nest mound and then very gracefully slips under the leaves. The mound pulses up and down a few times as the animal settles into its sleeping position and then all is quiet." (Rathbun, 1973.)

Breeding records from East Africa are scattered and a lactating female from Bwamba contained an embryo, suggesting a general agreement with Rathbun's *R. c. chrysopygus* which produce a single offspring at any season, giving birth four or five times a year. The young are born fully haired and probably stay in the nest for about three weeks, after which they follow the mother for a week. For a further six weeks they remain in the parental territory and then, almost adult-sized, leave to face the hazards of finding a vacant range. Rathbun has noticed an interesting difference between adult and young *chrysopygus* in the carriage of the tail. The former drag and soil the white tip, while the latter keep it raised and clean.

Rathbun observed the robin chat, *Cossypha natalensis*, associate with *R. c. chrysopygus* at Gedi. As the elephant shrew leaves a feeding spot the bird

"hops in and carefully picks out any small invertebrates that have not been eaten or (since the elephant shrew is a sloppy eater) it may pick up bits and pieces left behind. The robin chat also hawks insects flushed by the elephant shrew as it forages through the leaf litter. It is not uncommon to see an elephant shrew slowly moving along and close behind it in the undergrowth a robin chat hopping from branch to branch." (Rathbun, 1973.)

Large snakes are probably the main predators and *Rhynchocyon* have been found killed by cobra, *Naja melanoleuca*, and viper, *Bitis gabonica* (Ionides, personal communication). Hawks, small carnivores and perhaps owls are other predators. Rathbun found no trace of *Rhynchocyon* in *Tyto* pellets at Gedi. In some areas *Rhynchocyon* are eaten by local people and hunted with dogs to which they are peculiarly vulnerable. In southern and southwestern Tanzania they are often included in the bag of hunters who fire the savanna and woodland at the beginning of the dry season.

Rathbun sees the homeless, newly independent subadults as the most vulnerable age class but suggests that predation, disease and old age may take territorial animals of three or four years age.

Longevity is not known nor has the animal been kept in captivity successfully. When caught they never attempt to bite but kick energetically with the hindlegs.

The behaviour of captive animals in pens is reminiscent of that of duikers, headlong panic is succeeded by immobility or busy foraging. One impulse follows another in rapid succession and the general impression is that most of their behaviour is very highly stereotyped.

R. c. chrysopygus.

P. t. sultan.

Four-toed Elephant Shrew (Petrodromus tetradactylus)

Family Macroscelididae
Order Insectivora
Local names
Isanje (Kiswahili), Mudoro (Kibondo), Nyenge (Kinyaturu), Dongi (Kihehe), Sangi (Kiduruma), Mwonungu omballa (Kitaita)

Measurements
head and body
165—220 mm
tail
130—180 mm
weight
155—220 g

Four-toed Elephant Shrew (Petrodromus tetradactylus)

Races

Petrodromus tetradactylus tetradactylus	Western half of Tanzania
*Petrodromus tetradactylus rovumae**	Eastern and southern Tanzania
Petrodromus tetradactylus sultan	East Kenya
Petrodromus tetradactylus zanzibaricus	Zanzibar Island

In the field the four-toed elephant shrew is distinguished from *Rhynchocyon* by its sandy brown colouring, compact form with large head and well-defined "spectacle" of white round the eye. *Petrodromus* is considerably larger than the *Elephantulus* species, but, like the latter, it has very soft fur. The race *P. t. sultan*—formerly regarded as a separate species—has bristles on the undersurface of the tail, the tips of which are expanded into distinct knobs. These hairs are absent in *P. t. tetradactylus* and are less well developed in *P. t. rovumae*. Females are generally more warmly tinted than the males.

Petrodromus ranges over southeastern Africa, with an isolated population in the Congo basin forest south of the Congo River. The peculiarity of the latter distribution was discussed in Chapter IV (Volume I). Except for this population, the species is not characteristic of true forest and, over the rest

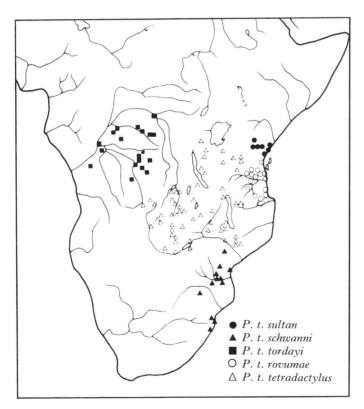

- ● *P. t. sultan*
- ▲ *P. t. schwanni*
- ■ *P. t. tordayi*
- ○ *P. t. rovumae*
- △ *P. t. tetradactylus*

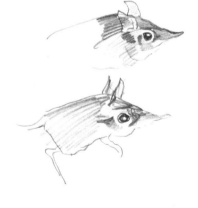

P. t. sultan.

* In the coastal region immediately south of the Pangani River, elephant shrews have various combinations of the characteristics of *P. t. rovumae* and *P. t. sultan*.

of its range, it is found in various types of dense vegetation: in thicket around rocky outcrops, in the moister *Brachystegia* woodland and in riverine strips as well as in the drier types of montane and coastal forest. These southern habitats generally suffer a distinct annual dry season between July and October.

Petrodromus makes very characteristic runs marked by bare patches in the litter which are at about 40 to 100 cm apart. Ansell (1969) has illustrated the plan of one of these path complexes, patrolled regularly each day. Ant-hills, *Tatera* burrows, hollow trees, logs or rocky outcrops are common features in their habitats and they often resort to the shelter of natural holes when they are pursued, but it is uncertain how frequently these natural shelters are used other than in emergency. They do not appear to make nests.

The principal foods of four-toed elephant shrews are termites and ants—including soldier ants—Hemiptera, Heteroptera and Coleoptera. Crickets are a favourite food. *Petrodromus* may benefit from a peculiar habit of many termite species; while these insects are foraging they occasionally make what appears to be an alarm signal by hammering their heads against the dead leaves, grass or wood on which they are feeding; the signal is then taken up by all the termites in the vicinity, thereby revealing what may be very large numbers of insects on the surface of the leaf litter.

Petrodromus is distinguished by long delicate legs on which it trots with the body held 3 or 4 cm clear of the ground. The gait is entirely quadrupedal, but it is capable of long leaps with the tail held high. While hunting, it uses

its hindlegs somewhat after the fashion of a chicken to scuff away debris, and the animal leaves patches of turned-over litter as evidence of its foraging.

A common noise made by this animal is a shrill squeak, which can often be heard at night. When relaxed, it makes a soft purr, but this is only audible within a few centimetres and its function probably resembles that of cat's purring. *Petrodromus* has also been heard to make a clucking, chirping sound and a cat-like mew when captured. It also frequently drums or raps with the hind legs and tail. Duff McKay (personal communication) has described hearing these rapping sounds being "answered" by shrews scattered in the Sokoke Forest and he thinks that the tapping may have a communicative function.

This species is noticeably active during the morning and evening and also at night, but detailed activity patterns under natural conditions have not been recorded to date. *Petrodromus* are found singly or in pairs. Very little is known about their social behaviour and the role of scent, which is obviously of central importance, has not been investigated. The peculiar knobs under the tail of some races may act as scent dispensers for glandular secretions.

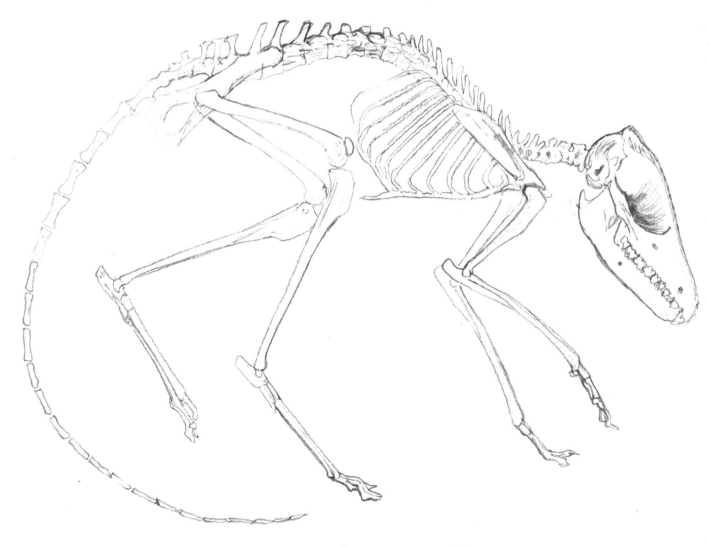

The species is probably preyed upon by a variety of small carnivores, by hawks, possibly owls and large snakes. Several coastal tribes eat this animal and in Mozambique local people catch it in a sisal net placed on a run with an anchored mouth string; the impact of the animal closes the mouth of the bag or net. In other areas simple snarcs are used.

The incidence of parasites in this species may be both a seasonally and a regionally determined occurrence. Brown (1964) found no parasites on animals from the Shimba Hills and from near Malindi, except for one specimen caught in June. Loveridge (1922) found that animals from Morogoro were heavily parasitized and noticed that ticks were plentiful on the glandular undersurface of the tail and on the ears.

Petrodromus probably breed throughout the year; Brown (1964) noted that out of 14 females collected in June, 8 were pregnant and that the embryos were of various sizes. Hollister recorded 3 pregnancies out of 10 females from the same area in December. Further south at Morogoro, Loveridge (1922) collected 2 females with large foetuses—64 mm head and body length— and 2 recently born young measuring 76 mm head and body length. In Zambia, pregnancies are recorded for the months of January, April and July. Pregnant females abort readily in captivity but the species has been kept successfully in zoos for several years.

Spectacled or Long-eared Elephant Shrew (Elephantulus rufescens)

Family Macroscelididae
Order Insectivora
Local names
Sange mdogo (Swahili), Ayole ngayolei (Karamojong), Mbulu (Kinyaturu and Kisandawe)

Measurements
head and body
103—150 mm
tail
100—165 mm
weight
25—50 g

Spectacled or Long-eared Elephant Shrew (Elephantulus rufescens)

The spectacled elephant shrew is immediately distinguishable by its small size and white ocular ring, which contrasts strongly with the black eye. It has a relatively long tail with a rudimentary subcaudal gland; the pectoral gland is well developed. Its colour tends to vary regionally and appears to be correlated with the dominant soils. Colour, therefore, is probably determined by predator selection, operating principally in the dry season. Animals from the red latosolic soil of southeastern Kenya tend to be the reddest, and it is from this area that the type "rufescens" was named. On the yellow sandy, clay loams that are the dominant soil types in parts of central Tanzania, the coat colour is yellowish. To the South-east of Lake Victoria, the loamy sands of that area are grey and the elephant shrews too are greyish. Similar local correlations can be found in other parts of its range. Heim de Balsac (1936) has demonstrated how almost all small mammals in the Sahara match the desert shade. Benson (1939) found white *Perognathus* mice on a white sand habitat in an isolated Mexican valley, while *Perognathus* (of another species) living on black lava flows in the same valley were quite black.

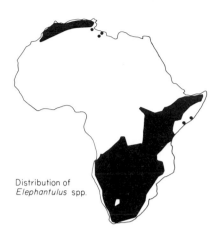

Distribution of *Elephantulus* spp.

Such dramatic contrasts are only possible with a considerable degree of isolation and interruption of gene-flow. The populations of spectacled elephant shrews mentioned above occupy peripheral areas of the species' range, and Corbet and Hanks (1968) note that animals collected between the grey populations near Lake Victoria and the yellow ones of central Tanzania are rather variable in colour and intermediate in type, probably reflecting greater genetic circulation.

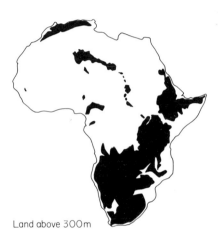

Land above 300 m

This species is restricted to dry acacia bush and thicket in northeastern Africa. Apart from some low-lying localities in Somalia and along the Tana River, the species is generally found above 300 m and it prefers well-drained soils. As fires and predators are major hazards in their dry habitat, they are very dependent on the presence of clumps of thick fire-resistant vegetation. Such vegetation foci may occur around termitaries, in gullies, or near cattle bomas (corrals) and in the hedges that are common in settled areas.

Aloe and *Sanseviera* species are conspicuous ground level plants in thicket clumps that are formed in heavily grazed and fired areas. These indigenous succulents and the introduced sisal, prickly pear and euphorbia (which are used as hedges or boundary lines) provide the elephant shrews with shelter during the dry season, and their shallow scooped "nests" are often found among the roots of these plants or among the crevices of rock outcrops and termite hills. A small clump of vegetation will shelter a pair of, or more, elephant shrews. Their runs, which are often shared with rats, connect with other clumps and radiate out to open feeding areas. Brown (1964) found that the principal food in Karamoja during December and January was harvester ants; numerous remains of these insects were found together with the dung of elephant shrews round the entrances of ant's nests. Watson (1951) also singled out harvester ants for mention. Termites, small grasshoppers and some vegetable matter are other foods that have been recorded. They do not

61

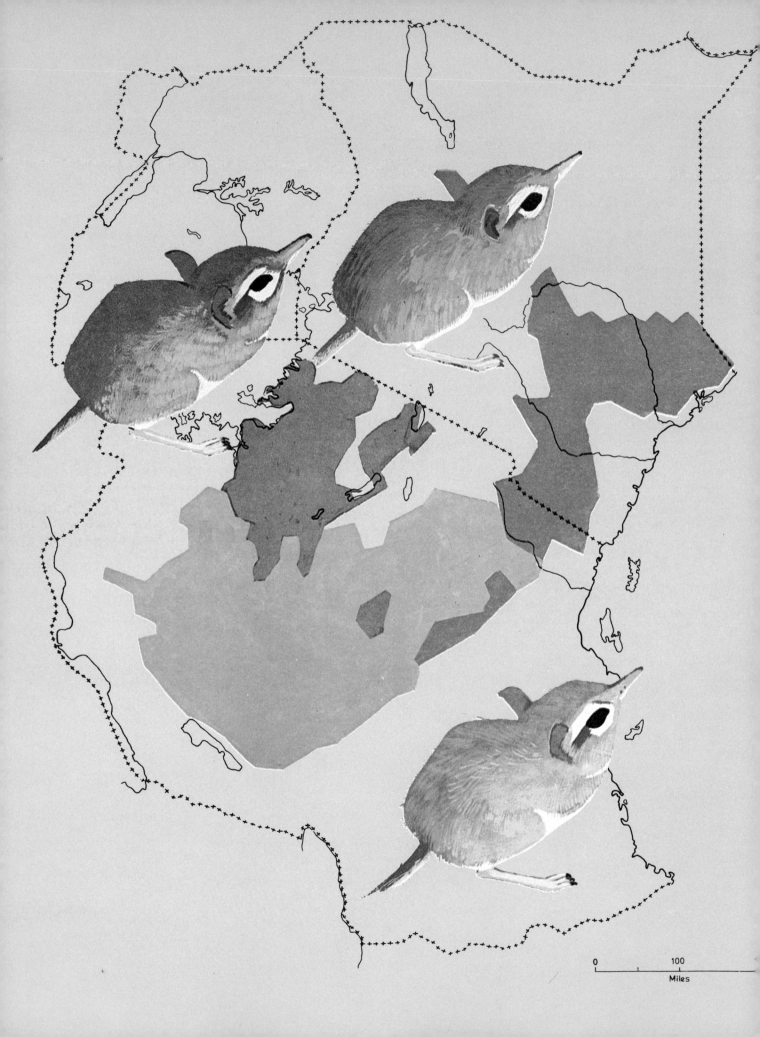

usually drink but will lap milk in captivity, having some difficulty at first keeping the tip of the nose out of the liquid.

They are very alert and take fright at every sudden movement or noise. They run on all fours and can make long quadrupedal leaps. Pocock (1912) described the locomotion and dispelled the belief that they hopped like a jerboa.

Rapping the hind legs is a common habit and is apparently an alarm signal; a shrill squeal is also made when frightened. In captivity squabbling is common when several animals are kept together at close quarters, and various chittering squeaks are associated with snapping and chasing. When handled, however, they never attempt to bite.

Pairs or small groups are the usual social unit. In suitable habitats where the animals are numerous, social interactions are probably most likely when the animals are feeding or foraging. Judging from spoor and dung, a concentrated food supply is often shared by several pairs, each centred on a home shelter area in some neighbouring thicket. Casual observations suggest that the total area normally used by a pair might be less than one quarter hectare. The area of sheltering thicket or thickets in which the animals spend the greater part of their lives in very small, sometimes less than 25 sq. m.

They have been found in close association with field rats, *Arvicanthis*, hyraxes (in Somalia) and gerbils. When natural shelter is hard to come by, they are sometimes found in holes dug by small rodents, for they are unable to dig themselves.

They are generally diurnal but like *Petrodromus* are probably active for part of the night as well.

In view of regional colour variations and the role of predator selection which was discussed earlier, predation is likely to play an important part. "Visual" hunters, effectively birds of prey and perhaps some snake species, may, therefore, be the elephant shrew's most important enemies. Laurie (1971) found remains of three elephant shrews in owl pellets collected in Serengeti National Park; this represented 0·8% of the pellets collected at this site. As the young are known not to exceed two, and one young is more usual, the level of predation seems to be appropriately low in relation to the more numerous rodents and shrews and to the mainly diurnal habits of *E. rufescens*.

A high proportion of females collected in April and May have been found pregnant. In December and in January, however, no pregnancies have been recorded and seasonal breeding seems to be most likely. Tripp (personal communication) saw frequent post-partum ovulation, inconspicuous sexual behaviour and very brief coitus. He found a 50-day gestation and spontaneous ovulation. Young are weaned at one month and are adult sized at two months. Mothers carry infants in the mouth.

Hoogstraal *et al.* (1950) discovered a malarial parasite (not affecting humans) in *Elephantulus rufescens* and the association has subsequently led to the establishment of *Elephantulus* species as laboratory animals for malarial research.

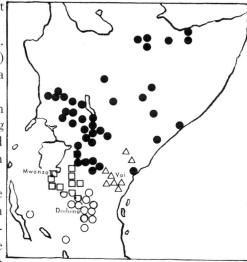

Elephantulus rufescens.
△ red population
□ grey population
⊔ intermediates
○ yellow population
● other populations

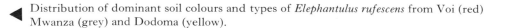

Distribution of dominant soil colours and types of *Elephantulus rufescens* from Voi (red) Mwanza (grey) and Dodoma (yellow).

Short-nosed Elephant Shrews (Elephantulus (Nasilio))

Family Macroscelididae
Order Insectivora
Local names
Sengi (Swahili)

Measurements
head and body
110—138 mm
tail
85—124 mm
weight
40 g (approx.)

Short-nosed Elephant Shrews (Elephantulus (Nasilio))

Elephantulus brachyrhynchus has a reddish-brown back, a grey-white belly, and a short simple supertragus in the ear. It lacks a pectoral gland. It is found in Kenya, Tanzania and southwestern Uganda. *Elephantulus fuscipes* is dark brown with a greyish-white belly. It has a long twisted supertragus in the ear and the pectoral gland is present. It is found in northern Uganda. Although apparently allopatric there are distinct structural differences between *E. brachyrhynchus* and *E. fuscipes*. The former species can be recognized from *E. rufescens* by the shorter thicker tail, by long guard hairs in a generally darker coat and by the lack of a black stripe behind the eye.

The short-nosed elephant shrews range over a variety of major vegetation zones but favour dry woodland, thicket or the denser vegetation around termite mounds or along water courses or gullies and, while preferring more open country than *Rhynchocyon* and *Petrodomus*, they do not venture into the open like *E. rufescens*. They appear to need shelters such as those provided by termitaries, gerbil warrens, other animals' burrows or rocky outcrops.

They have been found to eat termites, grasshoppers, cicadas and grubs from decaying logs but, in common with other elephant shrews, leaf-litter insects seem to form the greater part of their diet. They also eat vegetable matter; grain and the fruits of *Strychnos* and *Flacourtia indica* have been recorded.

They are generally very localized, feeding within twenty metres or so of their shelter. They are diurnal and feed during the hottest part of the day, between 10 a.m. and 4 p.m., moving between the feeding area and the home shelter in as direct a line as possible. In a restricted habitat in which these animals are well established, this may lead to the formation of small pathways, but the making of paths is not a conspicuous habit in this group as it is in some other elephant shrew species.

They are generally found as solitary animals or as "family" groups, never in colonies, although several families may live in some proximity to one another. Rankin reported that males displayed aggressive behaviour in defence of a territory and that sharp high-pitched squeals were uttered while fighting; another call is described as shrill and penetrating.

These small elephant shrews are the least conspicuous members of the family and consequently both locality records and data on their biology are scarce.

Hollister (1918) records three pregnant females from Kenya in November and two in June. In Rhodesia most births have been recorded during May. The young are born well developed and are able to run almost immediately after birth; they grow rapidly. Twins appear to be more frequent than single births.

An imperfectly pickled specimen of *E. brachyrhynchus* in my collection grew a thick white mould on a well-defined area of the rump and on the glandular area beneath the tail. As no other part of the body was affected, it looked as though some secretion was providing a suitable culture medium for the growth of this mould.

Species

Elephantulus brachyrhynchus
Elephantulus fuscipes

● *E. brachyrhynchus*;   ▲ *E. fuscipes*.

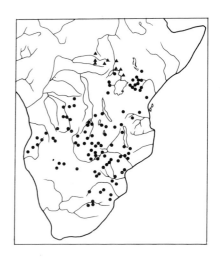

KEY TO SORICIDAE

Myosorex

$$\frac{3-1-2-3}{2-1-1-3} = 32$$

tooth formula

Sylvisorex, Scutisorex and Suncus

$$\frac{3-1-2-3}{2-0-1-3} = 30$$

tooth formula

Crocidura and Paracrocidura

$$\frac{3-1-1-3}{2-0-1-3} = 28$$

tooth formula

Shrews

Soricidae

Genera

Myosorex
Myosorex (*Surdisorex*)
Sylvisorex
Scutisorex
Suncus
Paracrocidura
Crocidura

Shrews are small, mouse-sized animals with long mobile noses and stout cylindrical skulls. They are specialized animals, living for the most part on insects by bulldozing them out of their well-hidden retreats in vegetable debris. To this end their bodies are powerful and tubular and they are provided with extraordinarily sensitive vibrissae with which to detect their prey. Various anatomical specializations have occurred in the course of evolution; the molars are well adapted to chewing insects, the middle incisors have acquired a peculiar form and prominence functioning as a sort of pincers, and most of the premolars have disappeared together with the lower canine; the zygomatic arch has been lost.

Shrews are known from the Oligocene in Europe and America. Still older fossil fragments are thought to represent links with the earliest ancestral insectivores (see margin drawing).

Living genera, *Sorex* and *Crocidura*, are known from the European Miocene, a *Crocidura* species (Butler and Hopwood, 1957) is known from the African Miocene and living species, *Crocidura hindei* and *Suncus lixus*, are found in the Lower Pleistocene beds at Olduvai.

Three African shrews have developed striking specializations: *Scutisorex* an extraordinarily complex backbone, *Surdisorex* a mole-like reduction of the eyes and ears with elongation of the digging claws and *Paracrocidura* a chisel surface to its incisors, rather like a rodent.

Within the seven East African groups a variety of phylogenetic levels appear to be represented. Heim de Balsac and Lamotte (1956—1957), correlating living with fossil forms, concluded that in the evolution of shrews there has been a progressive reduction of the back molar and of the premolars, excepting the fourth which has acquired a molar-like form. There has also been a simplification of the cusp pattern and a general flattening of the cranium with loss of the interparietal.

Myosorex, with the largest number of teeth, represents the most ancient of living groups. *Suncus, Sylvisorex* and *Scutisorex* probably represent descendants of this stock, retaining the upper P. 3, but loosing the lower P. 3.

Crocidura has reduced its dentition most and is a widely distributed genus with many species. It is the most advanced and at the same time the most adaptable and generalized genus in Africa.

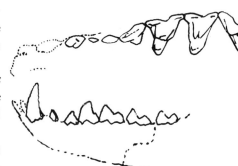

Saturnina,
upper Eocene Soricid.

tooth formula $\dfrac{3.1.4.3}{3.1.4.3}$

Deltatheridium,
Cretaceous Soricomorph.

tooth formula $\dfrac{?.1.3.3}{?.1.3.3}$

69

Shrews are found in a wide range of habitats but are generally commonest in relatively moist conditions. Some species are good climbers. One type, *Surdisorex*, habitually burrows underground, and many species nuzzle into soft soil or leaf litter to detect their prey. Many use rat runs or follow habitual paths; these are not usually obvious to the human eye but they are followed by a shrew with confidence. The scent glands on the shrews' flanks probably rescent the path with the animal's every passage and allow it to traverse its territory with ease. The scent may also repel other shrews and assist the spacing of individuals.

Food seems to be found principally by means of the long whiskers in the shrew's muzzle. Observers have agreed that the role of the sense of smell in finding food is not so important. However, the reactions of hunting shrews are generally very fast and the relative importance of tactile or olfactory stimulation is difficult to determine with certainty. Insects, arachnids, molluscs, small amphibians and even small mammals are eaten. Some species also feed on vegetable matter.

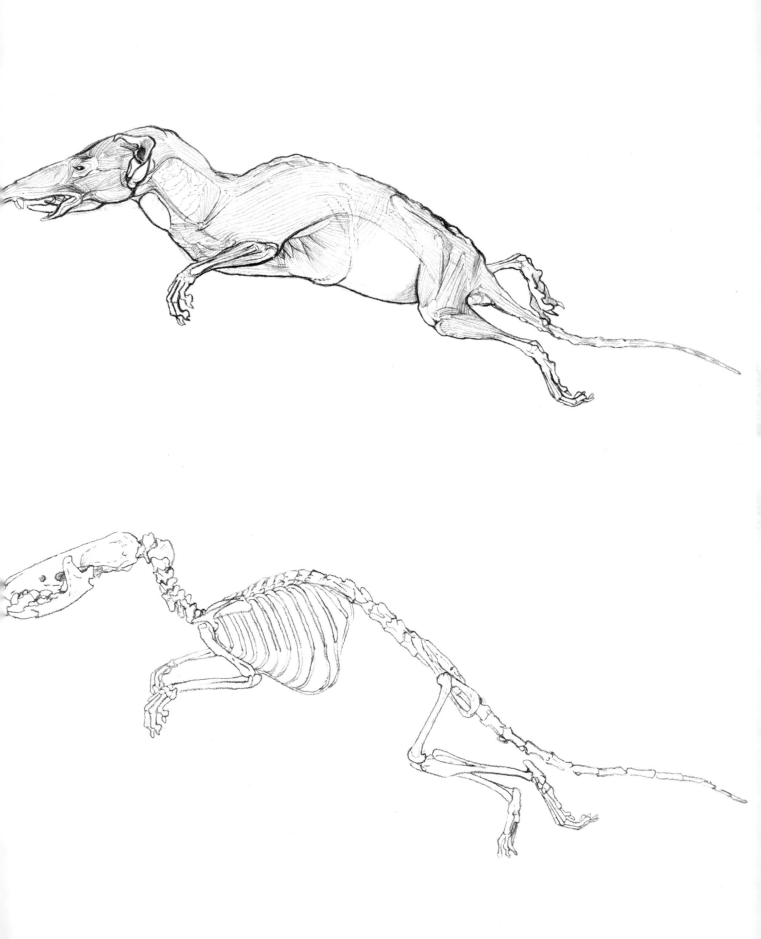

Individuals may eat from three-quarters to over three times their own weight in food per day, and some species die quickly if starved. Several species of *Crocidura* have been observed to eat their own faeces. This habit might assist survival during periods of food shortage, or it might be a physiological necessity as in the Lagomorpha. Dung deposits or latrines are known for several species of shrews. Food is stored by some species in or near the nest, which is generally built in some natural crevice, vegetation clump, termitary or burrow. The nest may be fairly large and round or rudimentary, depending on the species.

The shrews' dependence on water varies from species to species. Some arid-adapted types may do without drinking for many months, while forest species die quickly if they are confined without water.

The hearing of shrews is acute and the variety of noises shrews utter suggest that vocal communication is well-developed. While exploring, shrews twitter and their threats are punctuated by sharp explosive squeaks, they also crack their teeth together and make a number of other squeaks and chatters. They are more often heard than seen in the field. The vision of shrews appears to function at close quarters only, but Crowcroft (1957) remarks that the posture of rival shrews plays an important part in fights or threats.

Shrews are clean animals, grooming themselves by means of their teeth and scratching with their paws. The habit of worming along through vegetation and leaf litter and that of forcing the nose and sometimes the body down into loose soil probably assists cleaning and is also an effective way of depositing scent from the lateral glands. The fur moults from time to time and the new fur is generally rather darker and shorter.

Most species are intermittently active during both day and night, but there is probably some variation from species to species and from season to season. Crowcroft gives several examples of the shrews stereotyped behaviour. For instance, if a shrew carrying food to the nest drops it on the way, it will continue to the nest and only then return, as if on another foraging expedition, to find the lost food. Likewise, shrews collide with objects placed on their paths. Small obstacles on their paths may be removed with the teeth, whereupon the instinct to carry may take over and the object, irrespective of its nature, may be carried to the nest or food store. Crowcroft observed that the act of picking up a stone while digging seemed to trigger off a food-storing drive, and create a conflict between continued digging and storing behaviour.

Nest-building is also a highly stereotyped activity that consists of dumping materials and then turning round and round in the loose ring, grasping the material, usually grass or plant fibres, in the teeth. When the nest becomes more elaborate or covered in, the same behaviour pattern is followed; the shrew dumps the newly collected material outside the nest, and then entering it, thrusts its nose through the nest wall to seize the material and draw it round and round inside. It never carries the material directly into the nest.

Shrews are generally solitary, but captive groups and trapping records suggest that several African species, notably *Scutisorex*, may be more sociable than had formerly been thought.

The legendary shrewishness of shrews is well known and both sexes tend to fight when they encounter another shrew on their feeding forays. The scenting of pathways probably advertises the shrew's sexual condition as well as its actual presence on or possession of the path. Shrews, therefore, tend to avoid the central part of another shrew's territory, except when a female is in oestrus.

Aggressive encounters involve screaming, sparring with the teeth and forelegs and, sometimes, fighting with snapping and kicking. They seldom give chase, although Crowcroft saw lactating females pursue intruders.

The principal enemies of shrews are birds and snakes; mammalian predators are often put off by their strong smell. Owls, hawks and shrikes are known bird predators. It is probably unusual for most species to live much longer than two years in the wild, although captive shrews have been known to live up to almost four years.

Shrews have not been implicated in the transmission of human disease but they are infested by the fleas *Dinopsyllus* spp. and *Xenopsylla brasiliensis*, which are vectors of plague and, as some species of *Crocidura* take up residence in huts, it is possible that they could act as carriers. Ticks have been found on shrews and one species, *Ixodes alluaudi*, is specific to shrews of all genera in a wide range of habitats. The louse *Polyplax reclinata* is common on *Crocidura*, and mites and endoparasites have also been found. Parasites are listed in detail in Meester (1963).

A female shrew will generally attack any other shrew, and is only receptive to the male for about half a day during oestrus. At this time her genital tract gives out a scented secretion and, when a male approaches, the female will raise her tail and hunch her back. The female of an American species, *Blarina brevicauda*, shows less development of the lateral scent glands during oestrus (Pearson, 1946). This may also apply to other species of shrews, especially since in most African shrews the males' lateral glands are more developed than those of the females. The reduction of the female's lateral scent gland during oestrus could therefore play an important role in timing the reproductive behaviour and might offset the male's tendency to fight, display aggressive behaviour or even to run away. The genital secretion is probably the major stimulus attracting the male, as he will smell and lick the female before seizing her by the scruff of her neck with his teeth and copulating.

Once pregant, the female again drives off all other shrews. The gestation periods of shrew species are thought to range between 16 and 35 days. The young are born in the nest small, blind and weak. They first open their eyes after a period of about 12 days. They suckle for three weeks, growing with great rapidity and acquiring a nearly adult weight in five to six weeks. Weaning onto solid foods starts after 16 to 20 days. Caravans, in which the mother shrew is followed by her young gripping onto one another's rumps, are a common feature of *Crocidura* shrews and possibly other genera as well. This ritual can be induced by the mother seizing a young one by the nose and pulling it before presenting the rump to be bitten. The young soon learn to caravan in single file, although they sometimes attach themselves in clusters behind her. This behaviour is common between the eighth and the eighteenth day of life, after which the young find their own way about. A single female may rear two litters in quick succession.

Mouse-shrews (Myosorex)

Species

Myosorex : short tail without long hairs. Upper and lower P. 3 both present, velvety fur, small, forward folded ears. Claws and forelegs range between 2—3 mm and 4·1 mm.

Myosorex blarina
Myosorex zinki
Myosorex geata

Mouse-shrews (Myosorex)

Family Soricidae
Order Insectivora

**Measurements
head and body**
50—105 mm
tail
29—56 mm

Myosorex blarina are dark coloured shrews with greatly reduced ears and well-clawed forelegs. *M. b. blarina* from Ruwenzori is a dark brown shrew with longish tail and long claws. *M. b. babaulti* from Kigezi is blackish with a shorter tail and weaker claws. *M. zinki* is a paler form, endemic to the upper reaches of Mt Kilimanjaro. *M. geata* from the Uluguru Mountains is smaller with relatively longer tail; it is speckled dark brown, the fur is short revealing the ear and it has the least developed claws on the forefeet.

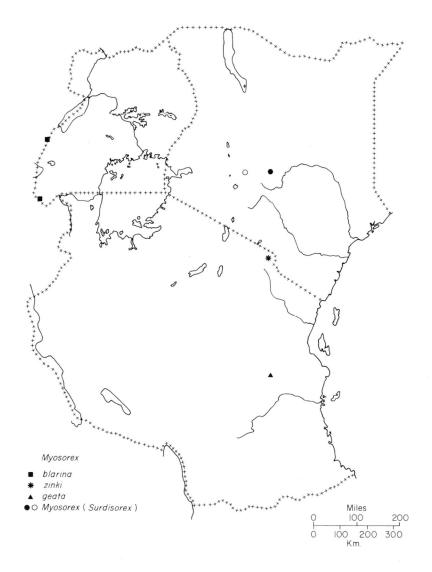

Very little is known of the habits of mouse-shrews, but they have been the subject of an interesting zoogeographic and phylogenetic discussion (Heim de Balsac, 1957a, 1966a, b, 1967) and (Heim de Balsac and Lamotte, 1956—1957).

Myosorex is considered to be the most primitive genus of shrew on the basis of its numerous teeth and the presence of an interparietal bone. The last feature is most marked in a form from the lowland forest of the eastern Congo, *Myosorex schalleri*. This species is also primitive in other respects. The third upper premolar is perfectly formed, rather than having the appearance of a vestigial tooth. Furthermore, the form of the cranium is rather simple and rectilinear. These features reveal not only that all other species of *Myosorex* are relatively more evolved, but that a specialized development in *Myosorex* spp. can be traced, step by step, to its culmination in *Myosorex* (*Surdisorex*), as a truly fossorial animal. The most striking evidence of this evolutionary trend towards a subterranean existence is the acquisition of a mole-like, wedge-shaped skull together with more heavily clawed forefeet and a reduction in the eyes and external ears. Further study of the biology of *Myosorex* will undoubtedly reveal a similar graduation reflected also in the habits of the various forms.

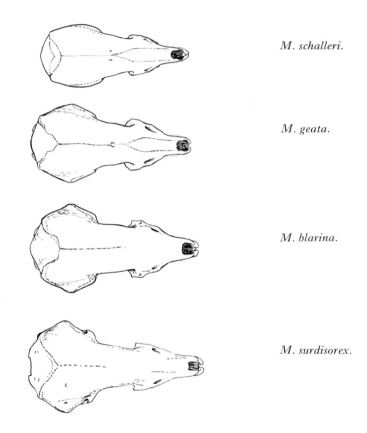

M. schalleri.

M. geata.

M. blarina.

M. surdisorex.

Myosorex geata is known, at present, from two specimens, one from the cultivated foothills and the other from the montane forest of the Uluguru Mountains. This species is the least specialized of East African forms. The most closely related types are found in South-central Africa, and there is good reason to suppose that similar forms may turn up in the Southern Highlands when more collecting has been done. It is interesting that the less evolved types of *Myosorex* have a lowland or more southerly distribution, while the more specialized species are on the equatorial mountains.

The mountain forests of East Africa are the home of relic faunas primarily because of their persistence as relatively stable and moist habitats in the face of climatic fluctuations in the ancient past (in spite of extensive forest destruction by man in the most recent past). They, therefore, provide refuges for ancient forms of life, but they may also provide conditions suited to adaptive change. Mountains isolate populations and offer a range of altitudinal zones and temperature clines. As was pointed out in Volume I, the dual role of mountains as specialized habitats and as preservers of relic species is inseparable in the case of *Myosorex*.

The range of the genus as a whole is restricted to moist habitats at various altitudes. This may reflect physiological limitations in a primitive group, but modern *Myosorex* may survive in moist habitats for another reason. Most shrews have feeble limbs, and the initial development of burrowing claws would depend on soils being singularly soft and moist, so that a capacity to dig, however inefficiently, might well have saved *Myosorex* from the competition of more advanced shrews.

The occurrence of the relatively unspecialized *M. geata* to the south of Mt Kilimanjaro, the home of the highly specialized *M. zinki*, shows that Kilimanjaro must have been colonized from the North-west, by some route that linked this mountain with the Central Refuge (see Volume I, pp. 69—76). As the Aberdare Mountains and Mt Kenya are close to this route, very interesting questions can be raised about the nature of evolution in these shrews; these are discussed in the following profile of the subgenus *Surdisorex*.

The little that is known of *Myosorex* habits derives from the South African species and does not add much to the general knowledge of shrews. They make nests, eat insects and small vertebrates and have been recorded as bearing four young.

Myosorex with hair shaved off ear to show folded structure open and closed.

Kenya Mole-shrews (Myosorex (Surdisorex))

Species

Myosorex (Surdisorex) norae   Aberdare Mountains
Myosorex (Surdisorex) polulus   Mt Kenya

These shrews are only known from 2,800—3,600 m on the Aberdare Mountains and Mt Kenya, in the upper reaches of the montane forest and in the Afro-alpine zone. They are commonest in grassy areas in the alpine and bamboo zones.

They are trapped in the runways made by *Otomys* species but they have also been caught alive, by means of buckets buried beneath the level of their burrows, which run just below the surface. It appears that they travel along their tunnels regularly and this implies the acquisition of a similar method of harvesting insects to that found in the true moles, Talpidae, and golden moles, Chrysochloridae.

Kenya Mole-shrews (Myosorex (Surdisorex))

Family	Soricidae
Order	Insectivora

Measurements
head and body
96—108 mm *M. norae*
89—100 mm *M. polulus*
tail
24—34 mm (both species)

The convergence with these subterranean insectivores is striking and the disappearance of the eyes and ears into the furry cone of the head and body betrays an important change. Shrews are surface dwellers "hunting" food with all their senses; moles and presumably mole-shrews too, are subterranean animals "gathering" food in the safety of their tunnels or along covered runways under grass, where eyes tend to lose their usefulness and the sense of touch becomes paramount.

The peculiarity of *Surdisorex* raises a number of questions. Firstly, it is the only endemic mammal on Mt Kenya (there is also only one endemic bird on the mountain), the implication being that isolation is not very ancient. Secondly, the related *Myosorex blarina* is found on the Ruwenzori Mountains and *Myosorex zinki* on Mt Kilimanjaro at equivalent altitudes and in very similar habitats. Although montane *Myosorex* species exhibit a distinct gradient in their adaptation towards fossorial habits, with the Kivu-Bufumbira populations the least developed and *M. zinki* the most advanced, it is evident that this evolutionary specialization does not follow a simple linear progression. It is not possible to postulate an uninterrupted ortho-selective trend either in a temporal or in a spatial dimension.

In the mountains of the Central Refuge *M. blarina* co-exists with *Chlorotalpa stuhlmanni* and it is possible that the abundant presence of the latter has inhibited *M. blarina* from becoming truly fossorial. However, this is not true for Kilimanjaro; furthermore this mountain is very much closer to Mt Kenya than Ruwenzori, which makes the extreme specialization of *Myosorex* on Mt Kenya and the Aberdare Mountains more puzzling.

Mayr (1954) observed that peripheral and isolated populations may very occasionally find a vacant ecological niche and become genetically and ecologically transformed. It is possible that such a transformation occurred when an original population of *Myosorex blarina* type became isolated on the Aberdares and Mt Kenya. Meanwhile the other populations have retained a more conservative form, apparently lagging behind in the evolutionary race towards a "mole" niche. A tantalizing field of enquiry.

The Cherengani Mountains and Mt Elgon may also harbour forms of *Myosorex*, and trapping in these localities may turn up new information on this problem. However, it is perhaps from a detailed comparison of the ecological needs and habits of the known montane *Myosorex* species and *M. (Surdisorex)* species that a better understanding may emerge.

These mole-shrews are both nocturnal and diurnal. Both species are quite numerous within their habitat, for instance, the Smithsonian expedition of 1909 collected 34 *M. polulus* on the west side of Mt Kenya in 16 days, and subsequent collectors have not found *M. polulus* difficult to catch. This series revealed that moulting was in progress during September and October and that two of the females were pregnant, likewise a pregnant female of *M. norae* was found on the Aberdare Mountains in October. One or two young are recorded.

S. megalura.

| **Long-tailed
Forest Shrews**
(Sylvisorex) | **Family**
Order | Soricidae
Insectivora | **Measurements**
head and body
48—70 mm *S. granti*
tail 47—64 mm
Hindfoot 10 mm
weight 3—5 g

head and body
52—70 mm *S. megalura*
tail 79—90 mm
hindfoot 14 mm
weight 4—6 g

head and body
72—83 mm *S. lunaris*
tail 50—56 mm
hindfoot 15 mm
weight 12 g |

Long-tailed Forest Shrews (Sylvisorex)

Sylvisorex has a long tail without long hairs, P. 3 present, P. 3 absent, soft, short fur.

Sylvisorex lunaris is a large black shrew with a grey belly, found on Ruwenzori and in Kigezi from 1,800 to 4,000 m.

Sylvisorex granti is a small black or blackish-brown shrew with the tail as long as the body, found on Uganda and Kenya mountains and on Mt Kilimanjaro between 1,800 and 3,500 m.

Sylvisorex suncoides is a well differentiated species known from the western side of Ruwenzori.

Sylvisorex megalura is a small brown shrew with pale belly. The tail is longer than the body. It is found at medium-high altitudes and in moist localities, from north of the Limpopo to as far west as the Niger.

Species

Sylvisorex lunaris
Sylvisorex granti
Sylvisorex suncoides
Sylvisorex megalura

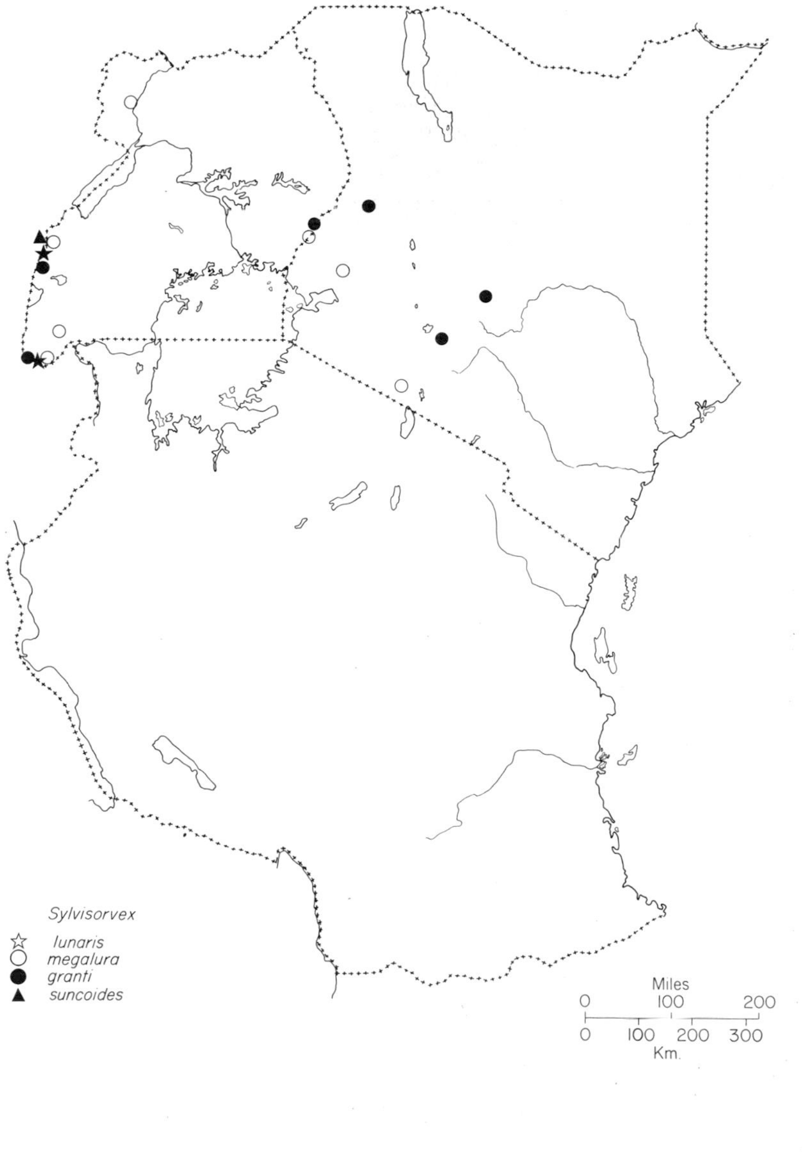

Sylvisorvex
☆ lunaris
○ megalura
● granti
▲ suncoides

Miles
0   100   200

0   100 200 300
Km.

S. granti.

S. lunaris.

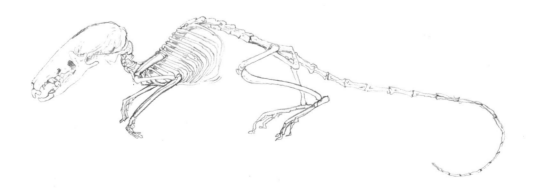

S. granti, skeleton.

Sylvisorex is sometimes regarded as a subgenus of *Suncus*. It has lost a lower premolar that is retained by the more primitive *Myosorex*, but retains three upper premolars and, therefore, probably resembles the ancestral stock of *Crocidura* and *Suncus*. Considering the similarity of their teeth this genus also shares a common ancestor with *Scutisorex*.

Members of this genus are sometimes arboreal, but the name "forest shrew" does not often apply, as all species can be trapped in grassy localities.

Sylvisorex lunaris is restricted to the Ruwenzori Mountains and the Kivu-Bufumbira area, where it is sympatric with both *S. granti* and *S. megalura*. It is found over a considerable range of altitudes and lives both in forest and grasslands.

Heim de Balsac regards *Sylvisorex granti* as a less evolved species, an interesting observation when one considers its discontinuous distribution from the Cameroons to Kenya. *S. granti* lives in montane forest, high altitude grassland and Afro-alpine habitats.

S. suncoides can be expected to turn up on the eastern side of Ruwenzori, although, at present, it is only recorded from the western side.

S. megalura is the most widely distributed and ecologically the most adaptable species, having been found in montane habitats, lowland forest, cultivation, grassland and riverine vegetation in semi-arid northern Kenya.

Virtually nothing is known of the biology of *Sylvisorex*.

S. granti.

Hero Shrew
(Scutisorex somereni)

Family Soricidae
Order Insectivora
Local names
Omusonso (Luganda)

Measurements
head and body
140 (105—150) mm
tail
87 (70—109) mm
hindfoot
20—23 mm
weight
70—113 g

Hero Shrew (Scutisorex somereni)

Races

Scutisorex somereni somereni   South Uganda
Scutisorex somereni congicus   Western Uganda

The hero shrew is a large grey shrew with a thick woolly fur, liberally interspersed with projecting guard hairs. Alive, it can be readily distinguished from other shrews by its "trotting" rather than "crawling" gait, a locomotion that is similar to that of the swamp rats *Malacomys*, *Deomys* and *Colomys*, with which it shares its swamp-forest environment. It is restricted to Uganda and eastern Zaire (Congo) within the forest belt.

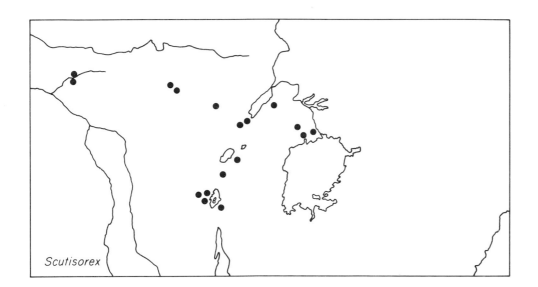

The English name of this shrew has its origin with the Mangbetu people of the Congo for whom this animal was a talisman bestowing invincibility in war, its charred remains serving as a medicine or passport to heroism. Although this species was first discovered by Van Someren near Kampala (Thomas, 1910) the extraordinary structure of the backbone was overlooked as it was the custom then only to take the skull and skin for scientific study. It was the Mangbetu, in fact, that brought the most extraordinary qualities of this shrew to the attention of the naturalists Herbert Lang and James Chapin—

"The Mangbetu gave it a name meaning 'hero shrew'. Those engaging in warfare or setting out upon an equally dangerous enterprise such as hunting elephants are anxious to carry along even a fraction of the ashes of this shrew. Though only worn somewhere about their body, they believe that neither spears nor arrows, nor any kind of an attack can seriously injure them, much less bear them down. One can easily imagine by the removal of the inhibitory influence of fear their courage, cunning and cleverness are set free for the best possible achievement. Whenever they have a chance they take great delight in showing to the easily fascinated crowd its extraordinary resistance to weight and pressure. After the usual hubbub of various invocations, a full-grown man weighing some 160 pounds steps barefooted upon the shrew. Steadily trying to balance himself upon one leg, he continues to vociferate several minutes. The poor creature seems certainly to be doomed. But as soon as his tormentors jump off, the shrew after a few shivering movements tries to escape, none the worse for this mad experience and apparently in no need of the wild applause and exhortations of the throng". (Allen, 1917.)

The backbone of this species is without parallel anywhere in the animal kingdom and there is no other shrew giving us any indication of a similar structure. Its evolution is therefore a tantalizing problem.

I believe that the explanation for the peculiar backbone of *Scutisorex* is to be found in the need for a swamp-dwelling animal to get its body well clear of the ground.

As Simpson (1940) remarks, the specialization of the vertebral column was probably a rapid development as the animal is obviously derived from a fairly typical crocidurine shrew. If we accept that *Scutisorex* has originated from a typical shrew, the backbone problem can be simplified by first considering the postural differences between a typical shrew and *Scutisorex*. The most obvious structural difference is that the typical shrew has a greatly over-extended body and head relative to the legs. This characteristic is such a familiar feature of shrews that it tends to go unremarked. Most naturalists are acquainted with its low-slung "front-heavy" form running in erratic spurts on widely spaced rather bent legs. Although well-suited to bulldozing through litter this build of body imposes some anatomical limitations on shrews, one of which is their inability to sustain a simple quadrupedal standing posture. As can be seen in the briefest observation of shrews (reference to the life sketches, p. 102) shrews drop their bodies when not moving, resting their chests and hindquarters on the ground.

When we consider the form of *Scutisorex* we find that the over-extension which places such limitations on other shrews has been counteracted by the simple contrivance of doubling up the forequarters into an "S" bend. In the normal standing posture the thoracic vertebrae are at ninety degrees to the ground—so that the rib cage forms a cup. The neck is also very upright so that there is a very sharp angle between the cervical and thoracic vertebrae. Incidentally, the sharpness of this angle, both at the neck and at the thoracic kyphosis is compensated for mainly by the intervertebral discs and not so much by the structure of the bones.

If a common shrew with the usual five or six lumbar vertebrae were up-ended in this way, it would be obliged either to walk on its hands or assume a most extraordinary posture. To escape this predicament it would seem that the ancestors of *Scutisorex* coordinated the changing orientation of their chest with an extension of the lumbar region. This was achieved by the simple expedient of sliding the pelvis down the vertebral column. The migration of the pelvis backwards is the only explanation for *Scutisorex* possessing twice as many lumbar vertebrae as other shrews. Variations in the position of the pelvis *vis-à-vis* the spinal column are in fact well known in many mammals.

Diagram of *Scutisorex* showing positions of pelvis and hindlegs if lumbar region were not elongated.

86

In *Homo sapiens* for example, the first sacral vertebra can be either the 24th, 25th or 26th in the spinal column. Extension of the lumbar region brings its own problems. As a slender bridge between two relatively large differentiated masses it is susceptible to considerable strain, particularly in the lateral plane. A simple multiplication of vertebrae in a hypothetical ancestor of *Scutisorex* would have introduced problems of instability and structural weakness. The solution seems to have been a general broadening and thickening of the vertebrae together with the formation of extraordinarily complex spicules along the outer margins of the lumbar vertebrae, these preclude any rotation and place some limit on movement in the lateral plane and would seem expressly designed to encourage stability and give structural strength to the spine. Discussing what elements of the vertebrae were involved in this spicular transformation, Schulte (1917) concluded that the

> "ankylosed costal element in the lumbar region becomes transformed into spicules. This deduction is strengthened by the position of the nervous foramina and by the contours of the rostral and caudal surfaces and the bones which are curiously like sacral vertebrae. Not only has the costo-transverse space been filled up as usual, but the costal process has expanded laterad and ventrad quite apart from this secondary equipment with spicules. The ventral tips of these modified ribs form the edges of the gutter which characterizes this portion of the spine. These lumbar vertebrae have therefore undergone a change in general form such as is usual only in the sacral segment."

Correlated with the increased breadth, bulk and strength of this bony arch is a decline in the bulk of the lumbar muscles compared with those of other shrews, or indeed, of other mammals.

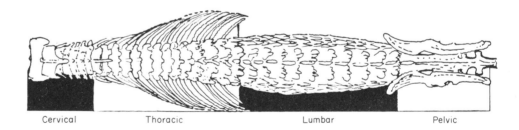

| Cervical | Thoracic | Lumbar | Pelvic |

From this the impression might be gained that the back is rather rigid. In fact, living *Scutisorex* flex the back a great deal and my captive assumed extraordinary postures while it rubbed its rump and sides on objects. The greatest mobility is of course dorso-ventral and a strong curvature is possible in this plane. The lumbar arch in the normal standing posture assumes an unexpectedly perfect geometric pattern. This can be illustrated by superimposing a quarter circle on a cross section drawing of the animal in this position. It will be seen that the lumbar arch follows the arc, while the centre of the circle falls at the point on the ground between the forelegs where most of the animal's weight is taken.

The compulsive rubbing of the body seems to have the object of scenting what is rubbed. In most shrews this is achieved by squirms that scarcely differ

from the normal movements of a shrew progressing through leaf litter. *Scuti-sorex* has very pungent scent glands and strong yellow stains are visible on the sides and chest of some individuals, where the secretion has discoloured the grey woolly fur. It is possible that the peculiar texture of the fur and the long guard hairs are related to scent dispensing. Apparent marking behaviour in my captive involved rubbing the hindquarters on objects that were both on the ground and also raised above it, rubbing the chest on the ground and arching and twisting the back against objects at any level it could reach. This rhythmic rubbing was most conspicuous just before the animal settled down to rest, the body assuming various lopsided positions as it pushed here and there using its back to tamp a nook for itself. Finally the animal would lie down curled on its side. When this animal was placed in its vivarium for the first

time after being trapped, it was quite uninterested in food until it has thoroughly "marked" its chosen resting corner, after which it sallied forth and started to eat voraciously.

Other more forceful movements of the shoulders and head were also employed to push into crevices and vegetable debris, but these were probably connected with feeding behaviour. My animal would trot slowly forward with the head held distinctly lower than the body, sweeping its muzzle from side to side with twitching whiskers. When it found live prey it champed it along its length rather tentatively at first but did not "savage" it. I did not see it use its forelegs to hold prey. Sometimes pieces of wood or vegetable matter were bitten and shredded and the animal would occasionally rub itself on the piece of remaining bark or wood before leaving. In general its search for food and feeding behaviour were more reminiscent of the rat *Malacomys* than a shrew. When this animal drank, its nose was lifted clear of the surface. Lang in Allen (1917) described one drinking dew:

> "One let loose in the early morning would busily lick off dewdrops from the margin of the leaves. But whenever it came to small tufts of grass it would press down the blades with the forelimbs starting near the base until it could easily reach the glittering drops that had gathered at the tips . . . the deeply grooved nose is moved in every direction and continually quivering, it explores actively the objects in view. The undersides of leaves and even stones are thus inspected. Fair-sized pebbles, pieces of bark and decayed wood are turned over or pulled away with the assistance of the incisors. When looking for insects or worms they squat resting the sole of the hindfoot on the ground."

Lang recorded insects, caterpillars, earthworms and the remains of frogs in the stomachs of trapped *Scutisorex*. My captive ate grasshoppers and birds' meat without hesitation but rejected beef; slugs were eaten with noticeably more champing than other prey and with the head held down very low.

When frightened it ran to a dark corner to hide, but it was less easily upset than most shrews and an approaching hand was never bitten nor did it give rise to snaps and shrieks. Indeed, the only noise I heard from it was a modest twitter when it fell into the water. The animal was able to swim but did not give the impression that it actively liked water. It spent much time grooming, using the forepaws to clean its face, teeth and tongue to comb its body fur; it scratched with the hindfeet. It was also an active but cautious climber, keeping very flat and close to the branches and not attempting to climb on twigs.

The mild disposition of these shrews could be indicative of more social instincts than those that seem to govern the behaviour of other shrews and Lang observed that *Scutisorex* never killed other species of shrews put in with it as *Crocidura nyansae* (*flavescens*) invariably did.

The largest series of *Scutisorex* caught to date in Uganda are the nine animals trapped by Delany in his study of rodent ecology in Mayanja Forest (1971). Several of these were caught in roughly the same spot, suggesting that the traps might have been near a scented pathway.

The scent of the lateral glands seems to act as an intra-specific repellent in some species of shrew (see Crowcroft, 1957; Pearson, 1946). It would be interesting to know what role it plays in *Scutisorex* since the activity of the glands seems to vary, a male caught in Bwamba in March and another in June had conspicuous yellow stains, immense fat deposits and large active testes (see drawing). This fatty condition can be contrasted with that of the lean subadult, which had a grey unstained coat.

Two lactating females have been collected in May and three sexually inactive animals were also caught in the same month.

Musk Shrews (Suncus)

Suncus has a tail with long hairs, upper P. 3 present, lower P. 3 absent, soft, short fur.

Species

Suncus murinus
Suncus lixus
Suncus varilla

    Suncus murinus (*S. leucura*) is a greyish-brown shrew. It was introduced to Zanzibar and Pemba Islands from the East Indies. *Suncus lixus* is a grey shrew with a brownish tinge, a dark grey belly and pale feet, found in Kenya and Tanzania. *Suncus varilla* is a very small shrew, with a finely variegated grey colouring. It is a southeastern African species, only known in East Africa from Ufipa in Tanzania.

    Suncus resembles *Crocidura* closely in external features and is very rarely trapped although widely distributed. It is possible that the more highly evolved *Crocidura* has largely replaced it over much of its African range.

    The genus has been studied in other parts of the world and *S. murinus* has been introduced to Guam and Madagascar as well as Zanzibar.

Musk Shrews (Suncus)

Family Soricidae
Order Insectivora
Local names
Kirukanjia (Swahili)

Measurements
head and body
110—115 mm   *S. murinus*
tail
70—84 mm
head and body
70—121 mm   *S. lixus*
tail
48—74 mm
head and body
52 mm   *S. varilla*
tail
32 mm
hindfoot
9 mm

Rodent Shrew
(Paracrocidura schoutedeni)

This close-furred black shrew has a tail about half the length of the body and a relatively short, robust snout. It has some resemblance to *Crocidura*, but is most remarkable for the development of a laterally flattened cutting edge to the incisors, which are very similar to those of a rodent.

This species was first described by Heim de Balsac (1956b) from Luboudaie in eastern Zaire (Congo). It has since been found in Kivu, on the Virunga Volcanoes and on Ruwenzori, so that it is probably only a matter of time before it is recorded from the montane forests of western Uganda.

Heim de Balsac has described this as a distinct radiation deriving from an ancestral stock common to both *Crocidura* and *Sylvisorex* but evolving in a different direction.

Records and observations would be most interesting in view of the absence of any information on the biology of this species.

 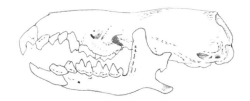 

Rodent Shrew
(Paracrocidura
schoutedeni)

Family	Soricidae	
Order	Insectivora	

Measurements
head and body
65 mm
tail
34·5 mm
hindfoot
12 mm

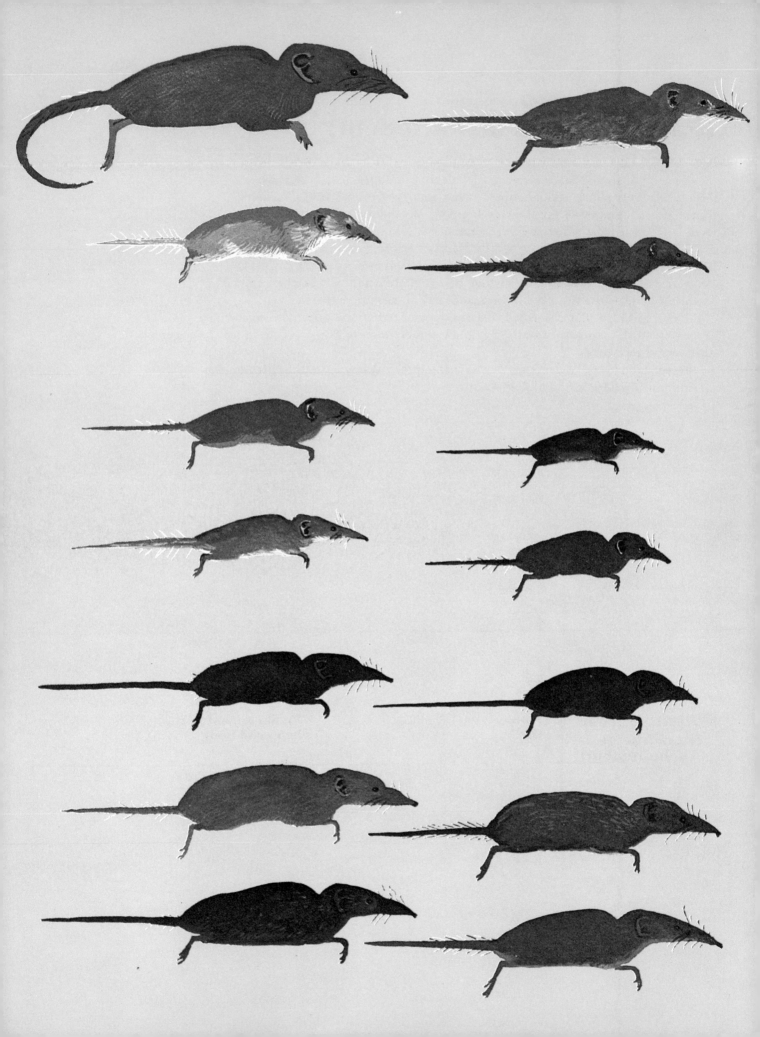

White-toothed Shrews, Crocidures (Crocidura)

One hundred and twenty-five forms of African *Crocidura* were described by Dollman (1915—1916), Cabrera (1925) listed 186 species in the genus the greater part of which were African and Allen (1939) recognized 162 African forms, of which 50 were described from East Africa and many more are actually recorded from this area. This proliferation of names is intimidating. Identification requires an examination of tooth and skull structures that are only clearly visible under a lens or microscope, so that the group is widely ignored because of the taxonomic and practical difficulties.

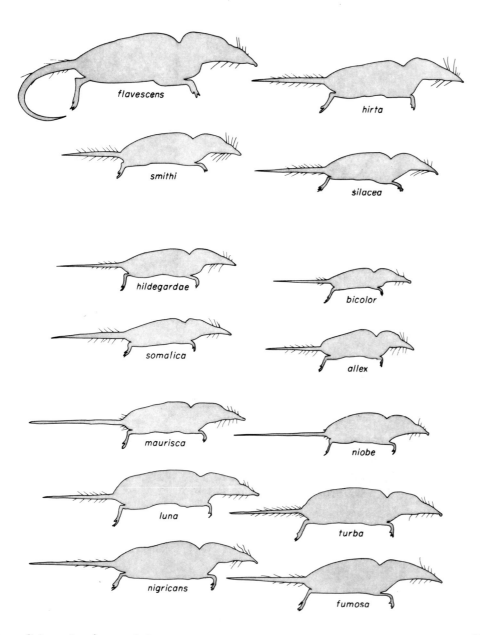

Family   Soricidae
Order   Insectivora
Local names
Kirukanjia (Kiswahili), Omusonso (Luganda), Omususu (Lunyoro), Oyo obile, Oyo ture (Lwo), Asurianya (Teso), Nyunga (Kihehe), Msusukwe (Kikerewe), Akasene (Kinyakyusa), Keke (Kisambaa), Munyunhe (Kikami), Cheptyoy (Sebei), Ntzeki (Kikinga), Munyongi (Kitaita), Tungu (Kipokomo), Chanyunga (Kimwera), Namyunga (Kikindani), Ntawara (Makonde), Keke (Sambara), Etutwi (Karamojong), Guchuru (Sebei), Namagaba (Lugisu), Lunihi (Ragoli), Iuhui (Lutereki)

Colour plate ⅔ natural size.

Notes

Very dark West Uganda montane species, tail without vibrissae

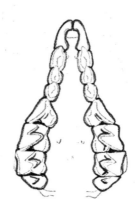
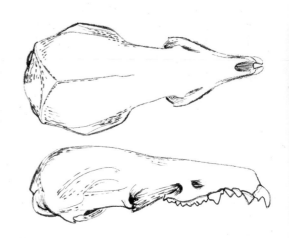

C. maurisca

Dark and small, tail without vibrissae

Uganda
Kilimanjaro
Jombenis
Ruwenzori

Closely related group, showing sympatry in some areas. Widely distributed in the moister parts of East Africa

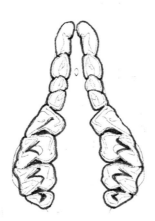
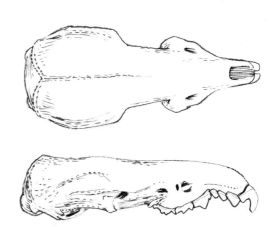

C. turba

CROCIDURA

Measurements	Well defined and/or commonly recognized species	Possible synonyms and/or subspecies	Less well known forms
H. & B. 63—95 mm T. 55—73 mm Hf. 13 mm Wt. 12—16 g	C. maurisca		
H. & B. 80—96 mm T. 68—85 mm Skull 19—21 mm Wt. 11—21 g		oritis neavei?	C. littoralis C. monax C. ultima C. niobe (Thomas)
H. & B. 80—107 mm T. 46—64 mm Hf. 13—16 mm	C. fumosa	raineyi, montis, schistacea, beta?	
	C. luna	electa	
	C. turba		C. nilotica (Heller)
		A giant turba?	C. zimmeri (Osgood)
		neavei?	C. nigricans

Notes

Widely distributed
in moister parts
of East Africa

Long tail and
very large canine,
West Nile district

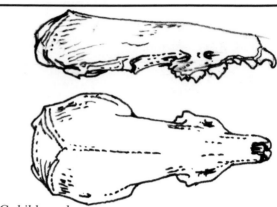

C. hildegardeae

Kenya and
North Tanzania

North Kenya and
North Uganda

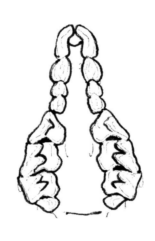
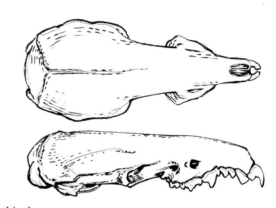

C. bicolor

Endemic to
Horn of Africa

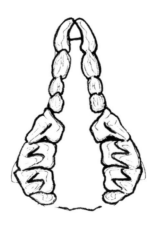
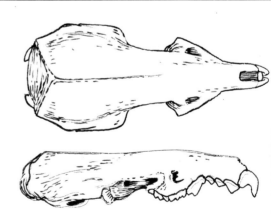

C. somalica

MIXED FEATURES

Measurements	Well defined and/or commonly recognized species	Possible synonyms and/or subspecies	Less well known forms
H. & B. 55—84 mm T. 38—67 mm Wt. 8—11 g	C. hildegardeae	gracilipes, lutreola maanjae, procera macowi, ibeana	
			C. roosevelti
H. & B. 50—60 mm T. 35—45 mm Wt. 3—10 g		hendersoni, elgonius zanzibarica, cunninghami	C. bicolor C. planiceps
H. & B. 57—66 mm T. 34—45 mm Wt. 3—11 g		religiosa alpina, zinki	C. nana C. allex
H. & B. 46—55 mm T. 31—36 mm Hf. 9—10 mm			C. bottegi
		rudolphi	C. nanilla C. pasha
H. & B. 55—80 mm T. 40—60 mm	C. somalica		

Notes

**Largest
Crocidura
species**

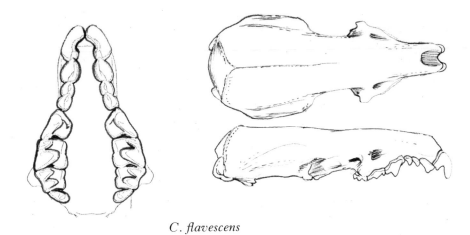

C. flavescens

**Group with
very short
foot and
laminated M3**

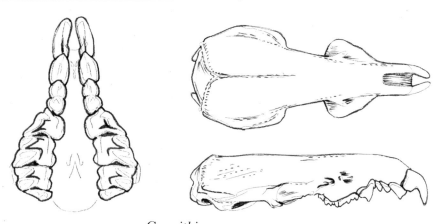

C. smithi

**Very local
form
(Nyanza District)**

CROCIDURA

Measurements	Well defined and/or commonly recognized species	Possible synonyms and/or subspecies	Less well known forms
H. & B. 75—140 mm T. 57—90 mm Skull 25—34 mm Wt. 15—40 g	*C. flavescens*	*nyansae occidentalis,* *martiensseni,* *kijabae,* *dapnia,* *doriana*	
H. & B. 62—110 mm T. 35—65 mm Hf. 125—12 mm Wt. 10—20 g	*C. hirta*	*sacralis* *velutina* *xantippe* *simuolus*	*C. canescens* (Peters) *C. annulata* (Peters) *C. hindei* (Thomas)
	C. silacea	*holobrunneus* closely related	
H. & B. 60—86 mm T. 43—57 mm			*C. jacksoni* (Thomas) *C. denti* (Thomas)
H. & B. 71—88 mm T. 34—42 mm Hf. 11—12 mm			*C. smithi* (Thomas) *C. boydi* (Dollman) *C. katherina* (Kershaw) *C. voi* (Osgood) *C. butleri* (Thomas) *C. macartheri* (St Ledger)
H. & B. 99—129 mm T. 53—70 mm Hf. 15·5—18 mm			*C. sururae* (Heller)
H. & B. 100—129 mm T. 66—80 mm Hf. 16—17 mm			*C. zaphiri* (Dollman)

Various efforts have been made at simplifying and grouping *Crocidura*. Swynnerton (1959) used simple metric criteria, tail to body proportions and colouring, and came up with 15 groups for the African continent. This scheme, however, took no account of the vitally important skull and tooth structures and is consequently of little value. Heim de Balsac and Lamotte (1956—1957) have been the first authors to define characteristics that allow the various forms to be grouped, irrespective of size, into phylogenetic categories. The criteria used are based on tooth and skull structure.

According to these criteria there are:

    1. advanced species;

    2. more primitive species;

    3. species with a mixture of advanced and primitive characters.

The more primitive species have sparsely-haired tails, somewhat bulbous braincases, a large third premolar and generally more complex cuspidation of the cheek teeth.

The advanced forms exhibit a general simplification of the teeth. The third molar decreases in size in the upper jaw, while the talonid on the lower third molar becomes simplified in the more highly evolved groups. The third upper unicuspid also decreases in size, the hypocone on the upper P 4 disappears and so do the cusps on the lower incisor and the lower P 4. There is also a tendency towards larger size and towards the development of long vibrissae on the tail. The skull becomes flatter.

In the key the East African *Crocidura* are grouped within these three broad phylogenetic categories, a task in which I am indebted to Professor Heim de Balsac for the benefit of his advice and numerous papers which are listed in the bibliography.

The key should enable the field-worker to place a shrew within its major grouping. However, the original description may have to be consulted for some of the little known forms.

In spite of very close resemblances within the groupings listed, detailed collecting has revealed that many of these closely related shrews are sympatric in some localities. There is little doubt that chromosomal studies may eventually assist in sorting out these difficult problems, but there is a most important role for the field naturalist to help define the ecological and geographic ranges of many of the forms of *Crocidura*.

Any worker seeking detailed information on *Crocidura* is referred to the many papers by Professor Heim de Balsac and Dr J. Meester that are listed in the bibliography.

C. maurisca.

"Primitive" Crocidura

Typically small and dark and lacking vibrissae on the tail, the more primitive species are distinguished principally by the large well-formed back molars, in both upper and lower jaw, and a highly domed braincase.

Numerous forms are known, some of which are isolated on mountain blocks.

Crocidura monax	On Mt Kilimanjaro.
Crocidura ultima	On the Jombeni Mountains.
Crocidura niobe	On the Ruwenzori Mountains.
Crocidura maurisca	In Bufumbira.

The upper molars of *C. maurisca* are peculiarly complex, and it may represent a particularly ancient relic population, with its closest relatives in the South African *C. pilosa* and *C. sylvia*. This species is common in the Echuya Swamp in Kigezi where it is readily caught with vegetable baits, a curious proclivity compared to most shrews which are generally insectivorous. *C. maurisca* shares this habitat with other species that have discontinuous or relic distributions, *Sylvisorex granti*, *Sylvisorex lunaris* and *Myosorex*. The last genus is found in relic pockets on East African mountains and has a more extensive distribution in South Africa. (Echuya is also the type locality for the relic *Delanymys brooksi*.) This species has been confused with *Crocidura littoralis* from Butiaba but the latter is a slightly larger form with more simplified molars and brighter colouring.

Crocidura fumosa is a dark grey-brown species largely confined to highland areas of eastern Africa. The nominate race is from Mt Kenya and the Aberdare Mountains and another race, *C. f. raineyi*, has been collected from 1,500—2,100 m on Mt Gargues.

Crocidura luna is paler, grey-brown dorsally with a grey belly and is distinguished by a shallow pit beside the last cusp (hypoconid) of the third lower molar. It is a central African species, ranging through upland areas of Rhodesia, Malawi, Zambia, parts of the Congo (Zaire) and Tanzania, where it is probably sympatric with *C. fumosa* in some localities. Meester (1963) found the size of this species subject to marked geographic variation. Four

Zambian breeding records are for March and April, and I have collected a pregnant female from southwestern Tanzania in mid-December.

A typical habitat for all the "primitive" *Crocidura* is tussock grass in swampy areas, and *C. luna* has been commonly caught in such places. Vesey-FitzGerald (1962) also captured this species from river and lakeside, among boulders, in grass and sedge and on the forest floor, in wood piles, in holes, in woodland, in termitaries and he also found it common in the runs of *Otomys*.

Crocidura turba is an equatorial species found in the high rainfall belt of eastern Africa. In the Albert Park it has been found to co-exist with *Crocidura nilotica*, a black species formerly regarded as a race of *C. turba*. *Crocidura zimmeri* is likewise very similar to *turba* but for its larger size.

C. luna.

Crocidura nigricans is another eastern Africa highland species, known from Kenya, Tanzania and Zambia. It occurs up to 2,400 m on Mt Kenya. Its habitat preferences resemble those of *C. luna*. Vesey-FitzGerald found *C. nigricans* frequent among debris on the floor of evergreen forest at high altitudes but also on flood plains, where there was a tendency for this species to be driven into narrow drainage lines and sump areas during the dry season, because of grass burning. In such sites he found it associated with *C. hirta*, *C. bicolor* and *Otomys*.

The overall distribution pattern of the more primitive forms of *Crocidura* is very interesting and some of them have been described as "moisture restricted". As was remarked in Volume I, mountains provide moist and special habitats while at the same time acting as refuges for older, formerly widespread, species and the two factors are often inseparable. In addition to this, mountains encourage the isolation of populations. With the superimposition of climatic changes over long periods of time this can lead to the evolution and proliferation of numerous forms. Indeed this seems to be the case particularly among the older, more primitive *Crocidura*. In addition to the isolated *C. monax* on Mt Kilimanjaro and *C. niobe* on the Ruwenzori Mountains and *C. maurisca* in Bufumbira there are several related and extralimital forms, the most noteworthy being *Crocidura lanosa* and *Crocidura kivuana*, which are restricted to a belt between 1,800 and 2,200 m in the Kivu area. The limited ranges of these small forms contrasts with the extensive ranges of the larger, advanced species.

"Mixed" Crocidura

Crocidura hildegardeae is generally brownish on the back, with a grey-tinged belly but the somewhat variable colouring seems to be responsive to the relative humidity of the habitat. An intermediate condition between the more primitive and advanced forms is suggested by the rather sparse vibrissae on the anterior part of the tail. The third incisor and the canine are of similar size and the latter is spaced away from the premolar.

The species is widely distributed in eastern Africa, particularly favouring the moister areas, but it extends into Somalia and Ethiopia in the north-eastern frontiers of its range.

C. hildegardae.

This species is common in secondary growth, cultivation and forest. Hollister (1918) recorded five pregnant females from central Kenya with two to four foetuses. I have caught a lactating female in early August, climbing a tree near Kampala.

Crocidura roosvelti is only known from West Nile and is distinguished from *C. hildegardae* by having a longer tail and a greatly enlarged canine.

Crocidura bicolor is one of half a dozen very closely related shrews. These species are easily confused and their status and distribution are still in need of detailed definition. Several forms that have been treated as subspecies in the past have since been found to be sympatric. Heim de Balsac (1966c, 1971) has been the first to bring some order to the shrews named the "bicolor group" or the "religiosa group" in the past, both rather artificial groupings that have been dismantled by Heim de Balsac.

C. bicolor is a widely distributed form, ranging from the Sudan and Somalia to the Transvaal throughout the eastern half of the continent. Its short broad cranium is thought to be relatively primitive, while the well-haired tail, small canine and simplified last molar of the lower jaw are instead advanced features.

C. planiceps.

Crocidura planiceps from Uganda and *Crocidura nana* have both been treated as subspecies of *C. bicolor*, but both are now thought to be broadly sympatric with the latter. *C. nana* has been identified positively only from Somalia, Rhodesia and Egypt (where it occurs as a mummy in the tombs at Thebes). Heim de Balsac believes that there has been much confusion between species from the intervening area and that *C. nana* is found over a great area between the Nile delta and Rhodesia.

Crocidura nanilla, a savanna species ranging from the Kenya Somalia border as far south as Katanga, has been found to co-exist with *C. bicolor* and *Crocidura pasha* in the Garamba Park.

Crocidura bottegi is a northern savanna form ranging from West Africa to Lake Rudolph.

Crocidura allex is a highland species found in Kenya and on Mt Kilimanjaro. It is principally remarkable for the large talonid cusp on the last lower molar, which suggests an archaic condition. An ecologically isolated alpine race occurs on Mt Kilimanjaro and another on Mt Kenya.

Crocidura somalica is a species endemic to the Horn of Africa. It is a pale grey species with very short feet and a thick tail with vibrissae along its entire length. The skull is very flattened. Outside Somalia it is only likely to occur in northeastern Kenya.

C. flavescens.

"Advanced" Crocidura

Crocidura flavescens is a large shrew of which various races have been described. The thick tail always carries numerous long vibrissae. Colouring is subject to some variation both regionally and individually but the back is usually brown, its tone apparently responding to the relative humidity of the environment. Size also varies but in all areas where it is sympatric with *C. hirta* it is always decidedly larger (Meester, 1963). This author found that the skulls of males were significantly larger than those of females.

The species is widely distributed throughout subsaharan Africa, but it is absent from the driest and the coldest high-altitude habitats in East Africa.

The diet is very catholic. According to Marlow (1955) it includes mammals, birds, molluscs and millipedes (presumably some of these can only be scavenged accidentally in nature).

Young animals have been collected in most months of the year and, while breeding seasons might occur in local populations, no clearly defined season is apparent in East Africa as a whole. Moulting animals have also been collected throughout the year.

Crocidura hirta is a closely related species but is smaller than *C. flavescens* in all areas where their ranges overlap. *C. hirta* is commonly pale brown or fawn on the back with a light grey or creamy belly. This is a common species of the southern savannas and eastern half of Africa but it ranges through a wide range of wooded grassland and dry bush habitats. Vesey-FitzGerald (1962) has noted it particularly in piles of thatching grass in villages, in rat and gerbil holes, in termitaries and under the dense herb mat of flood plains. He pointed out that riverine vegetation, termitaries and holes probably serve as dry season concentration areas where they may associate with *Otomys*, *Pelomys*, *Mastomys*, *Arvicanthis* and *Tatera* as well as with other *Crocidura* species. Meester (1963) found a birth season in South Africa from December to April, with one to five young born at a time. He also found one female giving birth to two litters in a single breeding season; this is possible because of the eighteen-day gestation period, which is also the lactation period in captives.

It is possible that similar but distinct species may have tended to be lumped together under the name of *C. hirta*, notably *C. canescens* and *C. annulata*.

Crocidura silacea is a grizzled buff-brown shrew with a light grey belly. It is usually distinguished by an entoconid cusp on the last lower molar. This is a southern African species, entering East Africa in southern Tanzania.

Crocidura hindei belongs to the C. hirta group (and has been identified with a Pleistocene fossil). It is found in the vicinity of Nairobi.

At least six advanced species can be grouped together on the basis of their very short feet and the lamination of the last upper molar.

Crocidura smithi is a dry country species of northeastern Africa, coloured very pale blue-grey with a pure white underside and very pale feet. The canine is wedged within the incisor and premolar (see pictorial key).

Crocidura boydi probably embraces *C. lutrella* (Heller) and *C. parvipes* (Osgood). These shrews generally range throughout equatorial East Africa in relatively moist habitats.

Crocidura katherina is known from the savanna areas of the Congo (Zaire) and Zambia. It probably occurs in the Semliki valley as it has been recorded across the river. It has recently been recorded near Masindi, Uganda.

C. katherina

Crocidura voi is known in East Africa from the type locality, Voi; however, Heim de Balsac (1966c) believes that shrews belonging to this species occur in savannas of the Ivory Coast and Somalia, and he suggests that this may in fact be a northern savanna species.

Crocidura butleri is a pale yellowish species from the Sudan with a white belly and a very short, thick tail. This species is known from the southern Sudan through Kenya to Somalia.

Crocidura macarthuri is another shrew belonging to the short-footed group, known from near the mouth of the Tana River.

Crocidura maurisca, only known from the West Nile area, has been reported as *C. flavescens*. Heim de Balsac considers this to be a distinct species.

Crocidura zaphiri is another local form resembling *C. flavescens*, known from Kisumu.

C. katherina.

Mimetillus moloneyi.
Vespertilionidae.

Nycteris thebaica.
Nycteridae.

Cardioderma cor.
Megadermatidae.

Eidolon helvum.
Pteropodidae.

Taphozous peli.
Emballonuridae.

Tadarida midas.
Molossidae.

Bats

CHIROPTERA

Facts can never dispel the element of mystery that surrounds bats. Their lives are lived in a dimension so greatly removed from our own that an appreciation of even the simplest elements of their biology makes demands upon the imagination.

Those who have watched bats as they hunt in a clear sky may have experienced a slight sense of disorientation; the aerial acrobats cannot be related to the ground, that solid surface to which we unconsciously refer every experience. While they are in flight, the structure of their form and the movements of their limbs embody a spatial freedom that is almost inconceivable to us.

Throughout my study of these animals I have felt humbled and awed by what Wordsworth called "blank misgivings of a creature moving about in worlds not realized".

After rodents, bats with about 175 genera and 800 species are the most numerous mammals on earth. It is curious that so fecund and ubiquitous a form of life should escape the consciousness of so many people. It is hardly possible to walk into a tropical night without being surrounded by bats, even in the cities. Counting bats flying over the disc of the full moon and using a method devised for estimating the numbers of migrating birds, Tomkins (personal communication) has estimated that between 3,500 and 10,000 bats are theoretically "in view" at a time over Makerere Hill, Kampala. The bat also does not seem to have touched that aspect of human imagination that finds expression in art, unless one excepts the vulgar comic-book. One exception is in the art of ancient China, where the bat is represented as a symbol of long life and happiness in beautiful reliefs and paintings.

The scientific name, Chiroptera, is derived from two Greek words meaning hand-wing. This abbreviated name draws attention to that feature of the bats which distinguishes them most from other mammals. The elongated hand and arm has a minimum of eleven joints, giving them very refined control over their wing membrane and consequently great spatial dexterity. In this last aspect they are generally superior to birds, some species being able to negotiate tangled thickets which most birds would be obliged to hop through. However bats' wings and bodies are very varied and the shape of the wing is related to function; long, thin pointed wings belong to fast high altitude flyers, broad rounded wings to the skilled navigators of thickets and forest. The illustrations opposite compare the wing shapes of several bat species.

The most detailed account of the functional anatomy of the bats' wing and shoulder is that of Norberg (1970b) where the origin, insertion and action of every muscle is described and illustrated. The body of bats has also been modified in several respects. The ribcage is stout and rigid with a small keeled sternum from which arises a cartilagineous extension so that the pectoral muscles can be said to actually pull against one another (see Pennycuick, 1972). The pelvis has rotated so that the knees face in the opposite direction

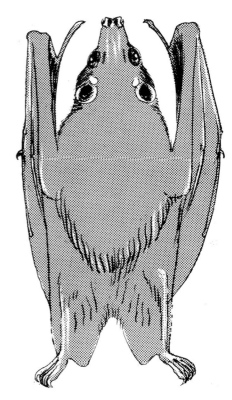

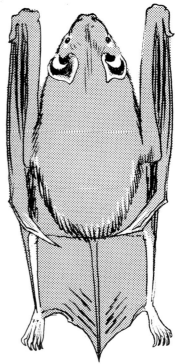

Insect bat Microchiroptera.

Eidolon fruit bat in flight (after Kulzer, 1966).

to that of other mammals. The membranes of the bats' wings are thin and naked, serving both as an aerofoil and to circulate cool blood which might otherwise overheat, for energy burnt in flight must be considerable. The legs are also involved in flight and serve as membrane struts as well as body braces.

Most important advances have been made by bats in the refinement of their "flight guidance systems". There are two methods by which bats gain information about their environment and the line between the two suborders of bats starts with this division. The Megachiroptera, or fruit bats, have increased the sensitivity and acuity of their eyes so that they can see in the faintest of lights. Feeding principally on fruit, this sharpening of night vision is adequate for their way of life. The Microchiroptera instead must detect and catch fast moving prey in the dark and vision is found to be inadequate for this demanding task. The technique of echo-location has been developed to an astonishing degree of precision in the insect-eating bats and the major differences between them and the fruit bats can almost all be traced to the limitations or the advantages brought about by their divergent orientation systems. Echo-location is discussed further in the profile of the Microchiroptera.

The two suborders have much in common and must derive from a common ancestor that developed flight before diverging and specializing. There are no early fossil bats to give an idea of what this ancestor might have looked like, although the earliest known fossil bat, *Icaronycteris index*, of the Lower Eocene shows a few connecting features (see Jepsen, 1970).

Because of the absence of fossils the nature and the time of early bat evolution are completely unknown. All that can be said is that at some time in the later Mesozoic or in the Tertiary a primitive mammal took to the air.

There is no doubt that the radiation of bats is very ancient. That some of the insect bats perfected flight and echo-location at a very early date is testified by the presence of the contemporary genus *Rhinolophus* in the Eocene and *Taphozous* in the Oligocene. These and other genera must have acquired a remarkable equilibrium and stability to have remained virtually unchanged for over forty million years.

So ancient and so well hidden is the bats' genealogy that discussion of their evolution is severely limited. In most cases living species and their "chrono-species" probably already occupied a definite niche long before other living mammals—ourselves included—were recognizable in their present form. It is therefore an imposing challenge merely to gather data and try to identify these niches and correlate the form of bat species with their particular way of life. Here lie open fields of endeavour for present and future generations of naturalists and scientists.

Many African bats have their closest relatives in the Oriental region rather than closer at hand, suggesting that primary differentiation may have taken place at a very early date. In the case of forest bats this connection dates back to the Oligocene.

Bats occupy all habitats and their range is almost world-wide. Various factors must combine to stop bats becoming diurnal. Competition from birds is probably the most important, but physiological limitations related to heat and water balance may also operate.

A dominant consideration for all bats is the need to escape the daylight world. Such retreats are not always easily available so that every possibility has had to be explored, particularly in those extensive areas where natural holes and crevices are rare. For instance, in the open flat valley of the Nile— in the Sudan—thousands of bats find refuge during the dry season in the cracks formed in dry alluvial mud that has been deposited on low-water islands by the river when it was in flood (J. Owen, personal communication). Roosts being limited, bats have had to adapt to particular situations and this is a major factor in their biology and in the analysis of form and function. The major division here is between "hangers" and "clingers" and a second less decisive line divides those that have to find capacious shelters because they must live in large colonies and those that can take advantage of more modest crevices for single bats or very small groups. The factors controlling the choice of a shelter vary for each species and much useful work could be accomplished by measuring the temperature, humidity, degree of light, air circulation, physical volume and structural formation of the various roosts used by bats. Briefly, the roofs and ceilings of caves, rock shelters, houses, culverts, bridges, animal burrows and the interiors of large hollow trees are used by "hanging" bats, and some species hang in the branches of heavily leafed trees and bushes. "Clinging" bats that are social, tend to congregate on the walls of caves and buildings while the less social species seek out crevices in trees, houses, rocks or in cracked soil, rock or bark.

The aspect of niche-filling about which least is known is food types and feeding behaviour. All bats are restricted to nutritionally concentrated and easily absorbed foods with a high energy return. Digestion in all bats is rapid and most species have relatively short guts. The advantages of such diets are obvious; weight is saved and the time spent feeding and digesting is also cut to the minimum. Fasting is generally better tolerated by insectivorous, carnivorous and frugivorous species than by herbivorous animals. Many Microchiroptera possess the capacity to live off fat deposits and this must enhance their unusual capacity to fast. The very different feeding behaviour of the two suborders is discussed in their respective profiles. Most species have to drink although fruit and nectar supply adequate moisture for many fruit bats. Nectar feeding is widespread amongst fruit bats and many species of forest trees, notably *Spathodea* and *Kigelia*, may rely primarily on bats for their pollination.

Specific patterns are apparent in the flight of bats to and from roosts and to feeding areas and also in the level at which they pursue insects or seek fruit. A few species are listed below in some sort of stratified order according to the heights at which they are commonly observed flying and hunting:

Tadarida	3,000 m—In U.S.A.
Otomops	200 m
Taphozous (*peli*)	150 m
Eidolon	100 m
Scotophilus	50 m
Epomops	20 m
Many Vespertilionids	10 m
Cardioderma, *Nycteris*	3 m
Myotis and some Pipistrelles	1 m

In recent years studies of bats in the laboratory have become more common but the ultrasonic sensitivity and the general dominance of hearing in Microchiroptera make the use of electronic equipment and plenty of carefully designed laboratory space particularly important prerequisites.

Field studies with fully mobile equipment have been pioneered by Pye and Flinn (1964) and Griffin (1958). A description of the instrumentation including diagrams of equipment used for generating imitations of bat sonar, recording, filtering and detecting bats' ultrasound is to be found in Pye (1968a). A commercial superhet bat detector is manufactured by Holgates of Totton Ltd. This instrument adds an entirely new dimension to bat studies as it allows the listener to eavesdrop on bats and even follow their movements to some extent as they fly over-head in darkness.

Ringing and netting in Europe and America have revealed many interesting data on migration, homing, orientation and range and ringed wild bats have been recovered 15—20 years later, showing that the Chinese chose an appropriate symbol for longevity.

Where there are very large colonies of bats their guano can be a real economic asset as the guano is rich in phosphorus, nitrogen and potash and where the roosts are in limestone caves, extensive phosphatization of the bedrock can occur. In Tanzania there are large deposits at Amboni and at Songwe, both of which have been exploited for fertilizer—some 3,200 tons having been extracted from the latter (Leedal, 1971).

These caves might harbour millions of bats but no attempt has been made to estimate numbers in East Africa. Some caverns in America have been estimated to shelter as many as 25 million bats.

Where there are many species in one cave they often occupy distinct zones. For instance, *Coleura* and *Cardioderma* may be near the entrance in weak light, with other species succeeding them until *Rhinolophus* is discovered in the deepest darkest recesses. Colonies may contain females with or without young, bachelor groups or mixed groups and the social structure may change with the seasons.

Their principal predators are small carnivores, snakes, owls and falcons, the most important being the bat hawk, *Machaerampus*.

Bats have excited interest as vectors of disease. They are known to carry rabies in America, where the vampire bats, *Desmodus*, may spread it to other bats. The disease is apparently not carried by African bats, perhaps because there are no vampires. *Trypanosomes*, *Plasmodium* and numerous viruses have been discovered in African bats. Current research on African bats is sometimes ethically dubious. I have heard a virologist working in West Africa boast of having killed ten thousand bats in an effort to turn up a new virus. To this worker academic novelty seemed an adequate excuse for inflicting the massacre and no other use was found for the destroyed animals. Bats are eaten in some areas.

Bats have a wide variety of breeding patterns. Gestation periods are long, the growth of the young is rapid and most juvenile bats can fly and fend for themselves before they are a month old. The females yield exceptionally rich milk. Some carry their young about, others park them in nurseries and most females have "false nipples" for the young to hang onto.

Many species exhibit sexual dimorphism which often involves glandular areas in the male sometimes associated with differentiated hairs, tracts, tufts, sacs or pouches. Some sites for glandular areas are listed below:

Epomophorus	shoulder pocket
Taphozous	gular pouch
Rhinolophus landeri	axillary tuft in "armpit"
Hipposideros commersoni	forehead pocket
Vespertilionids	muzzle and other facial glands
Tadarida cistura	anal sacs

Very little is known about sexual behaviour. Odd observations of aerial chasing, posturing at the roost and the utterance of unusual cries suggest that a study of courtship and other behaviour associated with breeding would be most interesting.

Literature on bats is scattered but extensive; the following works are particularly useful—Miller (1907), Andersen (1912), Verschuren (1957), Griffin (1958), Rosevear (1965), Wimsatt (1970), Hayman and Hill (1972).

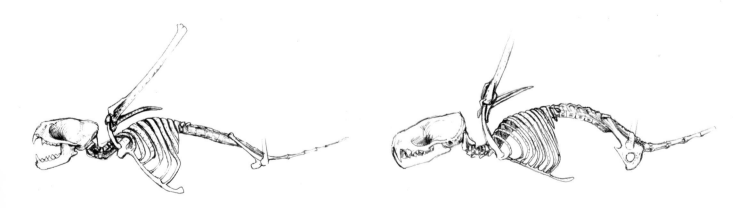

Taphozous peli.

Tadarida midas.

KEY TO MEGACHIROPTERA

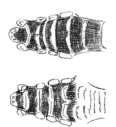

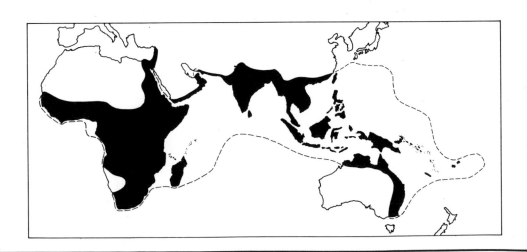

Fruit Bats

MEGACHIROPTERA

The large brown eyes, second claw on the wings and the funnel-shaped ears are the most obvious distinguishing features of the fruit bats. Powerfully clawed legs, broad wings and blunt simple teeth are others.

The second claw on the wing is frequently brought into play in those species that climb about in trees and handle fruit. The large papillated tongue, blunt, short teeth and deeply ridged palate of many species appear to act together, crushing, squeezing or rasping fruit, so that juice and pulp only are swallowed, and fibres, seeds and rind are spat out. Very elastic folded lips (see drawing overleaf) allow juice to be contained and controlled while the jaws chew the fruit.

The adaptive significance of the fruit bats' eyes has been discussed by Kolmer and Neuweiler in Russel Peterson (1964). The retina has a unique structure; along the outer layers of the retina are minute corrugated mounds containing many thousands of nerve endings connected to light-sensitive rod cells. The mounds undoubtedly increase the total surface area of the retina and it is this arrangement that is probably responsible for improving the night vision of fruit bats. The absence of cone cells suggests that they are insentitive to colour.

The skull and teeth of fruit bats have such a striking resemblance to those of lemurs that even the great taxonomist, G. G. Simpson, mistook a fossil fruit bat from Miocene Kenya for a lorisoid (Walker, 1969). Bats are as a whole close to insectivores and primates, but fossils of fruit bats are extremely rare and fragmentary. The earliest known is *Archaeopteropus* from the Italian Upper Oligocene.

Some of the most highly evolved types have reduced dentition, an example being the subfamily Macroglossinae, the members of which have become specialized nectar feeders. The least differentiated genera are *Rousettus*, *Stenonycteris*, *Lissonycteris*, *Eidolon* and *Pteropus*, which retain 34 teeth. The exclusively African epomophorine bats are generally advanced species. The name Megachiroptera is somewhat misleading for the suborder includes several very small types. Sexual dimorphism is marked in several species.

All Megachiroptera are restricted to the Old World tropics and subtropics (see map opposite) and are entirely dependent on a year-long supply of tree and shrub-borne fruit. Not unnaturally, the family is best represented in forest areas although some *Epomophorus* and *Rousettus* species are well-adapted to fairly dry savannas and woodlands. A number of species, notably *Eidolon* and *Hypsignathus*, are known to migrate. In many areas this would be the only means by which supplies of fruit could be found throughout the year. Andersen (1912) first pointed out that so far as the distribution of Megachiroptera is concerned Africa is divisible into two principal provinces, a western and a southeastern. In fact, the majority of Megachiroptera, in spite of their power of flight, have broadly similar distribution patterns to other mammals; the resemblance is particularly noticeable in forest species. Low-

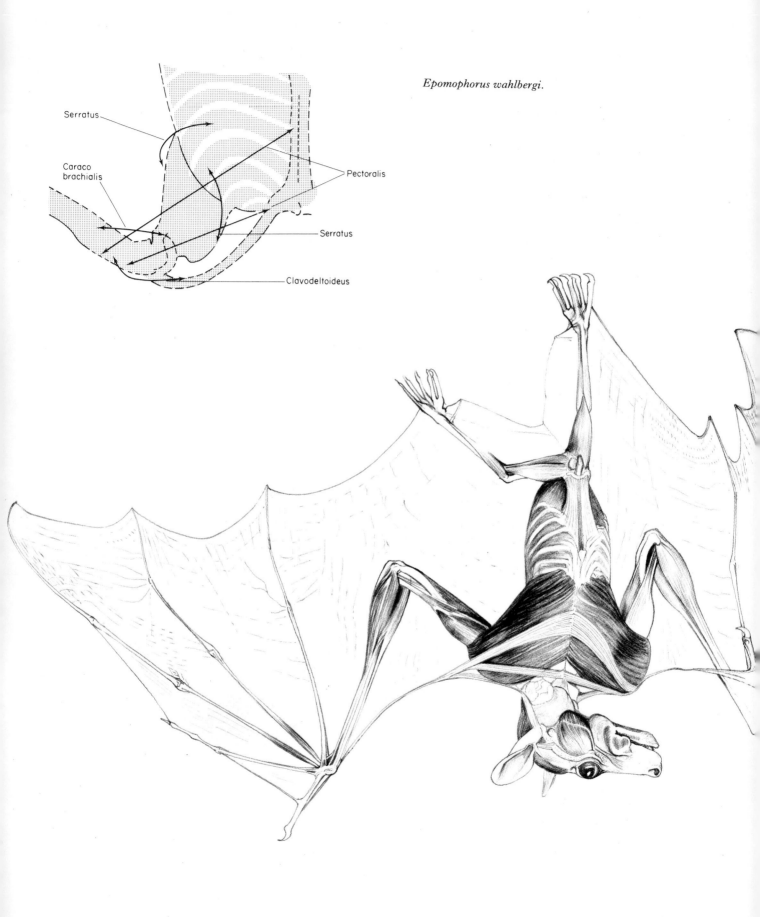

Serratus

Caraco
brachialis

Pectoralis

Serratus

Clavodeltoideus

Epomophorus wahlbergi.

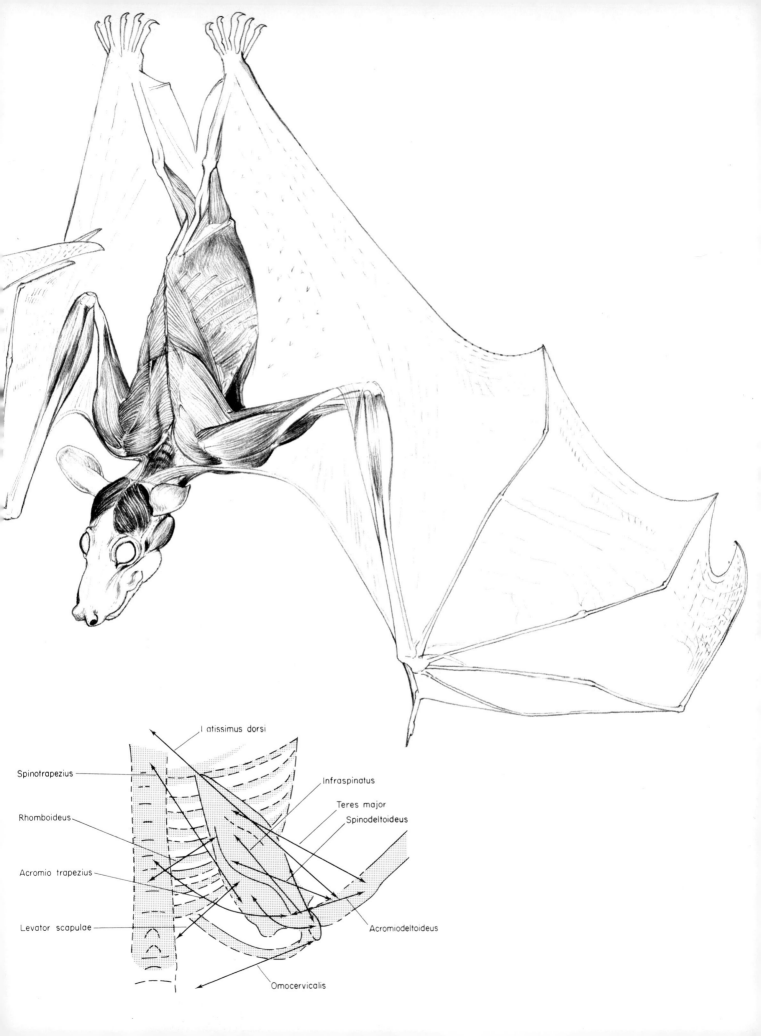

Latissimus dorsi

Spinotrapezius

Infraspinatus

Teres major

Rhomboideus

Spinodeltoideus

Acromio trapezius

Levator scapulae

Acromiodeltoideus

Omocervicalis

land forests shelter *Myonycteris*, *Megaloglossus*, *Hypsignathus* and *Epomops* and these genera tend to have similar distribution patterns to terrestrial forest mammals. *Stenonycteris lanosus* appears to be a relic montane forest species, scattered over the highlands of eastern Africa, but peculiar also in occurring on Madagascar; its distribution is discussed further on p. 134. *Eidolon helvum*, a long distance flyer, living in huge colonies and known to migrate, is a widely distributed species found from Senegal to Ethiopia and south to the Cape. By far the widest range belongs to *Rousettus aegyptiacus*, which inhabits all but the driest parts of Africa and extends down the Nile to the eastern Mediterranean and Arabia.

Apart from the cave-dwelling rousettines, fruit bats shelter in trees, where most species are moderately cryptic. However, *Eidolon* and *Pteropus* roost in vast, noisy, smelly and very conspicuous roosts high up in bare-branched trees. *Eidolon* exhibits a peculiar preference for towns. *Pteropus* are restricted to two islands off the East African coast and their peculiar distribution is discussed in the profile.

Most fruit bats feed on the juice and pulp of a variety of fruits. The juice diet requires a capacious stomach but digestion and excretion are accomplished in a few hours, and fruit bats flying off to feed in the evening are generally quite empty of food. Some fruit bats may eat leaves.

Juice extraction is achieved entirely by the action of the large tongue rasping against the corrugated palate, with some chewing by the teeth; the cheeks store and control the food. Fruit is detected primarily by smell and both the nasal and olfactory areas of the fruit bat's head are well developed. The eyes are similarly large. These specialized activities concentrated in the head give the fruit bats their peculiar proportions (see drawings, pages 118, 119), for the head is generally rather large in relation to the body.

Some fruit bats are thought to be important pollinating agents as they visit arboreal flowers for nectar, which is abundant in trees such as *Kigelia* species. Where the bats actually destroy flowers, they are usually borne on the tree in very great abundance and where such trees are indigenous they must be adapted to this situation and probably rely on the bat's clumsy and wasteful feeding habits to overlook an adequate residue of flowers for the trees' propagation. The specific role of bats in the pollination of such species awaits investigation. More certain is the role of bats in the dispersal of seeds. Most species pick the food amidst a squabbling flock and fly off with the fruit to eat it in another tree. Whether spat out or passed through the gut, the seed is generally dropped away from the parent tree. In the more open habitats, thicket foci such as those found around anthills provide the bats with a sheltered perch and the seeds with a bed protected from fire, sun and floods.

Occasionally where the fruit is very large, bats will eat it on the stem; *Eidolon* bats have been collected inside the football-sized fruit of the Borassus palm, having eaten their way in to get at the pulp surrounding the seeds. In many tropical areas palms are tapped for their sweet juice to make toddy and Eisentraut (1945) reported fruit bats drinking this. Fruit bats are easily attracted even to bare trees by the smell of ripe fruit and can be netted by artificially scattering sweet scented juice near a net. Whether fruit, flowers, nectar and pollen constitute their entire diet is not certain; one species, *Hypsignathus*, is known to scavenge for meat and to attack chickens (Van

Deusen, 1968).

Smell is clearly the primary means of finding fruit but the sight of other bats converging on a tree probably plays an important part. Well-established migratory habits similar to those of some birds may also carry the bats to distant areas where there is a seasonal abundance of wild fruit. The huge changes in natural vegetation patterns that are now taking place through the expansion of agriculture will probably affect fruit bats fundamentally. It is a pity that so little is known, as interesting comparisons with birds' migrations might emerge and the opportunities to learn about them may diminish very soon.

All species have pigmented membranes which can be drawn over their faces in sun and rain. Likewise all conduct lengthy self-grooming sessions. Muffled snuffling noises seem to be associated with sociable feeding activity in many species and loud shrieks and cackles with aggressive sparring. When these bats fight or defend themselves they bite and also strike out with their clawed thumb, holding their wings half open.

Rousettus, *Eidolon* and *Pteropus* have a strong scent, which may be both a protection against predators and also a strong bonding device for these closely packed colonial bats. The scent may be principally due to the glandular areas on the males' neck and shoulders, but both sexes seem to smell strongly. The role of epomophorine scent seems to be more definitely linked with the males' secondary sexual characteristics. In this group males have white "epaulettes" on the shoulder, which are reversible glandular pockets lined with long white hairs. These white hairs appear to advertize the males' presence in a manner similar to that of a bright scented blossom on a tree. The opening of these shoulder pockets is synchronous with the utterance of loud repetitive calls.

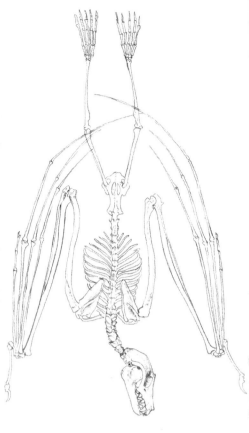

Among epomophorine bats the function of the voice seems to differ from species to species. For instance, the species with the loudest calls, *Hypsignathus* and *Epomops*, have different patterns. Several males of the former species frequently honk together in a group, whereas *Epomops franqueti* males always seem to call alone. Neither of these species seems to have a seasonal breeding pattern. What does seem certain is that for the majority of African Megachiroptera sexual behaviour is initiated by the male, who attracts the female by a battery of auditory, olfactory and visual signals.

In all fruit bats the voice obviously serves as a very important "social adhesive" but, as with scent, there appears to be an interesting difference in role between the epomophorines and the highly social species. The very mobile epomophorines have developed the voice particularly as a means of establishing contact in heavily vegetated country and over relatively long ranges; they are generally rather quiet and still during the day. By contrast the colonial species seem to chatter and cackle the daylight hours away in an almost continuous clamour. Much of this might appear to be bickering, but aggressive incidents are not common and the noise seems instead to be aimed at reinforcing cohesion within the roost and may actually maintain, or at least encourage, contacts between individuals. This rather undifferentiated chattering may only differ qualitatively from the epomophorine long-range calls—which are specific and distinctive. In this connection there is some evidence suggesting that the voice assists in the cohesive process; for instance,

Pteropus, which does not roost as densely as *Eidolon* or *Rousettus*, draws into closer social formation during the mating season, this is accompanied by the making of loud penetrating croaks (Pollen). On the other hand, *Eidolon*, which have been kept as isolated laboratory animals, sleep quietly throughout the day rather like typical epomophorines. The function of the voice seems to differ from species to species, from season to season and it also differs with the condition of the individual bat.

The peculiar range of social organizations found among fruit bats and their ramifications are central to any discussion of megachiropteran biology. Why are some species relatively solitary, while others form vast roosts? Why are some species widely distributed while others are intensely local?

For *Rousettus* the problem may be relatively simple; by relying on caves it limits its local range to areas within easy flying distance of caves; but since new caves are hard to find, it is limited in turn by the food available within range of the home shelter. However it gains several important advantages. The most obvious is the diurnal security the cave offers to a vulnerable nocturnal animal and its young. Depending of course on the size of the cave, such a secure refuge allows large numbers to gather together. This in turn increases the effectiveness of breeding, as no fertile females will find it difficult to meet a mate.

Some solitary or semi-solitary species of the epomophorine group seem instead to depend for their security on looking like dead leaves in the trees and on keeping quiet during the day. When the season changes they can freely seek food in other areas, but this mobility must make contact and mating more difficult. For many species of this group of fruit bats the solution has been the development of loud contact or mating calls and, in some cases, continuous breeding. Between these extremes are a variety of interesting conditions.

Pteropus may form large colonies and like *Rousettus* are extremely conservative about the locality of their roost. Unlike them, however, they hang in tall exposed trees where, for much of the year, they live in a rather loose social pattern. Like many other fruit bats they have a loud call but this seems to be restricted to a short mating season and may be less for contact than for conditioning females to accept the males' close approach.

Eidolon seems to follow another and quite unique pattern, roosting in bigger numbers than either *Rousettus* or *Pteropus* but with similar advantages for their security and breeding pattern—which is strongly seasonal. There is no need here for a loud meeting or mating call, since these bats are already densely gathered in noisy abundance. *Eidolon* appears to occupy its principal roost for many years and may be as attached to a locality as *Pteropus* or *Rousettus*. But it is capable of leaving for definite periods of time and of dispersing into smaller mobile units. Nevertheless, these units are small only in relation to the main colony and they often contain over 1,000 individuals.

Several points need to be stressed in discussing *Rousettus*, *Pteropus* and *Eidolon*: their need for numbers, their conservative attachment to a roost over many generations and the apparent equilibrium they have acquired in relation to their environment.

Once animals have become as highly social as these bat genera, they become completely dependent on being surrounded by their fellows for survival. The vast flocks of American passenger pigeons were exterminated very easily and very rapidly because they were unable to breed in anything less than vast concentrations. Similar limitations probably apply to these bats.

The formation or colonization of a permanent new roost appears to be a rare phenomenon in the genera *Pteropus* and *Eidolon* and "cloning" of surplus populations has not been reported. This could be due to genetically fixed inhibitions about breeding in small groups. The choice of villages as sites for roosts is often cited as a strategy to avoid predators but it is possible that the activity and bustle of a village could act as a direct stimulus to the formation of an *Eidolon* colony, simulating the noise and movement of a fully fledged colony. Alternative sites are sea-side trees above the waves, or above torrents, sites which have noisiness and water as their only common denominators. Furthermore, principal roosts such as that of *Eidolon* in Kampala and of *Pteropus* at Matanga are well placed, for these strong flying bats are within easy reach of year-long fruit supplies.

The rarity of "cloning" mentioned earlier may be due to what appears to be a remarkable ecological equilibrium. The main colony can presumably accommodate normal population fluctuations and it is possible that we are seeing in these colonies a very stable, ancient and well-balanced relationship between these bats and their guaranteed fruit supply.

The distribution of *Pteropus*, which is peculiar to Indian and Pacific Ocean islands and to parts of South-east Asia and Australia has been the subject of much zoogeographic discussion and is pursued further in the profile of that genus.

Continental forms display some interesting patterns in distribution, suggesting that certain forms are mutually exclusive, although in no case are there enough data to be certain that this is so throughout their range. *Epomophorus wahlbergi* is the typical large Epomophorine fruit bat throughout southeastern Africa. Its niche is taken in West Africa by *E. gambianus* and the western rift, from Lake Albert to Lake Tanganyika, seems to define the boundary between forms. Other mutually exclusive forms may be *Epomophorus crypturus*, a southern species, and the tropical *E. anurus*.

Fruit bats are known to be eaten by snakes, genets, owls and hawks. In some areas the spotted eagle-owl, *Bubo africanus*, feeds regularly on fruit bats. It is possible that crows and hornbills may occasionally harry fruit bats but I have seen the tables turned when an *Eidolon* returning to roost at dawn suddenly dive-bombed an early flying pied crow. The bird hastily evaded its attacker by swooping low to the ground and made a rapid getaway while the bat resumed its course toward the roost.

Fruit bats feed exclusively on ripe fruit and are a limited nuisance in orchards as they are generally anticipated by birds or man. Some people, however, object to their presence when they roost near or in houses, mainly on account of superstition, but sometimes because of the noise and musky odour.

Eidolon has become an important laboratory animal at Makerere University, as the vast Kampala roost is a ready source of supply. Its importance as an experimental animal will certainly grow, particularly as delayed implantation is a phenomenon of topical interest in relation to birth control. Furthermore, the bats show certain physiological similarities to primates which enhance their usefulness.

As was remarked earlier, the sexual behaviour of fruit bats is initiated by the male. Some species are seasonal—with a dramatic increase in the size of the testes—while others appear to breed throughout the year. There is marked sexual dimorphism in the majority of species. The phenomenon of delayed implantation in a Uganda population of *Eidolon* may turn up in other species. It is even possible that this mechanism may be absent in other populations of the same species. This would open an important field of research.

Actual gestation periods probably range between about 100—125 days. Copulation has been seen to be performed both belly to belly and with the male on the female's back. The young are born blind but well developed, with disproportionately huge feet, a hairless belly but hairy back. They fasten onto their mother's nipples or "false nipples" with such tenacity that the nipple may bleed if the infant is pulled off. Rosenberg (1840) describes seeing an infant *Pteropus* fall off its flying mother but before the infant hit the ground the mother had successfully retrieved it in mid-air, catching it in her teeth. Young are carried for 5 or 6 weeks and suckled for about 4 months. Judging from arm measurements and sexual development, some fruit bats—e.g. *Rousettus*—are not ready to breed until they are two years old.

Rousettine Bats

Rousettus aegyptiacus : Heavy cheek teeth. Flight membrane attached to first toe.

Stenonycteris lanosus : Very narrow cheek teeth. Shaggy fur. Flight membrane attached to second toe.

Lissonycteris angolensis : Short broad cheek teeth. Flight membrane attached to second toe. Peculiar hairy collar in males.

Conspicuous and well-marked genera like *Pteropus* or *Eidolon* or the epomophorine bats are not difficult to distinguish from one another. However there are three fruit bats, *Rousettus*, *Stenonycteris* and *Lissonycteris*, which lack obvious distinguishing features; their very close superficial resemblance—particularly as dry museum specimens—led to their being lumped as *Rousettus* or at most regarded as subgenera. The resemblances are reinforced by an identical tooth count, so the situation stood for many years. However, as more has become known about their habits, fundamental ecological, behavioural and anatomical differences have appeared and the single genus has had to be abandoned (see Lawrence and Novick, 1963). The recognition of three genera is necessary, because the separation between them may very well be more ancient than the radiation of other fruit bat genera. *Lissonycteris* for instance, may be near the stem of the epomophorine radiation. Furthermore, although *Rousettus*, *Stenonycteris* and *Lissonycteris* may occasionally occur together in a single roost, they occupy distinct ecological niches—another important criterion for the recognition of a genus.

Andersen (1912) lumped the three forms in *Rousettus*; he regarded this group as being close to the ancestral stem of the Pteropodidae and he pointed

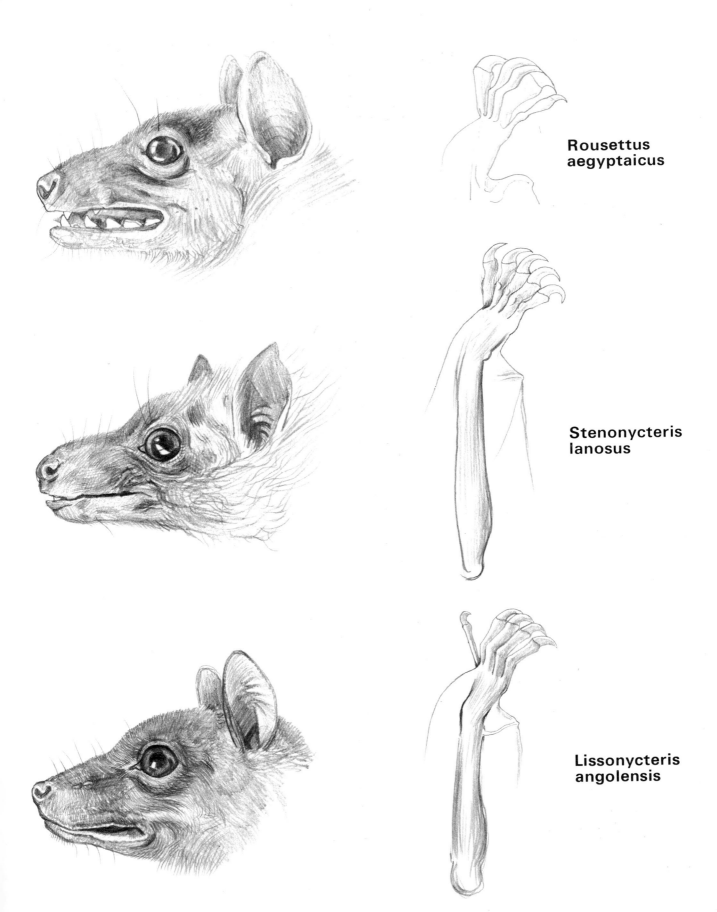

Rousettus
aegyptaicus

Stenonycteris
lanosus

Lissonycteris
angolensis

out their relatively unmodified skull, full dental formula and the possession of a tail. If one attempts to reconstruct the situation that faced ancestral fruit bats, it seems possible that they might have been heavily dependent on well-shaded shelters, such as the mouth of caves and large hollow trees. The peculiar social patterns of *Pteropus* and of the epomophorine bats suggest that special adaptations are necessary for survival on open branches or among the foliage of trees. The security of deep caves was likely to be denied to the ancestral fruit bats, because of their inability to see in total darkness. *Rousettus* and *Stenonycteris* had probably remained less specialized than other fruit bats and a wholesale shift deeper into caves was made possible by the acquisition of echo-location. The independent evolution of this device gives them some of the advantages enjoyed by insect bats and there has been little need to acquire any other specialization (see margin diagram). A conservative form is not necessarily an unsuccessful one and this applies to *Rousettus* (*sensu stricto*), which is a widespread and most successful genus with a pan-African, Australasian and Asian range. Its distribution over three continents and some Pacific islands suggests that its adaptation to total darkness has been of very long standing. By contrast *Stenonycteris* is restricted to some East African highlands and to Madagascar. Although it has several peculiarities it is, in some respects, intermediate between *Rousettus* and *Lissonycteris*.

It is possible that *Stenonycteris* represents an early population of cave-dwelling bats that has been displaced in all but the coldest habitats by the superior *Rousettus aegyptiacus*. If we consider the presence of several species of *Rousettus* in the tropical Orient and in Australasia it is permissible to suggest that this genus may originate in that region.

Lissonycteris is completely unable to echo-locate but it does seem to occupy the original niche that has been suggested here for the ancestral fruit bats and, perhaps more than any other of the rousettine bats, it represents a relatively unchanged form. It is to be expected that at least one successful contemporary species should maintain the tenancy of a convenient niche, but it is interesting and in some ways surprising, that in this case the tenant seems to be close to the stem of an important and exclusively African radiation, that of the epomophorines. This radiation is discussed later on p. 152.

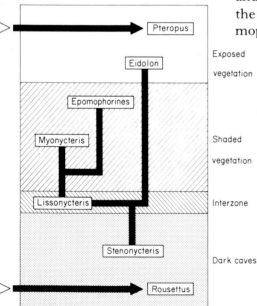

Radiation of African Fruit Bats into various niches.

Egyptian Rousette Bat
(Rousettus aegyptiacus)

Family Pteropodidae
Order Chiroptera

Measurements
head and body
130—155 mm
tail
16—20 mm
forearm
85—106 mm (ave. 95 mm)

Egyptian Rousette Bat (Rousettus aegyptiacus)

This species is the most widespread and best known of the rousettine bats. Geoffroy (1818), writing on the natural history of Egypt during the Napoleonic campaign found these bats in the Great Pyramid at Gizeh and first described them as *Pteropus aegyptiacus*. They are recognizably portrayed by the Ancient Egyptians in the wall-paintings of the tombs.

Andersen (1912) regarded this rousette as one of the least specialized bats among all living Megachiroptera; he described the braincase as perfectly un-modified in general shape.

These bats' capacity to echo-locate is probably the key to their success and their principal specialization. This ability must have evolved independently of that of the Microchiroptera.

The species ranges across the whole of sub-Saharan Africa and up the Nile Valley to Egypt, Arabia and the eastern Mediterranean. In East Africa, the Egyptian rousette can be found in almost all habitats and from sea level to the montane forest of the Ruwenzori Mountains, but they are commonest in forested areas.

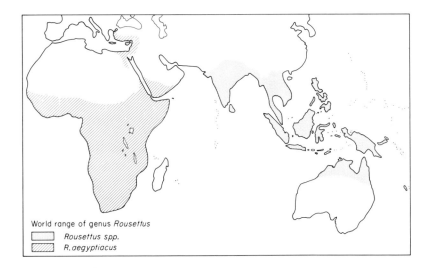

World range of genus *Rousettus*
Rousettus spp.
R. aegyptiacus

Almost exclusively a bat of dark caves, this species dislikes being exposed to daylight, and captives always seek the darkest corner they can find. Even in caves they crawl into crevices to roost, a habit which is not at all typical of fruit bats. When man-made structures offer favourable conditions these bats may colonize them.

Their food includes a wide range of fruit. In drier regions, dates and figs may be their staple food for a large part of the year. In Egypt, the nectar of *Bombax malabaricum* attracts them, and it has been suggested that they are the principal pollinators of this tree. They have been seen to chew soft stems, possibly for the sap. Cunningham van Someren (1972) has found this species feeding on the rough leathery leaves of *Erythrina* which it rasps away with its tongue, spitting out the skeleton leaf in a rolled-up ball. He suggests that the bat may be seeking chlorophyl or a specific alkaloid. After chewing, they spit out the fibres of all but the softest fruit. They usually eat the fruit directly off the stem without carrying it away, although nursing mothers may carry a mouthful of food to their young in the roost. They have been seen to drink by dunking their breast into the water while flying low over a surface and then licking the water out of the fur.

When they are climbing about in their food trees, these bats are reminiscent of *Eidolon* or *Pteropus*, as they travel along under a branch with a hand-over-hand motion. They also resemble these genera in the way they fold their wings against their sides in the normal resting posture. When their roosting place forces them to hang against a vertical surface, they often prefer to put their backs against the wall.

They are attracted to their food by smell, and I have captured these bats by scattering mango peel on the grass beside a net. Within their caves they chatter noisily for much of the day and they have a wide range of squawks and cries. Cries clearly play a large part in communication, but Kulzer (1969b) has shown that recognition does not depend on hearing but on scent:

"Two young bats (37 and 25 days old) were taken from their mothers and put down on a landing stage six metres away. Each one chirped immediately. After a few minutes the first mother arrived, but landed next to the baby which was not hers. She smelled it and flew to the other one, her own. It was recognised immediately and together they returned to their cage. The initial error shows that the call notes are not sufficiently distinctive to allow each mother to recognise her own offspring. In this instance either the youngster was recognised by its size—it was older and bigger than the other—or it was identified by scent. The test was repeated, except that this time the two young were placed in gauze bags. Each chirped again. Now they could not be seen but they could be scented. The result was clear-cut, for however often the positions of the bags were interchanged, the mothers never made a mistake. After scenting her baby the mother frantically tried to free it. This shows that the sense of smell plays an important role during the growing period of the young. It is the final proof of identity in social contact."

The noise emitted to echo-locate in darkness is a rhythmic click made by the tongue and sent out through the corner of the mouth, which is drawn back into a slit and resembles an artificial grin. Each click lasts up to 6 millisecs and many of the upper frequency waves reach 70 kilocycles, the lowest audible ones being about 5 kilocycles. This noise is seldom uttered in an adequately lit situation, so that an Egyptian rousette released in a blackened room may stop clicking the instant a light is turned on, and start again the moment it is extinguished; the electric switch can become an exact way of controlling the emission. Young bats use echo-location from their first flight.

Each night's activity is preceded by a long toilet session with much scratching of the body and ears by the claws, which are also used to clean the spaces between the teeth. The wings and face are very thoroughly licked by the long tongue.

Unlike *Pteropus*, *R. aegyptiacus* is a very agile flyer and can negotiate narrow passages with ease. Kulzer (1969b) describes them flying easily through a 25 by 20 cm opening and they can fly at speeds of 16 km an hour.

Their colonies, which in numbers of individuals range between a hundred

129

or so to many thousands, are characterized by very densely packed clusters of hundreds forming wherever there are suitable footholds.

The sexes segregate during part of their breeding cycle, which is very clearly defined and seasonal in East Africa. Females form "nurseries" of maternal colonies, while the males segregate in their own bachelor groups. Their shelter may be shared by insectivorous bats, such as hipposiderids, emballonurids, *Rhinolophus* and *Miniopterus*, but these are often in the deeper recesses.

In southern Africa the numbers in a population have been known to fluctuate (Lombard, 1968). However, in a large colony in Uganda numbers appear to be stable. Seasonal changes in the abundance of fruit outside the tropical belt might enforce some local migrations. Where this happens there is perhaps a well-established movement from one reliable feeding area to another. Because of its need for caves, this species is perhaps even more dependent than other fruit bats on well-established if locally restricted ranges. This would apply to migration beyond the range of the home cave; such behaviour

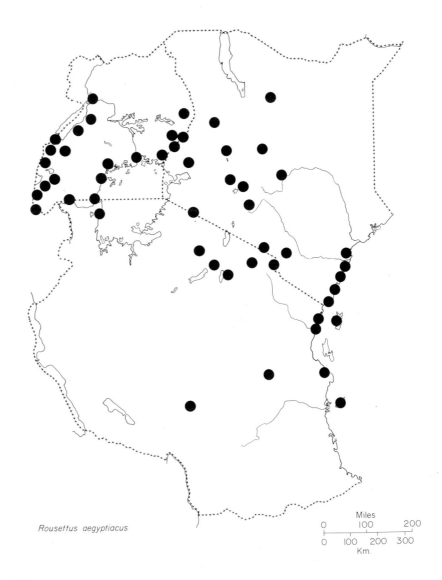

Rousettus aegyptiacus

would probably be maintained within a local population by the young learning from and following the older bats.

In East Africa the Egyptian rousettes breed biannually, bats from Masaka and Mt Elgon bearing their young in March and then again in September. Around Lake Baringo there is a birth season in July—August (Start, personal communication). In more northerly latitudes the pattern may be different and in Egypt they breed throughout the year. To the south, in the southern Congo and in Zambia—about 10° to 16° south—births are reported to occur in October. These births would be the product of a mating season that coincides with the end of the southern single rain season. The presence or absence of a second season at these latitudes awaits investigation.

As with other fruit bats, the male initiates sexual activity. Mutere (1968) found bats from Masaka, on the Equator, with testes of maximum size in April and then again in September. This period coincides with that of the birth or nursing of the young, at which time separate sexual groups have been observed. The weaning of the young follows, which is probably coincidental with mating: an affair that Kulzer (1958) has likened to rape. The female gives every appearance of being very unwilling, resisting and protesting, while the male climbs on her back, bites her neck and embraces her beating her with his wings. This struggle continues until her resistance subsides and she becomes excited, clinging to his belly with her claws while he copulates. The timing of the reproductive cycle would therefore appear to depend on the male but what might trigger the two phases of testes growth is not obvious, except that if we consider the bats from southern latitudes we see that having benefited from the rich harvest following a long wet season, they must be in a peak condition.

The gestation period has been estimated at 105—107 days. One young is born at a time, but twins have been recorded by Kulzer and others. Walker (1968) records twins occurring "at about every fourth birth" in London Zoo. Captive females abort readily if subjected to stress. Kulzer (1966b) has described parturition. When there is one infant only, its weight is about $\frac{1}{5}$ of that of the mother. Kulzer saw the first of a pair of twins participating actively in its own delivery, having ruptured the amnion within the mother before emerging. The second twin was born enclosed in the membrane and fell to the ground when the placenta appeared, whereas the first one quickly attached itself to the mother with its teeth and claws. The mother ate the placenta.

The young are born with their eyes closed, their ears folded and with some hair on the back of the head. At about 10 days of age they open their eyes and their ears become erect and mobile shortly after. They soon acquire the appearance of small adults. They cling to the mother for about 6 weeks, which is the period of lactation. At the time of weaning the infant may have almost doubled its birthweight and has a forearm length of about 60 mm. They begin to fly at about 9 or 10 weeks. While they are being weaned the mother brings back mouthfuls of food for them and they are often seen licking the mother's mouth. Over this period the young can probably survive without much food and judging from Mutere's weights of juvenile bats (1968), there is very little gain and sometimes some weight loss between the drying up of the mother's milk and the time the young start joining the adults. Thereafter weight-gain is rapid but at the same time limb growth—of which the forearm measurements are a good indication—slows down and levels out (see diagram below). The mother continues to be closely associated with her young. I have seen single adult bats, probably mothers, flying in an agitated manner around juveniles (forearm measurements 84—85 mm) shrieking in a net. Kulzer has illustrated the role of calls and of scent in the relationship of mother and young:

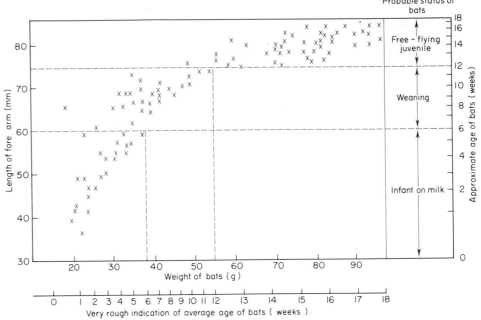

"When I held a 37 day old bat in my open hand about six metres from the cage it chirped incessantly. In a few seconds the mother left the cage, circled round me and finally settled on my hand. She grabbed the youngster in her mouth pushed it onto her breast and returned to the cage with it. After 70 days the youngster can leave and return to its cage. It flies near its mother and starts from and lands back at the same spot. After a 100 days it remains within reach of its mother during the daytime rest period only. Its behaviour now resembles that of the adults, whose movements it watches carefully. Each animal that comes near is smelled all over. This suggests how the bats recognize one another, as is shown by the following experiment. Two breeding pairs and their young lived in two identical cages. Soon after the young had left I interchanged the position of the two cages, so that on their return the young landed with the wrong mothers. Almost immediately each one noticed its mistake and somewhat upset, left again. I was never able to deceive them."

Mutere found the minimum forearm length for a pregnant female was 90 mm; males of the same size had minute testes. Since older bats of the same sex are roughly of similar size, it is probable that males mature more slowly than females. Sexual maturity is probably not reached until females are one year old, males perhaps a little older.

A captive in the Gizeh Zoo lived 20 years. A captive female I kept had severely damaged the phalanges and web of one wing at the time of capture. This wound healed rapidly but it was unable to fly properly. In the process of healing, the broken first phalanges of the third and fourth fingers formed a cartilaginous connection (see margin drawing).

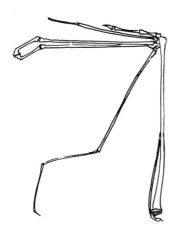

Mountain Fruit Bat, Long-haired Fruit Bat
(Stenonycteris lanosus)

Family Pteropodidae
Order Chiroptera

Measurements
head and body
140 (130—155) mm
tail
15—25 mm
forearm
90 (85—95) mm
tibia
40 mm
weight
145 (120—165) g

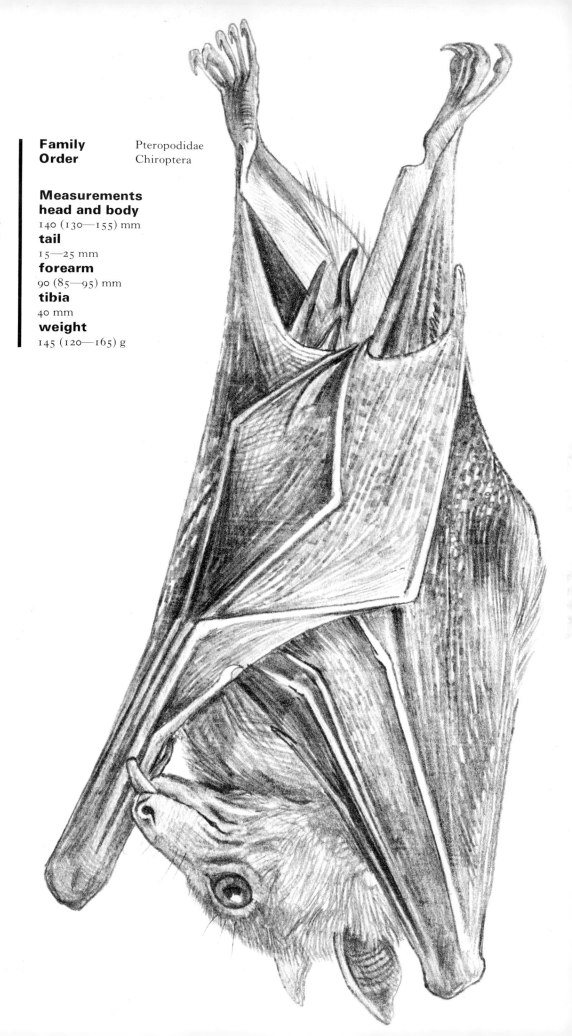

Mountain Fruit Bat, Long-haired Fruit Bat (Stenonycteris lanosus)

Young *Stenonycteris*.

In several respects this bat is intermediate in form between *Rousettus* and *Lissonycteris*; it wraps the wings round the body when roosting like *Lissonycteris* but it clambers about using the foreclaws in a manner unlike that of *Lissonycteris*, although it is neither as adept nor as fast as *Rousettus* and it makes shorter "strides" when clambering. In its dependence on caves and in its capacity to echo-locate it resembles *Rousettus*. In the attachment of the wing membrane to the foot there is an ambiguous connection between the first and the second toe.

Stenonycteris bats are also peculiar in a number of other respects. They have narrow, weak cheek teeth and a peculiarly tilted skull. The angle and shape of the ears are also odd and give the impression of being set to a special position, although they are capable of some movement. *Stenonycteris* tend to keep their ears relatively still, whereas the ears of *Rousettus* are extremely mobile and twitch constantly. The tibiae are elongated and, at the slightest alarm, the animals assume a conspicuous posture by raising the body and bending the knees. Their long hair with fine, woolly underfur is probably an adaptation to the cold regions in which they live.

They have a very odd distribution; they are restricted to a few mountain massifs in eastern Africa and to the east coast of Madagascar. This is essentially a relict type distribution, but it might also be a specialist's distribution, if more were known about their biology. These bats are locally very abundant in the areas of montane forest in which they live. I once found one impaled upon a barbed wire fence on the cold and treeless Elton Plateau in southwestern Tanzania, at an altitude of 2,700 m. The Plateau is surrounded by relict gallery forests, all of rather small extent, so that in order to feed the bats must fly over an extensive area and the bat I found was at least 4 or 5 km from the nearest trees.

The food of *Stenonycteris* probably resembles that of other fruit bats but their weak teeth are suggestive of a more delicately textured diet; possibly very soft fruits and nectar are the preferred foods. The dung of these bats is generally rather liquid and collects on the ground under their roosts in a mushy ooze that is attacked by the larvae of various insects. I have found young plants germinating well in this mush.

They roost in dark, damp caves, where they assemble in hundreds and may occasionally share their roost with either *Rousettus* or *Lissonycteris* or even both. One such site shared by all three genera is a cave hidden behind a screen of water, at the celebrated Sipi waterfall on Mt Elgon. They are able to fly in utter darkness, uttering a rapid succession of chinks, that seem to me to be quieter and more "musical" than the clicks of *Rousettus*. I found that, if there was only the faintest glimmer of reflected light, they were able to fly in a darkened room without any sound emission.

In a colony that I have kept under observation down a mine adit in western Uganda, I found that pregnant and nursing females formed their own cluster in the twilight zone near the entrance, while the rest of the colony, which numbered several hundreds, roosted deeper in the mine and in total darkness. I also found that some of the females that were still nursing were already pregnant again and that breeding was certainly continuous throughout the year. The number of females and young in the "nursery" cluster varied from month to month and sexually active males were also present sometimes. A lactating female collected beside an unusually large cluster of males with active testes was found to have her uterus filled with semen. The figure below illustrates the allometric growth rates of *Stenonycteris* from a pre-natal stage to adulthood.

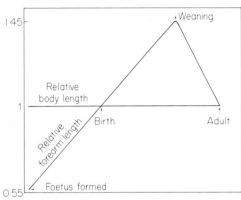

Above: ratio of forearm length to body length from formation of foetus to adulthood.

Six growth stages drawn to approximately same size.
Top row: foetuses
Bottom row: (left) at birth; (centre) weaned juvenile; (right) adult.

This species adapts to captivity rather less successfully than *Rousettus*; however, once settled it is not difficult to keep. A newly captured female I kept had lost her young and was "adopted" by a motherless infant; this foster child died a few days later, although the bat did not reject it and appeared to nurse it. A month later the same bat was introduced to another infant but refused to associate with it at all. The distress cries of this infant attracted the attention of an African goshawk, *Accipiter tochiro*, displaying thereby the vulnerability of this species outside its own habitat.

An interesting fact came to light when I caught a juvenile bat—weight 60 g, forearm 74 mm—and noticed that it had broken a forearm which had already knitted together and healed.

Angola Fruit Bat
(Lissonycteris angolensis)

Family Pteropodidae
Order Chiroptera
Local names
Ndema (Kisambaa), Bebea (Sebei),
Ebugut (Lugisu)

Measurements
head and body
105—135 mm
tail
9—25 mm
forearm
65—91 mm
weight
650—910 g

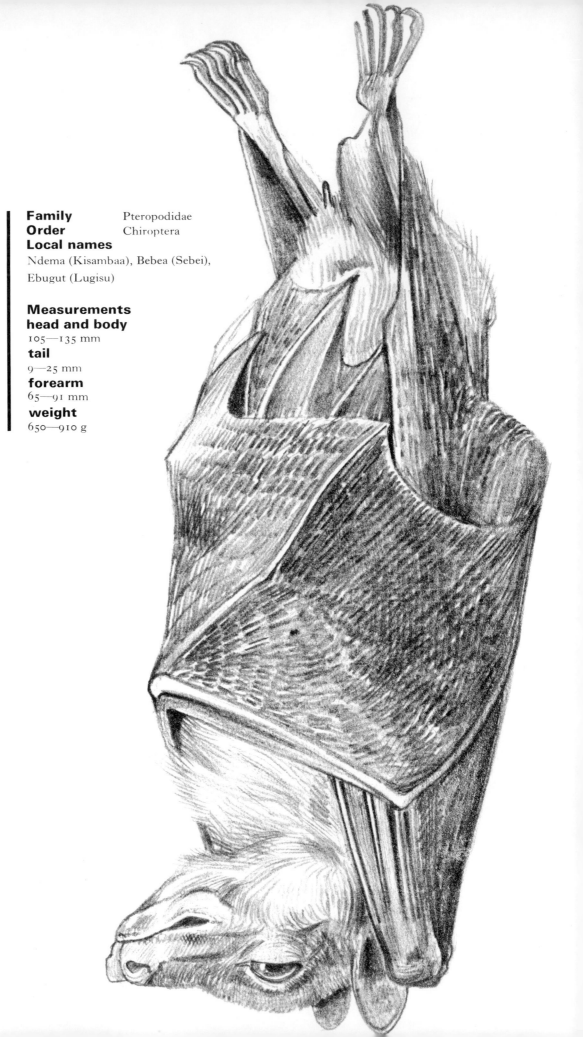

Angola Fruit Bat
(Lissonycteris angolensis)

Lissonycteris is a plain brown bat; the males of the species are distinguished by a "collar" of long sticky hairs. The Rousettine key (p. 125) illustrates some of the characteristics which distinguish *Lissonycteris* from *Rousettus* and *Stenonycteris*. It also resembles *Myonycteris*, but this genus is easily identified on the basis of its small size and lower tooth count.

Lissonycteris range over much of tropical Africa, but are commonest in forest or well-wooded areas. They are found from sea level up to 2,200 m. They roost in small parties, in hollow trees, in well-lit caves and I have captured a solitary individual roosting in well-shaded undergrowth in montane forest.

They feed on various soft fruits, which they chew slowly, spitting out the fibres and stones. When out foraging, they stuff their cheek pouches with food before taking off to a nearby perch to eat at leisure. Food is often manipulated with a foot.

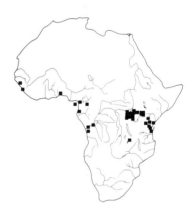

They fly by sight and do not penetrate very deeply when roosting in caves. They fly low and, judging from the places in which they are caught, they may be better able to fly among thick vegetation than many other fruit bats. Their normal sleeping posture is hanging free with the wings wrapped round the body. They assume this posture immediately on alighting and do not crawl about the branches like *Eidolon* or *Rousettus*; in this they resemble the epomophorine bats.

Little is known of their biology but the "stickiness" of the males' collar appears to increase and decrease from time to time and this may be due to glandular secretion connected with sexual behaviour. The size of the testes also appears to fluctuate. Although breeding seasons probably occur, their timing has not yet been recorded. In West Uganda—Budongo—I collected a male with large testes and a sticky collar in September. Eisentraut (1956b) found females forming nurseries in the Cameroons.

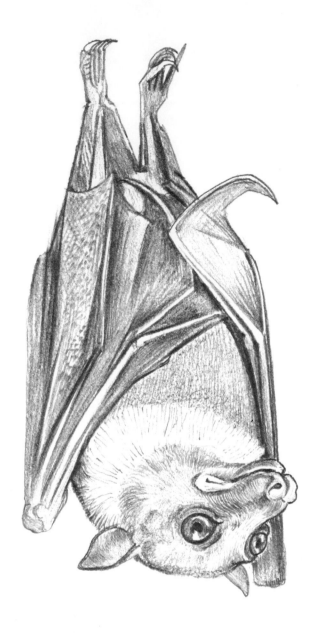

Little Collared Fruit Bat
(Myonycteris torquata)

Family	Pteropodidae
Order	Chiroptera

Measurements
head and body
87—110 mm
tail
7—9 mm
forearm
55—70 mm
tibia
19·5—23 mm
weight
30 g+

Little Collared Fruit Bat
(Myonycteris torquata)

Race
Myonycteris torquata wroughtoni.

This uniform buff or brown-coloured fruit bat resembles a smaller and shorter-faced *Lissonycteris*. The males have similar ruffs of coarse hair rooted in a glandular zone on the shoulder. The species was known from rather few specimens until very recently when large series were netted in West Africa.

On present evidence, western Uganda appears to be the eastern limit of its range. It is mainly a forest species but has been recorded from savannas, plantations and clearings from within or near the forest zone. It has been seen in exposed situations and it changes its roost readily. The flight has been described as slow and feeble. Three races are recognized, each occupying one of the three forest refuges.

The food of this bat appears to be the juices of soft fruit, but they are not thought to be specialized feeders. It is possible that they make local "migrations" after favourite fruits. In captivity they feed on various soft fruits and will take honey and butter.

Brosset (1966b) kept and bred the Gabon race *M. torquata torquata* successfully. He found the males to be sexually active throughout the year, but the ruff exhibited changes in colouring due to the waxing and waning of the sticky secretion which covers the hair and oxidizes on it. Females were polyoestrus in captivity but he records them breeding twice a year, in December—January and again in June. By contrast fifteen wild females collected between November and March were pregnant or lactating, but three collected in June were sexually inactive. A female of the eastern race *M. t. wroughtoni* has been collected in the eastern Congo with a half-grown young in September.

Flying Foxes (Pteropus)

Species

Pteropus comorensis : Black wings and back. Pointed ears and muzzle. Yellow mantle and belly. Occurs on Comoros and Mafia Islands.
Pteropus voeltzkowi : Black wings and back. Deep red head and orange belly. Short ears. Occurs on Pemba Island.

The genus *Pteropus* extends from the western Pacific through Australasia

Flying Foxes (Pteropus)

Family Pteropodidae
Order Chiroptera

Measurements
head and body *Pteropus comorensis*
224—231 mm
forearm
151—157 mm
ear
31—32 mm
head and body *Pteropus voeltzkowi*
240—265 mm
forearm
151—161 mm
ear
21—23 mm
weight
430—610 g

P. comorensis.

and South-east Asia to the tropical islands of the Indian Ocean. The East African species are restricted to two islands off the coast. The largest, *P. voeltzkowi*, occurs on Pemba at Matanga-Mgogoni and *P. comorensis* is found in the mangrove swamps of Mafia Island. *Pteropus* includes the largest bats known; some Oceanic species can weigh 900 g and have a wing span of 1,700 mm.

The genus has been the subject of much speculation because of its peculiar distribution. Moreau and Packenham (1940) suggest that the *Pteropus* stations on Pemba and Mafia may be the relics of a former extension of the genus up the East African coast and they quote Hesse's remark (1937) that broad stretches of Ocean do not seem to be crossed by bats on their own accord: an observation, I suppose, of the remarkable degree of endemism found in the *Pteropus* species of the Indian Ocean Islands where nine species occur on seven island groups (see margin map). In the instances where two *Pteropus* species inhabit the same island, each one belongs to a different sub-division of the four groupings described by Andersen (1912). Hesse's remark is of limited validity, however, because *Pteropus* is scattered over a vast range, indicating that at some time in the distant past it did indeed cross broad stretches of ocean. The occasional presence of sympatric species on islands implies an equally ancient primary differentiation within the genus.

However, the secondary speciation of *Pteropus* on tiny island groups supports Hesse's contention. For once established on an island, their isolation seems to be maintained. The suggestion of a relic status for *Pteropus* is a tempting corollary of the great age that is implied by their speciation. However, being a relic does not imply very much in this case and the rate at which speciation might proceed on isolated islands is unknown.

Andersen described 83 species and 103 forms of *Pteropus* in 1912; at present they are considered as numbering about 36 species (Walker, 1969). The most numerous variety of forms inhabits the "island world" lying between the Pacific and Indonesia; the Indian Ocean populations are probably peripheral to this evolutionary centre. Krzanowski (1967) has attempted to correlate size in *Pteropus* with the degree of isolation of the islands on which the species are found.

Asking why flying foxes have successfully colonized Madagascar and the Comoros Islands and yet not reached the African mainland, Novick (1969) suggests that they cannot compete with other mainland bats or other frugivores, or that they are exceptionally vulnerable to some continental tree-climbing carnivores. The suggestion of competition is vitiated by the presence on Pemba Island of three large fruit bat species, a monkey, *C. aethiops nesiotes*, and a large galago, *G. crassicaudatus*. It could be argued that a coastal frontier between competing groups might have been established during some hypothetical period in the past. Expanding westwards from their insular home, ancestral *Pteropus* met their African competitors and were found wanting. According to this argument, *Eidolon*, *Rousettus aegyptiacus* and *Epomophorus wahlbergi* between them should have ousted *Pteropus*. This is not the case, for both *P. voeltzkowi* on Pemba and *P. comorensis* on Mafia appear to flourish within their own small realms. One must conclude that on islands at least *Pteropus* can withstand competition from bats normally found on the mainland.

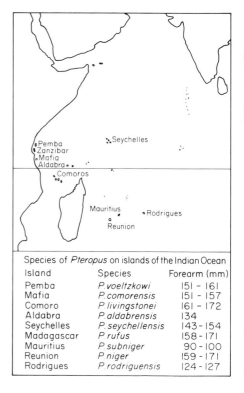

Species of *Pteropus* on islands of the Indian Ocean

Island	Species	Forearm (mm)
Pemba	P. voeltzkowi	151 – 161
Mafia	P. comorensis	151 – 157
Comoro	P. livingstonei	161 – 172
Aldabra	P. aldabrensis	134
Seychelles	P. seychellensis	143 – 154
Madagascar	P. rufus	158 – 171
Mauritius	P. subniger	90 – 100
Reunion	P. niger	159 – 171
Rodrigues	P. rodriguensis	124 – 127

P. voeltzkowi.

P. comorensis.

Some possible predators are also present on Pemba and Mafia; the python, *P. sebae*, and a variety of falcons and owls are sympatric with these bats. The arboreal mongoose, *Myonax sanguineus*, and the leopard are admittedly present on Zanzibar and not on Pemba and Mafia but it is most unlikely that these or any other carnivores should be a deterrent to *Pteropus* colonizing Zanzibar or even the mainland. Besides, mainland populations of *Eidolon* appear to be just as vulnerable and have not been seen to be attacked. It was remarked earlier that a love of "bustle" may play a large part, or even a more important part, in *Eidolon's* urbanization than predation, although town dwelling could be seen as an equivalent to roosting in "predator-free islands". The role of enemies and competitors is present, but it must be rather indirect. For instance, the initial advantage for ancestral fruit bats to roosting in caves may have included freedom from predation among other factors. Predation becomes a more remote factor once specific behavioural, physiological and even anatomical adaptations have evolved. *R. aegyptiacus* has become a "cave bat" rather than a refugee from the dangers of open air. It is therefore possible that *Pteropus* is an "island bat" having specialized in living on tropical oceanic islands much as *Rousettus* has in caves.

What is peculiar to islands that might eventually leave its mark on fruit bats that have adapted to living on them? For the bats that colonized them, the spatial and botanical resources of the primaeval islands were finite but their moist equatorial climate must have yielded a reliable and regular harvest of fruit. In any event, bat colonies would have achieved a perfect balance between their numbers and the food available for them. More bats than there is food for might cause some of them to fly away; but while this sort of migration presumably colonized the islands originally, the subsequent isolation of populations led to speciation on each island group. Once all the island populations were genetically distinct the "stay-at-homes" were favoured, as an already occupied island was unlikely to support a newly arrived group at the expense of the established species. Ultimately, selection for "home residence" may have made this trait characteristic of oceanic *Pteropus* species. As a result, *Pteropus* species evolving on isolated islands would seem to have lost any "wanderlust" they might originally have possessed. All movement was restricted to the island or island group, creating static populations in terms of both numbers and long-range movement.

Such a situation is not found on large continents and the evolutionary pressures on a continental bat must therefore differ profoundly. It would be surprising if along with their physical peculiarities *Pteropus* had not evolved subtle behavioural and other peculiarities that reflected the stability and static centripetal exclusiveness of island life. It seems that some productive research might be pursued along these lines, collating differences in biometrics, anatomy and behaviour of species and correlating this with the geography of their scattered insular range. There could, for instance, be more general implications for the biology of island communities in relation to continental ones.

These bats' preference for bare, upper branches in very tall trees might be influenced by their large size and the need for a free-fall take off. The great breadth of their wing-span (over 1,120 mm in *P. comorensis*) makes free manoeuvre difficult. Lorenz tells of exercising a boyhood pet *Pteropus* by

making it fly down a long passage. The flight was timed so that the bat, which was quite unable to turn in a narrow space, alighted upon the young Lorenz's governess.

The flapping flight of *Pteropus* is strong and high and they have been reported to take advantage of air currents to soar (Novick, 1969). Many of the birds that truly soar are raptors or scavengers, and the advantages of this habit for birds would seem to be associated with long-range visual searching. Several species of fruit bats can be seen planing down in fairly rapid glides, notably *Eidolon* returning to their roosts, but they are not aerodynamically stable at slower speeds. Soaring in this genus should therefore be investigated further, as it is difficult to see obvious advantages to a nocturnal animal unless its night vision is quite extraordinarily acute over long distances. It is, of course, also difficult to be sure about observations made in the dark.

The diet of *Pteropus* does not differ materially from that of other large fruit bats. They chew the pulp of various types of soft fruit, extracting and swallowing the juice. They also chew flowers that contain nectar and pollen.

Pteropus appear to live a long time and a captive has survived over 17 years. They are common in zoos but, according to Walker (1964) they are a prohibited import to the U.S.A. because of being a potential menace to orchards. The threat they present cannot be very real, when one considers the exceptional vulnerability of their colonies to armed men and the geographical and ecological limits within which they seem to be confined.

Colonies may number many thousands but individuals do not generally clump into the dense aggregations so typical of *Eidolon* and *Rousettus*. However, when the breeding season starts there is a change in the social pattern. This is marked by a reduction in the distance between individuals (Kulzer, 1969a). Pollen (in Eisentraut, 1945) reported that *P. comorensis* emit croaking cries during the mating season and Oceanic species have been seen to clap and rattle their wings before copulating.

Gestation is thought to be in the region of 6 months for the largest species. Several *Pteropus* species are known to have a very definite single annual season. Juveniles of *Pteropus voeltzkowi* have been collected in August and, as growth after birth is very rapid, it is possible that this species has a birth season about June.

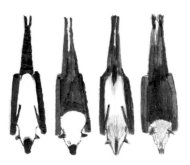

left: *P. comorensis*, right: *P. voeltzkowi*.

Straw-coloured Fruit Bat Eidolon Fruit Bat
(Eidolon helvum)

Family Pteropodidae
Order Chiroptera
Local names
Kakorokombi (Lukonjo),
Chagugu (Lutoro)

Measurements
head and body
150—195 mm
tail
6—20 mm
forearm
110—135 mm
weight
250—311 g

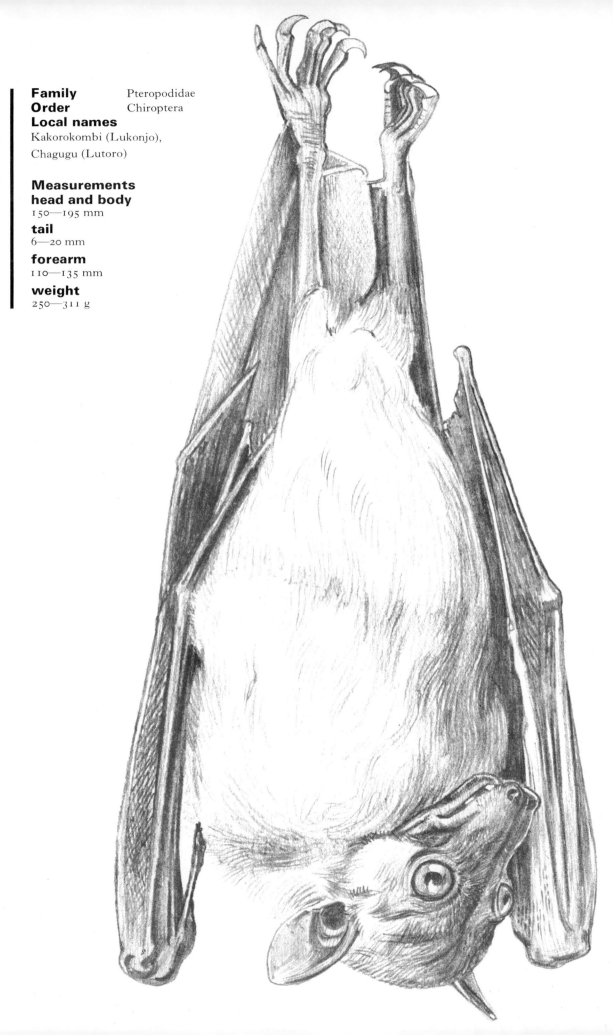

Straw-coloured Fruit Bat
Eidolon Fruit Bat (Eidolon helvum)

Eidolon helvum
Possible 'home' zone
Total migratory range

This very large bat is thought to have derived from a *Rousettus*-like ancestor. The most conspicuous differences between *Eidolon* and the living *Rousettus* species are its manner of flying—which is strong and direct—and the shape of the wings, which are long and have a tapered, pointed tip. Males generally have a bright orange ruff, and the pale tawny fur of the back makes a strongly patterned contrast with the black wings.

This bat is becoming an important experimental animal and has been used in the research on prostaglandins which is being developed by Professor Karim at Makerere University. The structure of *Eidolon*'s heart has been investigated (Rowlatt, 1967); this is of exceptional size—as in all bats— probably in relation to the physical challenge of sustaining flight. The heart appears to have been modified to receive sudden increases in venous blood; furthermore, in the phalanges there are short-circuit devices, which allow the peripheral capillaries to be by-passed while the bat is at rest, as the arteries and veins are linked while the anastamoses are open. These close when the bat flies and circulation becomes complete.

Although widely distributed over the whole of sub-Saharan Africa, *Eidolon* is very localized, forming vast colonies in those areas where there is a year-long abundance of fruit and smaller ones in less favoured areas, where it may only occur as a visitor for a few months of the year.

It appears to be primarily a bat of the forest belt, or of areas that would be forest were it not for man. It lives in a wide variety of roosts and at various altitudes; it has been collected at 2,000 m on Ruwenzori, and Osmaston (1965) records these bats flying to the upper forests on the slopes of Mt Elgon.

Eidolon choose to roost in peculiar situations: they have a marked preference for tall trees in busy villages or towns or for the vegetation on lake or river islands or along the sea shore. I have seen them in Dar-es-Salaam, temporarily roosting in casuarina trees along the esplanade, with the roar of the sea on one side and that of the main road traffic on the other, apparently an ideally noisy situation. They are very conservative about the locality in which they roost; the Kampala site was already occupied in pre-European days; at that time the bats roosted in a patch of indigenous swamp forest near a small village. This vegetation was cleared early in the present century to eradicate a malaria breeding site. Later, the exotic *Eucalyptus saligna* was planted for its special qualities as an absorber of ground water. Although when the forest was felled the bats left the area, they promptly recolonized their ancient site as soon as the Eucalyptus plantation was established; this occurred at some time in the twenties or early thirties.

They will roost in almost any tall tree that offers adequate purchase for the thick bunches of bats that will hang in it. Large roosts cause considerable damage to smaller twigs and branches and the favoured perching areas lose most of their leaves and small branches. The largest roosts may contain tens or hundreds of thousands of bats and the best known ones are scattered at widely spaced intervals across the entire African forest belt. Although almost all major sites are known to be abandoned for some period of the year—July and August in Uganda and the Congo—they all seem to be well placed, within easy reach of forests or, in West Africa, fruit plantations. Large roosts are seldom closer than 60 km to one another and often much further apart; this could imply that the foraging range of a large roost is at least 30 km, and observation has shown that the Kampala bats will fly over 20 km from the roost to feed. These are conservative records of their foraging range and the catching of *Eidolon* 200 km out at sea testifies to their flying capacities.

The published lists of fruit eaten by *Eidolon* tend to reflect the vegetation types from which they are recorded but, in captivity, the species eats any sweet juicy fruit and wild bats are known to raid exotic orchards when ripe fruit is on the trees. They also feed on the buds and young leaves of certain trees as well as on the flowers, nectar and pollen of many other species: *Ceiba, Bombax, Adansonia, Kigelia, Eucalyptus* and *Chlorophora* flowers are particularly favoured, but Morrison found the pollen of many other species present in the stomachs of Kampala bats. Wild figs appear to be a most important food, and the abundance of these bats at Kampala may be influenced by the Baganda's traditional use of barkcloth, which is manufactured from

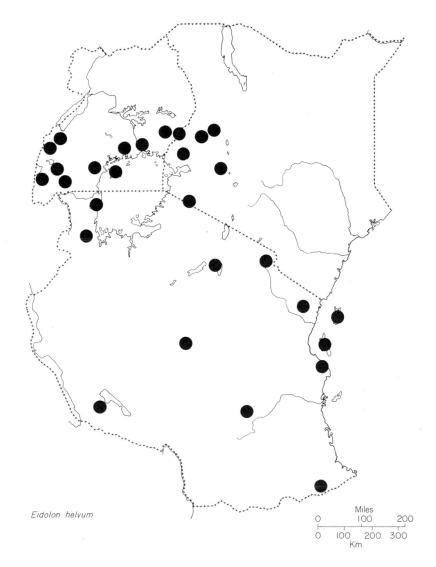

Eidolon helvum

Miles
0   100   200

0   100   200   300
Km.

Ficus natalensis. The tree is very widely used as a live fence and as shade for coffee and is the commonest indigenous tree in most cultivated areas of southern Uganda. Other favoured fruits are from *Celtis*, *Cordia*, *Pseudospondias* and *Syzygium* species, all common trees in the wetter parts of Uganda. *Parinari*, a dominant tree in some forests of western Uganda, yields fruits that attract large numbers of *Eidolon* (Osmaston, 1953).

Like other fruit bats, *Eidolon* spits out all but the smallest seeds as relatively dry pellets, having sucked all the juice out of the fruit and it excretes all food between one and three hours from the time of eating. Larger fruit may be chewed on the stem; Heuglin (1869) shot *Eidolon* bats inside the fruits of the Borassus palm—the animals had eaten their way into the fruit in order to suck the pulp. Large cheek pouches allow these bats to pick fruit and to take it to another perch to be chewed without disturbance. As most of the favourite trees bear their fruit in large quantity over a short season, and as several hundreds of bats may be in one tree at the same time, this habit is advantageous both for the individual animal and for the dispersal of the tree's seed. *Eidolon* is an active climber and a very noisy feeder, so much so that once I suffered a robbery while the bats were squawking in a fig tree outside my bedroom window.

Bats leave the main roost as darkness falls, flying high in broad streams that head in several directions. When a tree is in fruit, I suspect that it is actually advertised by a fine aerial display on the part of the bats. Some individuals plunge out of the high-flying main stream of bats and, coming down to the tree in a zig-zag, make long tacking swoops at great speed but stalling with each change of direction; those bats that follow seem to pursue a very similar aerial pathway. Once down at tree level, their flight does not repeat the splendid style of the arrival flight.

Brief fruiting seasons certainly draw these bats to areas from which they are otherwise absent, but how do they get to know about them? For example, Kock (1969) reports that, during the wet season, they now occur at El Obeid, a town of the Central Sudan in a semi-arid locality where they were completely unknown until recent years. The attraction is the fruit of an introduced tree, *Azadirachta indica*, first planted at El Obeid in 1951. Food supply, therefore, seems to be the primary incentive for seasonal movements in the peripheral or less equatorial parts of their range. Lang and Chapin (1917) recorded some scattered observations on the irregular appearances and disappearances of *Eidolon helvum* in the Congo, but they were unable to show any coordinated pattern of movement.

It may be necessary, however, to make a distinction between migrating after food and the well-known annual desertion of their favourite roosts by *Eidolon*; for instance, in Uganda the huge Kampala roost loses its bats in June and regains them in September. The calendar (diagram opposite) shows the known fruiting periods of some *Eidolon* favourite food plants in Buganda. It illustrates the fact that there is no shortage of food in June, the time of the bats' dramatic dispersal from the Kampala roost. Chapin noticed that *Eidolon* periodically absented themselves from a big roost at Avakubi (Congo) and the period seems to agree with that of the dispersal from the Kampala roost. Chapin remarks "Such movements might be necessary to find new fields of fruit supply, although throughout the equatorial forest fruits seem

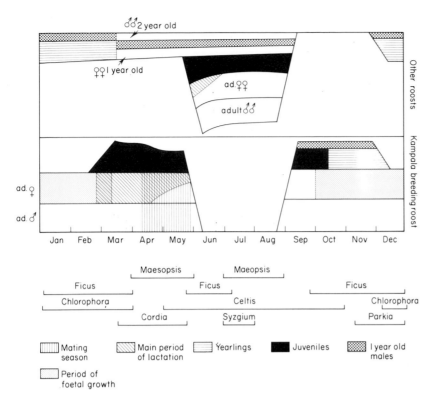

E. *helvum* : simplified calendar of changes in population structure over the year at the Kampala breeding roost. Fruiting periods of favoured wild fruits indicated below calendar.

to be available during the whole year". This puzzled qualification about the lack of any dietary incentive for the "migrations" of *Eidolon* is often echoed by observers of Uganda bats.

Although fruit is abundant in June, this month marks the end of the mating season and the desertion of the roost seems therefore to be controlled by the socio-sexual behaviour of the bats in the colony. Is breeding the major factor holding these bats together in such large numbers? Is it the decline of the males' sexual drive that loosens the bonds that tie these bats to their main roost—which probably ought to be termed the breeding roost? What exactly happens during the dispersal period is not known. A relatively small number of about 5,000 Kampala bats goes to Namulusi Island, a mere 10 km away from the city and this group is presumably typical of the numerous smaller colonies into which the main colony breaks up. The sexual composition of these sub-colonies remains unknown.

Mutere (1965) showed that throughout this period of the bats' dispersal, the fertilized ovum lies attached and quiescent in the lumen of the uterus. At this time the females are probably still in milk and are accompanied by their young. This is likely to be a crucial period for the young that are learning to fend for themselves. I agree with Kulzer (1969a) who thinks that learning by experience is an especially important part of the bats' life. The dispersal period may, therefore, be necessary to allow the young bat to learn how to find food and make foraging flights at its mother's side. In these small sub-colonies the possibility of getting separated from the mother would be reduced and foraging distances should also be reduced. If dispersal is a mechanism designed to favour the young, what triggers the return of the bats to the main roost remains a mystery, although the drying up of the milk and the decline of the mother-young link could conceivably be factors. Whatever the reason, September sees a great convergence of bats onto Kampala and, by October, there are a quarter of a million bats in the small wooded valley linking Kampala City with Makerere University. All ages and sexes are represented in the colony. Younger animals often form their own clusters within the colony, suggesting that the break in the mother-young link has indeed been achieved by this time. It may also anticipate a situation that occurs towards the end of the year when the total number of bats decline. Sample shooting of these bats suggests that most of the younger non-breeding animals depart at this time. The colony is diminished by perhaps as much as one third. This is a statistically appropriate proportion only if all the young have survived and also all leave; however, if males do not mature in less than two years, the ranks of the non-breeders might swell to this rather high proportion. The remaining bats are still formidably numerous and are now a true breeding colony, being mostly composed of gravid females and adult males. In October the quiescent ova implant in the uterus. The factors that lead to the implantation of the blastocyst over 4 months after fertilization remain obscure, but the final independence of the young from their mothers might be associated with changes in the female's physiology which could prepare her for implantation. The presence of a quarter of a million neighbours might also constitute quite a considerable stimulus to the physiology of the female and sheer numbers in the roost might be the final trigger for implantation. Various experiments might be done to explore this problem and these could have a special interest in a world beset with problems of population control.

From December to late February (when the young start being born) the size of the colony seems to remain stable. Pregnant females often form single very large clusters, usually together with males; these agglomerations may contain several hundreds of bats and are situated at low to medium heights, where there are branches capable of bearing the weight. These large clusters also tend to be at the heart of the colony. Occasionally there are all-female clusters and small all-male clusters are common, particularly on the higher branches. The young are born in the midst of the clusters and a few males continue to be intimately associated with the mothers and young, although some sexual re-grouping may take place. The young grow rapidly and at the age of 6 weeks can no longer be carried in flight by the mothers. Between April and June the nights are filled with urgent cries. These contact calls are not conspicuous at other times of the year and they probably serve to link mothers and young. Then the females mate again with the males that are sexually active from April to June. April is a period of abundant food supply for mammals of all kinds and this is true also for bats; Mutere has shown that both sexes of *Eidolon* are heaviest in April. This coincidence of peak condition and mating suggests that a high nutritive level is a likely factor in determining the time of mating. The concentration of all mature animals of both sexes in one main roost must surely facilitate maximum reproductive effectiveness.

Eidolon are often electrocuted on wires and their still, gaunt silhouettes against the sky are a familiar sight in Kampala, Jinja, Mbale, Kakamega and Dar-es-Salaam. They fly well in daylight, when disturbed, although they are

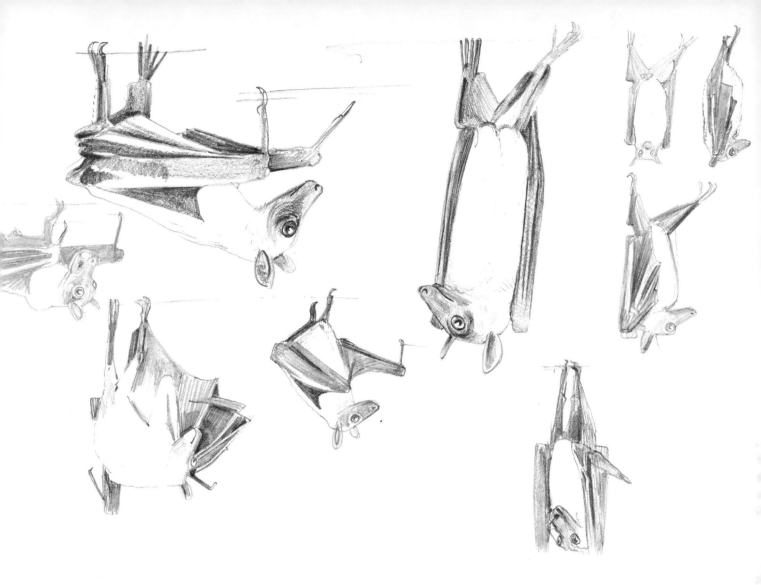

reluctant to leave the immediate vicinity of the roost. The clatter of hundreds of leathery wings and loud squawking probably deters most predators and their musky smell intensifies whenever they are disturbed. As far as is known wholesale attacks on the roost are made only by man and, considering the fecundity of these bats, predators can probably take quite a toll each year without greatly affecting numbers. I have seen and trapped genet cats in fig trees that were being visited by fruit bats, but it was difficult to know whether it was birds, bats, or both that attracted them there. The spotted eagle-owl, *Bubo africanus*, is known to eat fruit bats. Crows and steppe buzzards sometimes appear to be interested in them. An attack on a pied crow by *Eidolon* was mentioned earlier. A kite, *Milvus migrans*, has been seen to kill a subadult *Eidolon* during the day, about 1 km from the roost; the bat however, may have been wounded already for, in spite of being very common round the roost, these birds have never been seen to attack the colony.

The usefulness of this bat for scientific purposes has already been stressed. The Kampala roost is also a very famous spectacle at dusk and is a perennial source of wonder for residents and visitors alike; it would be a great pity if the city's expansion were to destroy this link with the past.

The breeding cycle of this bat has been described but it was not mentioned that their mating season from April to June is a prolonged affair. In

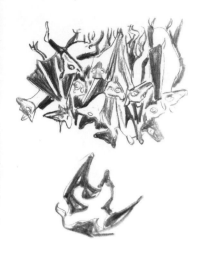

contrast to fertilization, implantation in October is amazingly closely timed; all foetuses are virtually at the same stage of development throughout the colony. Almost 100% of the adult females are pregnant. I only found one adult female in a hundred that was not pregnant at this time. This individual had small but apparently functional ovaries and uterus, but had male colouring and sported a sticky orange ruff. Presumably she suffered from some hormonal unbalance. Lang and Chapin found a female with two embryos and I came across one case of twins in over a hundred pregnancies.

The young weigh 40—50 g at birth and are typical of other fruit bats in having a bare belly, a hairy back and hugely disproportionate head and legs. At one year of age the juvenile's forearm measures about 110 mm and the juvenile's weight is 175—225 g. Adults seldom weigh less than 250 g and their forearms are over 115 mm.

Epauletted or Epomophorine Bats

The epomophorine bats are distinguished by buff to brown colouring with white tufts at the base of the ears and, in most cases, white shoulder pockets or "epaulettes" on the males (from which the group receives its name).

They are exclusively African and have radiated into eight genera, of which only four have been recorded to date from East Africa, which is indicative of their pre-eminently forest distribution. Most of these genera have specialized in some way and contain only one or two species but *Epomophorus*, the most generalized type, has several "size classes" which altogether range over most of sub-Saharan Africa.

Some species and genera of epomophorine bats form flocks but these are never as large or as conspicuous as rousettine or pteropidine bats. Several species are largely solitary.

Epomophorine bats have emancipated themselves to a great degree from dependence on a home shelter or traditional roosts and consequently can range more freely in search of food.

Of the many unknowns, migration or seasonal movement by epomophorine bats is among the most intriguing of possibilities. Periodical arrivals and departures of these bats have been noticed in several places, but no true migrations have been proved to occur. Observation coupled with ringing or banding programmes by local naturalists might help to answer some interesting questions.

Breeding patterns and sexual behaviour vary widely.

Epomophorus anurus (male).

African Epauletted Bats
Epomophorus gambianus complex

Family Pteropodidae
Order Chiroptera
Local names
Mipulumusi (Kinyakyusa)
Measurements
head and body

105—125 mm (males)
95—105 mm (females)
forearm 62—70 mm (males)
55—65 mm (females)
weight 48—65 g (males)
38—70 g (females)

Epomophorus labiatus
Smallest size-class

head and body
120—145 (males)
110—135 (females)
forearm 71—79 mm (males)
66—74 mm (females
weight 80 g (males)
73 g (females)

Epomophorus anurus
Lower middle size-class

head and body
135—160 mm (males)
130—145 (females)
forearm
81—85 mm (males)
79—80 mm (females)

Epomophorus crypturus
Upper middle size-class, restricted
to South-east Africa and only
recorded from South Tanzania

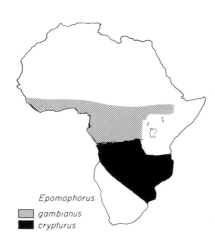

Epomophorus
anurus
labiatus

Epomophorus
gambianus
crypturus

African Epauletted Bats
Epomophorus gambius Complex

The *Epomophorus gambianus* complex contains some 5 or 6 "size-classes", ranging from small bats weighing as little as 40 g, with forearms of 55 mm to the large *Epomophorus gambianus* bat, a West African type not recorded in East Africa, which may weigh three times as much and has a forearm of over 90 mm.

The skull structure and the palatal ridges are identical in all forms, except that a positive allometric ratio exists between the rostrum and the cranial area; this has been studied by Lanza (1961) and by Kock (1969). In some areas there appears to be a size cline and there can also be replacement of one size class by another in other regions but there are also large areas of overlap where two forms co-exist. The maps indicate this situation as far as it is understood at present.

Kock collected a large series of *Epomophorus* in the South Sudan. He found that the males tended to tally with *anurus* and the females with *labiatus* and so synonymized *E. anurus* and *E. labiatus*. I have found populations from south-western Tanzania and coastal Kenya that are intermediate in size between *E. anurus* and *E. labiatus*, suggesting that there may be both an east to west, and a south to north clinal increase in size. A medium-large type, *E. crypturus* occurs in southern Africa. In West Africa, the largest form, *E. gambianus*, co-exists with the medium-small type *E. anurus*. In East Africa, the largest *Epomophorus* niche seems to be occupied by *E. wahlbergi* which co-exists with *E. anurus* and also with the small *E. labiatus*; in southern Africa it is sympatric with both *E. labiatus* and *E. crypturus* which is almost the same size.

Epomophorus bats range throughout the woodland and savanna zones of Africa but do not seem to be common at higher altitudes. They are not truly forest bats, but are common along forest edges and in the mosaic of cultivation and scrub that is typical of the former forest belts.

They roost in a variety of situations, sometimes in dense aggregations numbering up to 60 or 70, at other times more scattered or in smaller groups. Favourite perches are along the main rib of banana or palm fronds, in mango trees and other thick-foliaged trees and in tall buildings under the eaves of roofs; occasionally they may occur in hollow trees.

They feed on many wild fruits and also on some soft cultivated fruits such as mangoes, guavas, peaches, bananas, pawpaws. Fruit, however, is only eaten when ripe and the juice is swallowed while the fibres and pips are spat out. Flowers are visited for their nectar, particularly *Parkia*, *Adansonia* and *Kigelia* species. They prefer to alight beside the flower or fruit in order to feed, but they have been seen to hover clumsily in an attempt to pluck fruit and lap nectar. Fruit may also be taken to another tree to be eaten at leisure.

Epomophorus bats are alert during the day, and if disturbed, will fly away in spite of sunshine and glare. Their flight is fairly slow and usually at a low level since they seldom rise above the tree canopy. While feeding, they make various chuckles or squeaks and the males make a clinking note, which

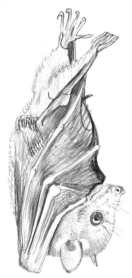

Epomophorus labiatus.

though being similar, does not carry as far as that of *Epomops* or *Hypsignathus.* Nonetheless, when numbers of males are scattered through wooded country, their chorus is reminiscent of that of tree frogs. Each monotonous chink is accompanied by a gentle fluttering of the wing tips, meanwhile the shoulder sacs are turned inside out to expose the long white hairs. This song attracts other bats of the same species.

In *E. labiatus* some breeding is probably going on all the year round, judging from the wide scatter of ages and conditions in netted samples. Fluctuations in the size of the testes have not been recorded. It is possible that male and female flocks may be formed, but contact, particularly sexual contact, may depend to a great degree on the males calling and displaying, thus attracting the individual females that are receptive. The call can be heard throughout many months of the year. Local movements of these bats after fruit and water might occur, particularly in areas subject to a fairly severe dry season.

Loveridge (1922) noticed that *E. labiatus minor* was absent from Morogoro in July. Ionides (personal communication) thought that there might be local fluctuations in the numbers of this species in southern Tanzania.

Lang and Chapin (1917) thought that *E. anurus* had no breeding season in the Congo, as they found young in various stages of development and embryos during the month of November. In southwestern Tanzania I have netted *E. labiatus* in December and January. Of 34 females caught, 15 were immature and of the remainder 3 were lactating, 7 were pregnant, and 9 were neither lactating nor pregnant; I also found a similar scatter in August and

September. In this area the average adult female weighs about 50 g (range 38—60) and has a forearm of 64 mm (range 60—65 mm). Animals with forearms less than 60 mm and weighing less than 38 g are immature and incapable of breeding. Loveridge (1933) was only able to capture males in March on the shores of Lake Malawi. He also caught no adult males in a series of *E. anurus* collected on Ukerewe Island in Lake Victoria. These scattered observations pose interesting questions about the social and sexual life of the *Epomophorus gambianus* complex. These very common and relatively tame bats would repay further study. In Moshi, northern Tanzania, a flock of 30 *E. labiatus* day-roosted for at least 6 weeks near some fruiting fig-trees. The flock deserted the site after being attacked one afternoon by two hornbills, *Bycanistes* spp.

Wahlberg's Fruit Bat
(Epomophorus wahlbergi)

Family Pteropodidae
Order Chiroptera
Local names
Nema (Swahili)

Measurements
head and body
115—175 mm (males)
110—150 mm (females)
forearm
77—98 mm (males)
72—86 mm (females)
weight
60—120 g (males)
54—97 g (females)

Wahlberg's Fruit Bat
(Epomophorus wahlbergi)

This large epauletted bat can be distinguished by its long, narrow snout and lappet folded lip (see drawing). There is only one simple palate ridge behind the toothrow, rather than two as in the *E. gambianus* group, or a series of serrated corrugations as in *Epomops*. Like that of other Epomophorine bats their colour varies a lot, ranging from pale yellow to very dark brown. Sometimes the males have almost black shoulders, creating a dramatic contrast with the white epaulettes.

This is a bat of southeastern Africa, ranging from Somalia through Kenya and eastern Uganda to South Africa and Angola, and along the southern margins of the Congo basin forests as far as the lower Cameroons. It is a savanna, woodland and forest margin species not found at altitudes above 2,000 m.

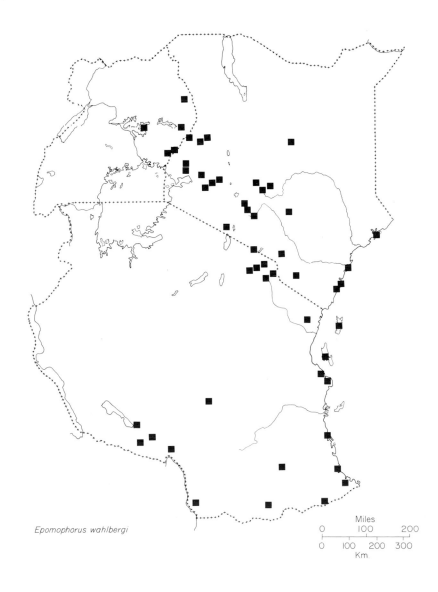

Epomophorus wahlbergi

Miles
0   100   200

0   100   200   300
Km.

157

Not infrequently it is associated with *E. labiatus*, roosting in separate groups but in the same tree and, at night, it will feed from the same tree at the same time. Flocks usually number between 20 and 50 individuals. Clusters of five or six individuals are sometimes to be found scattered about over a small area. The bats hang quietly on banana or palm fronds or along smaller branches bearing clumps of shady foliage.

They are not shy and can be seen in the suburbs of Dar-es-Salaam, Nairobi and other towns, roosting in palm trees, mango, banana, cypress or other farm and garden trees.

They eat ripe, sweet fruits of many kinds and are commonly seen eating wild figs; gardens and orchards attract these bats for as long as the fruit is ripe. They can be netted after scattering a little banana pulp or some mango skins near the net.

Like other fruit bats they seem to be guided principally by their sense of smell in finding food. They pluck the fruit by hanging onto the branch with one foot, while using the other to guide the fruit to the mouth. In some areas local movements after fruit are probably necessary during dry seasons. Ionides (personal communication) believed that these bats moved regularly between the Matengo Hills (about 1,600 m) and the shores of Lake Malawi, not a great distance away but nearly 1,000 m lower. This movement would be interesting to investigate in greater detail. They are not attached to a roost for very long, although, while fruit is abundant, a suitable spot may be used for a few weeks at a time.

The sexes of *E. wahlbergi* have not been noticed to be segregated. Of eight females caught in southwestern Tanzania in December none were pregnant.

Kulzer (1959) kept this species for up to three months and succeeded in getting these animals to Germany; however he remarks that they were much more difficult to keep than *Rousettus*.

Embryo male *E. wahlbergi.*

Dwarf Epauletted Bat
(Micropteropus pusillus)

Family Pteropodidae
Order Chiroptera

Measurements
head and body
75—105 mm
forearm
48—57 mm
tail
0—3 mm
weight
25—35 g

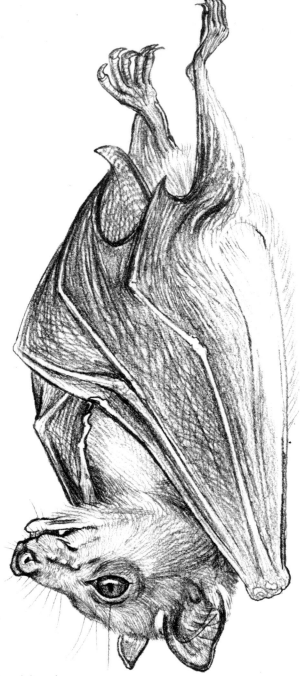

Dwarf Epauletted Bat
(Micropteropus pusillus)

The dwarf epauletted bat is a small tawny-coloured fruit bat, with colouring and markings similar to those of *Epomophorus*; white epaulettes in the male and small white tufts at the base of the ears in both sexes.

It can be distinguished by its short muzzle, small size and by the pattern of ridges on the palate. Andersen (1912) thought that *Micropteropus* probably originated from a type related to, but more primitive than, the living species of *Epomophorus*.

This species seems to belong broadly to the vegetation belts that fringe the forests and it is distributed along a broad swath around the forest zone of Africa, from Senegal to the southern Sudan, East Africa and Angola. It is probably much more widespread within these moist savannas and woodlands than the sparse and scattered records suggest, but it is easily overlooked.

This tame little bat roosts in a variety of situations, dense thickets, heavily foliaged trees, banana gardens and palm fronds and even in exposed,

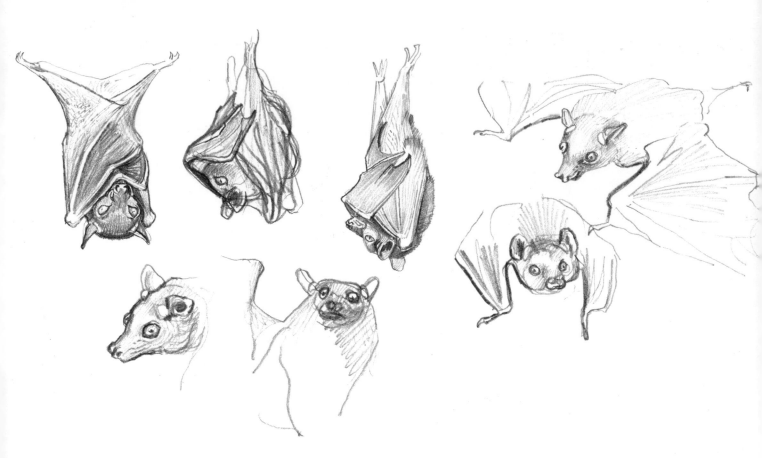

sunny bushes and trees. Its tameness may be linked with its cryptic appearance when hanging quietly from a branch, like a dead leaf. Kock (1969) found that they were not unduly disturbed at their feeding by gun-shots and torches.

It feeds on ripe fruit: bananas, mangoes, guavas, *Anona* spp. and wild figs have been reported. It has been caught in banana-baited rat traps set on the ground (Eisentraut, 1956b), so it probably feeds readily on fallen fruit. It has also been seen feeding on the nectar of *Kigelia aethiopica*, a very widespread tree in Africa and one specifically adapted to be pollinated by bats. The flowers open about dusk, after which time the bats arrive. They are presumably attracted by the scent, although this is not very strong and slightly unpleasant to the human nose. Likewise the nectar appears to be rather tasteless to our tongue. *Micropteropus* alights on the flowers which dangle in space, growing in series from long, hanging stems. Forcing its head into the tube of the corolla it laps the nectar at the base, meanwhile the anthers scatter pollen all over the bat's head and back. The flowers have generally closed at about 11 p.m. and the bats are no longer in evidence. *Kigelia* flowers irregularly but more commonly during the rains. Where it does flower during the dry season, it might provide adequate moisture as well as food.

I have netted *Micropteropus* over a small pond entirely surrounded by thicket, it sipped water from this in a light well-manoeuvred flight. This bat can hover well and negotiate quite tangled vegetation without difficulty.

They seem to be rather quiet bats, they make a little squawk when handled. The males make a shrill ringing note resembling that of other epomophorines.

160

This species often roosts singly, in twos or in small groups of up to a dozen. Pregnant females have been recorded roosting with immature animals and pairs. Social structure might, therefore, alter with the sexual cycles and conditions of bats. Kock, shooting at *Eidolon* in a *Kigelia* tree, brought down a number of *Micropteropus* that he had not noticed before firing.

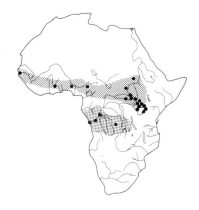

They do not seem to do well in captivity and those that I have attempted to keep soon died.

There are inadequate data on breeding, Verschuren (1957) found that most births occurred in the northeastern Congo during November and December and at the end of February. Some breeding probably continues throughout the year. Like other epomophorine bats the male can open his epaulettes and vibrate them; presumably this is part of the male's mechanism for attracting females.

Singing Fruit Bat, Franquet's Fruit Bat

(Epomops franqueti)

Family Pteropodidae
Order Chiroptera

Measurements
head and body
160—178 mm (males)
140—150 mm (females)
forearm
88—100 mm (males)
86—94 mm (females)
weight
123—158 g (males)
78—130 g (females)

Singing Fruit Bat, Franquet's Fruit Bat (Epomops franqueti)

This large fruit bat is distinguished by its broad muzzle and by the series of flat serrated ridges at the back of the palate. The colour ranges between tawny orange and dark brown, with very dark tones tending to surround the males' yellowish shoulder tufts. The belly is often white and the inner surfaces of the wing membranes may have a pale lemon-green bloom. The females are difficult to tell apart from *Epomophorus* unless the palate is examined.

This species is typical of lowland forest where it is very numerous. It is occasionally caught just outside forest areas, where it may be attracted out by fruit or water.

Franquet's bat ranges through the forest belt from Ghana to the Victoria Nile and to the western shores of Lake Victoria. It occupies all the forested areas of the Congo basin, the most southerly records being from Angola and northern Zambia.

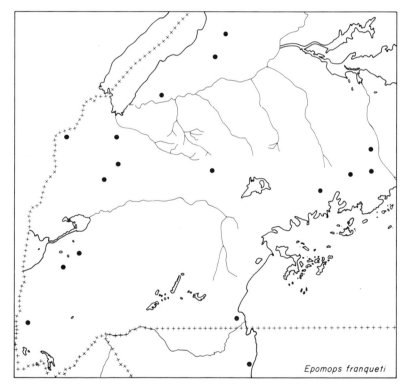

Epomops franqueti

The males' highly distinctive call is the most reliable indication of the species' presence, as these bats are shy and, during the day, will roost in very thick vegetation often among tangles of creepers in the forest canopy.

They feed on soft fruit and flowers. The wide maw and very elastic lips allow it to carry quite large fruit, which it can eat while in flight, as the squeezing action of the jaws is augmented by strong suction. Generally, however, the food is taken to a perch some distance from the point where it was plucked, and the seeds, pulp and sometimes also pieces of fruit are dropped

to the ground. The seeds of many species of forest trees are probably dispersed in this manner. This species will readily eat fallen fruit lying on the forest floor and it is quite adept at taking off from this position. It flies about through the forest with a slow dancing flight and avoids obstacles with remarkable ability. Brosset (1966b) had his net collapse under the weight of these bats that were caught while coming to a flowering bush to eat the flowers.

Their dung is very liquid; I have noticed that one favourite food is voided as a viscous liquid with the consistency of a pale yellow rubber glue, turning orange and then dark red within about one hour.

Like other fruit bats they are guided to food by scent. When in distress they utter a loud nasal shriek and they squawk when quarrelling. The most characteristic noise, however, is the males' ringing note, which can sometimes be heard from dusk to dawn without a break. This note has something in common with the much louder call of *Hypsignathus*, and consists of two component parts. It was well described by Lang and Chapin (1917): "when heard at short range might be written 'kurnk' or 'kyurnk' but at a distance it has a whistled effect almost musical, being repeated slowly with intervals of $\frac{3}{4}$ to 1 second and often for many minutes without a break". The call is made by the males only and they remain well spaced while singing; just before

164

breaking off the song often speeds up and the tone changes. As far as is known the bats do not take up special stations to call and it is assumed that the function is to advertise the sexual condition of the male and so attract oestrous females.

Although many of these bats may be attracted to a source of food or to a forest pool to drink, they seem to roost singly or in very small groups. They are probably residential as periodic appearances or disappearances have not been noted.

My experience of trying to keep this species in captivity has been similar to that of Brosset who found that they refused to eat and died after some days. Blackwell (1967), however was successful in keeping them in Ife Zoo, Nigeria, and even had a young one born and reared in captivity. The species is very shy and it may be necessary for captives to have adequate shelter and flying space before they will settle down.

Both Lang and Chapin (1917) and Brosset (1966b) agree that the species does not have a breeding season, as they found a very wide scatter of sizes among young animals and females in breeding condition throughout the year. In Uganda, scattered records tend to confirm the absence of a breeding season. The mother sometimes carries her young and sometimes "parks" it while foraging (Blackwell, 1967). As in other bat species the female remains capable of flight even when carrying a young weighing as much as two thirds of her own weight.

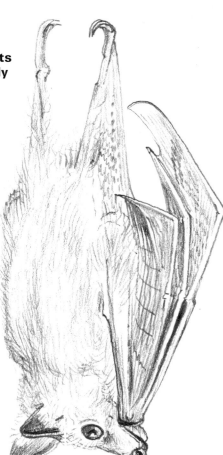

| **Broad-faced Fruit Bat** (Plerotes anchietae) | **Family** **Order** | Pteropidae Chiroptera | **Measurements** **head and body** 87 mm **forearm** 48—53 mm |

Broad-faced Fruit Bat (Plerotes anchietae)

This very rare bat, which is only known from the female, is related to *Epomops* and *Hypsignathus* but is much smaller than either of them. It is pale greyish brown and has the typical epomophorine tuft of white fur at the base of the ears. There is no cartilaginous calcar on the heel.

Plerotes has been recorded almost on the Tanzania border at Abercorn, and there can be little doubt that it occurs in Ufipa, although no specimen has been caught yet.

The very feeble dentition suggests that this species might feed on flowers and nectar, but it can probably manage soft fruit as well.

Hammer Bat, Hypsignathus Bat
(Hypsignathus monstrosus)

Family Pteropodidae
Order Chiroptera

**Measurements
head and body**
220—285 mm (males)
170—220 mm (females)
forearm
120—140 mm (males)
118—128 mm (females)
weight
228—450 g (males)
218—377 g (females)

Hammer Bat,
Hypsignathus Bat
(Hypsignathus monstrosus)

This is the largest bat occurring on the African continent and it is also one of the most peculiarly specialized animals known. It could almost be described as a flying loud-speaker or musical box on wings, so extensive are the structural modifications devoted to the production and resonance of noise. The most obvious external features of this bat are its large size and, in the male, its hugely inflated nose.

Detailed discussions and illustrations of the internal anatomy of this bat are available in Matschie (1899), Dobson (1881), Lang and Chapin (1917) and Schnieder *et al.* (1967). At an early date dissection revealed that a great part of the male's body is taken up by an enormous laryngeal structure. Dobson (1881) thought this might be a pumping apparatus devoted to extracting fruit juices, but more careful investigators showed that the mechanism is actually a huge voice box and an exaggeration of the condition found in *Epomops*. Indeed Andersen (1912) pointed out that this very specialized animal must derive from a form very similar to *Epomops*, which also exhibits an

167

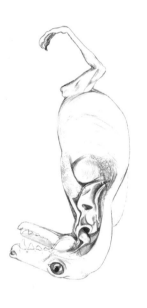

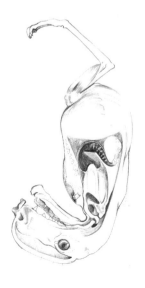

enlarged larynx in the male. The largest unit within this structure is the thryoid cartilage, the greater part of which is a chamber containing the broad, thick vocal chords. The anterior part is subdivided by the flanges of the false vocal chord just before the epiglottis; the soft palate and the base of the cranium form a greatly enlarged nasopharynx which must act as a resonating chamber for the voice. Vocal resonance must also be augmented by two pharyngeal sacs, which lie beneath the skin on each side of the neck and connect with the nasopharyngeal chamber through wide orifices, thus enlarging the total volume of this space several times. These are not, however, the only chambers for resonance; the whole head of the male has become modified to enclose still further space. The skull of a male *Hypsignathus* bat is distinguished by a wholesale elevation and thickening of the nasal bones. In effect the name *Hypsignathus* describes the prominence of the rostrum on the upper jaw. Such an enlargement of the rostrum might be taken to signify an enhanced capacity to smell but, in this instance, the enlargement is restricted to the male, and dissection of the male skull reveals that the bone above the orbits has become inflated by empty sinuses which are not connected with olfaction.

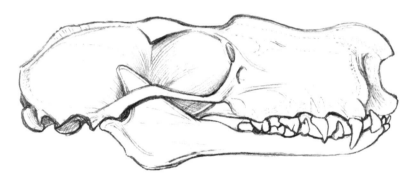

The peculiar architecture of the male skull probably improves the acoustic function of the huge cheek pouches, which have spread up to the rostrum and above the eyes, to be separated from each other by a thin membrane along the midline. This large pouch which encircles the whole head is one more of the male's secondary sexual characteristics and therefore suggestions that its main purpose is storing fruit can be dismissed. I can confirm Mertens' observation (1938) that this *vestibulum oris* on top of the head is distended with air while the male makes its call (drawing, p. 171). It seems that it is the last of a series of chambers built into the "vocal tract" of this extraordinary bat to enhance the resonance of its voice.

The peculiar development of this bat's vocal apparatus can be compared with that of *Epomops franqueti* and *Epomophorus wahlbergi* by means of a simple cross-section (see margin). These sections show that *E. wahlbergi* has a more or less normal larynx in the throat with a slender trachea leading straight into the lungs, while *E. franqueti* has an enlarged larynx which has invaded the chest cavity and forced the trachea into a bend. The larynx of *Hypsignathus* has displaced the heart, lungs and diaphragm and filled the entire chest, the trachea is also very broad and curves round on the cricoid cartilage like a tuba. To make a loud call air must gather in this cricoid arch before

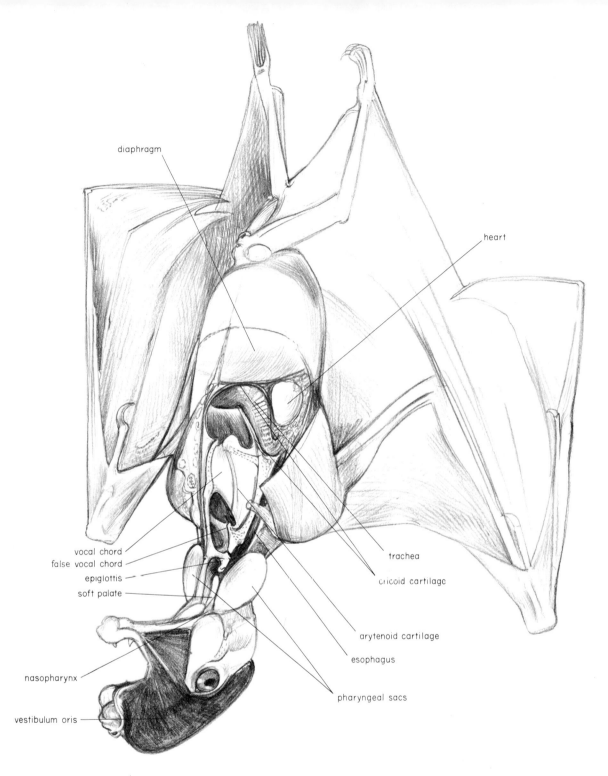

diaphragm

heart

vocal chord
false vocal chord
epiglottis
soft palate

trachea
cricoid cartilage

arytenoid cartilage

esophagus

nasopharynx

vestibulum oris

pharyngeal sacs

being forced past the huge vocal chords which are inside a rigid box. Compression of the air presumably comes from the *cricothyroideus* and other muscles acting on the flexible cricoid arch, possibly assisted by a tightening of the diaphragm.

The voice-box of *Epomops* on the other hand has not developed the cricoid arch and the vocal chords are very much smaller. Instead the false chords and the epiglottis are relatively much larger than they are in *Hypsignathus*. Could this signify that *Epomops* makes its call while it inhales? It would be very interesting to correlate these anatomical differences with the quality of the

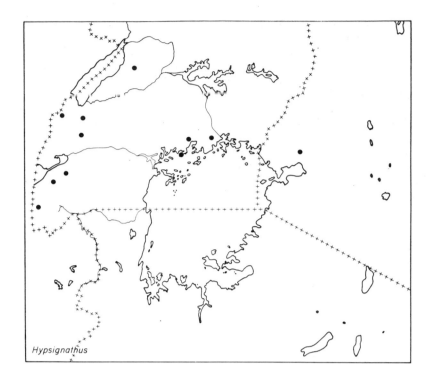

Hypsignathus

sounds emitted by these bats.

Hypsignathus bats range throughout the lowland forest belt from Senegal across the entire Congo basin to western Uganda. They visit the outlying forests of southern Uganda and western Kenya from time to time, but they are unlikely to be permanently resident there. They are commonly found in riverine strips and swamp forest and numerous observers have remarked on their association with forested rivers, mangroves and swampy, palmy areas. They have not been recorded above 1,800 m.

They roost in trees, sometimes quite low down, in the well shaded middle and lower storeys. Sanderson reported finding them among rocks in West Africa, and Walker (1964) mentions caves; however these do not appear to be common habitats for this species.

Like most other fruit bats this species also depends on the juice and soft pulp of various fruit and discards pips, rind and fibres. They have been recorded as eating various cultivated fruits, mangoes, bananas, guavas, *Anona* spp. and wild figs. Large fruit may be eaten *in situ*, the fruit being held by the forearm and clawed thumb while the juice is sucked out, but they prefer to pluck small fruit while hovering, seizing it in the mouth and then flying to a nearby perch where the fruit is cut up by the teeth and sucked; bits and pieces tend to fall to the ground in the process. Pulp and seeds are ejected after chewing.

Hypsignathus may also be somewhat of a "vampire", for it has been seen scavenging for the skinned bodies of small birds and attacking chickens in Gabon. In both cases the attraction was probably blood and other semi-liquids sucked and chewed out of the bodies. This extraordinary habit was observed on several occasions by a museum collector, H. A. Beatty, and was reported by Van Deusen (1968):

170

"a large bat would enter the field of light cast by an oil lantern shining through the doorway, bank sharply, brake to a fluttering stop in mid-air, reach down and seize a small bird carcass. Almost immediately a second bat would repeat the same routine. For the next two hours he watched these bats as they appeared at irregular intervals, always coming from the east, to pick up the scraps of meat. Not once did one alight to the ground. Late on the night of the 8th July, Beatty heard the squawking of one of the chickens tethered to a stick on the ground in the adjacent cook-shed and hurried there. In the glare of his jack light he saw a large bat clinging to a lacerated and bleeding chicken. Beatty prodded the *Hypsignathus* loose with a stick and the bat went hopping on its wings across the floor and under the siding of the shed. Unable to take flight from the flat ground, the bat started to clamber up a nearby tree stump where Beatty captured it. Twice in the succeeding weeks Beatty went out to the rescue of chicken now tethered in the main house attacked late at night by *Hypsignathus*".

The lack of fear evidenced in this report is typical of this bat; they are not easily flushed from their day-roosts and they sometimes seem to exhibit curiosity. While in the forest or wading along rivers at night, I have not infrequently noticed them low overhead, flying to and fro as if to investigate, and ignoring the light of my torch which picked up a vivid reflection from their eyes. When the males start calling they can be watched with ease, for they are oblivious of torches, voices or movement. Lang and Chapin (1917) remark: "During this performance their utter lack of fear was amazing. Neither talking, nor rapping on the trees, lighting a lantern, nor even firing a gun could induce them to cease their calling."

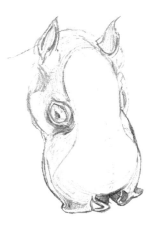

The male's call has been described as kwocking, kwacking, croaking or clanging. These words do not bring out a very curious quality of sound that the voice of *Hypsignathus* possesses, that of two distinct components: at close quarters it sounds guttural, explosive and blaring, somewhat like the burst from an electric alarm, but with increasing distance one only hears a ringing chink which has a similar quality to the "singing" of *Epomophorus*. Possibly the long distance component is the laryngeal sound, while the amplification gained from the series of air chambers and sacs has less range.

The call does not seem to be territorial, as males will sometimes gather and sing in "choirs", and the call is certainly very attractive to other bats of both sexes. One that I watched calling for about an hour attracted six other bats to him at one moment, and there was scarcely a time when at least one bat was not circling round the tree or occasionally alighting beside him. Each time a bat came in towards the "singer", he quickened the frequency of the note and vibrated his wings in a slow, characteristic flutter. On one occasion I saw a "singing" bat make a sudden movement towards another one hovering momentarily beside him; this appeared to drive the visitor off but, usually, the calling, wing-shivering bat appeared to provide a strongly attractive stimulus.

Calling may start shortly after dark, even while the bats are still in the day roosting site. Later they fly from tree to tree, often calling while they are far from sources of fruit. The most continuous noise is generally heard during the first three or four hours of the night, but intermittent calling may be heard throughout the night.

Brosset (1966b) heard *Hypsignathus* answering the cries of the bat *Epomops franqueti* and, more surprisingly, it also seemed to respond to the croaks of the

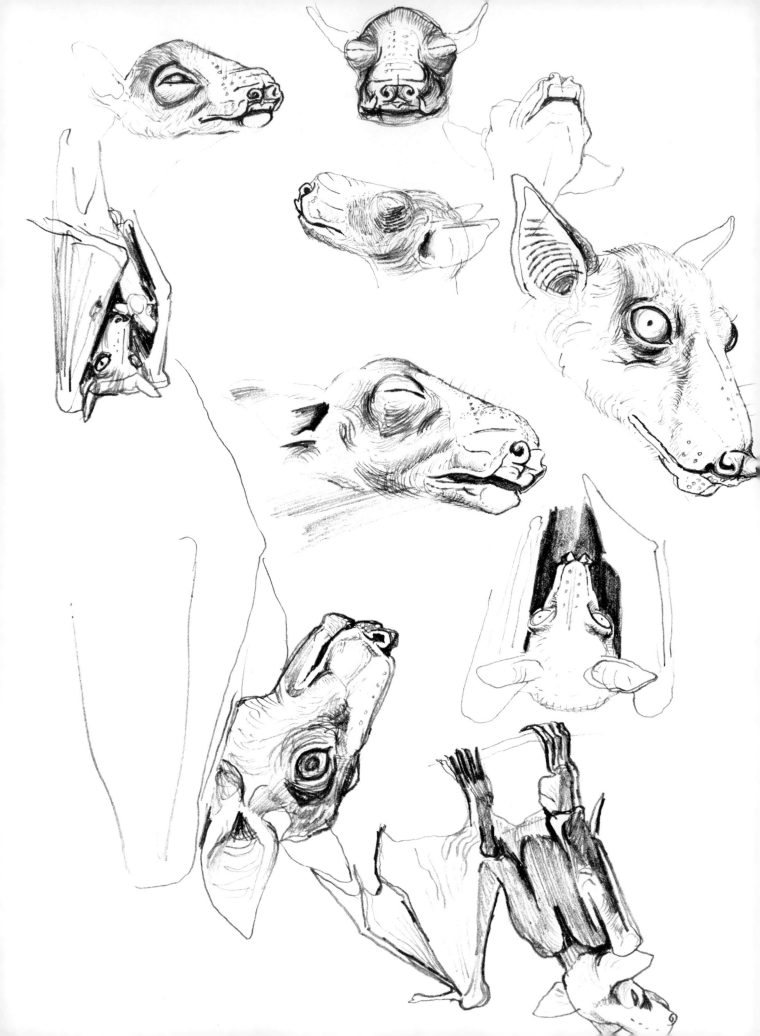

hyrax, *Dendrohyrax dorsalis*. He heard Hypsignathus calling in Gabon, from January to April and from June to September. In western Uganda I found them to be very noisy during February and March. Lang and Chapin also noted this to be a noisy period in the nearby Ituri (East Congo). Instead, during May and June they noticed mostly immature bats which were accompanied by a few adult males whose voices were sometimes heard. Their call was often heard at Kisangani from August to December. Hypsignathus is an occasional, vocal visitor of Kampala during July, August, September and March. They are numerous some years, and apparently absent during others. Since one largely depends upon hearing their call to notice their presence it is difficult to be certain that silence means absence.

It should be mentioned that July—August happens to be the period of dispersal for the big *Eidolon* roost in Kampala. I have noticed ripened fig trees that normally attract large numbers of *Eidolon*, being visited instead by *Hypsignathus* in August. Is it possible that competition for fruit resources could be a factor in their biology?

Ansell (1960) thought that a *Hypsignathus* collected at Mporokoso, Zambia, might have been a vagrant. The rare occurrence of this bat east of the Nile in Busoga and Kaimosi (West Kenya) suggests that it is an infrequent visitor. The bat is probably present throughout the year in the forests of western Uganda. Lang and Chapin were confident that it was likewise resident in northeastern Congo but added "yet there is no doubt that the ripening of certain fruits, wild and cultivated, influence its local occurrence". The vagrants recorded from the outlying forests of eastern and central Africa do imply that *Hypsignathus* can range over considerable distances. The male's voice could assist re-location and re-assembly during migration and the role of the voice as a social stimulus has already been mentioned.

The social groupings of this bat seem to vary, possibly with the seasons. Solitary roosting animals or pairs are not usual, but small parties are probably the most commonly observed. In Bwamba Forest I found a mixed party of males and females, some with young attached, during early March. Lang and Chapin found one male amongst a densely packed group of females and a few males with parties of immature animals. Brosset noted all-male groups in Gabon during May.

It is not known how the seasons impinge on *Hypsignathus* and it is also uncertain whether there is a breeding cycle. Lang and Chapin found foetuses in the eastern Congo during May and December. Brosset found five Gabon female *Hypsignathus* either pregnant or lactating in January—February and suggested a birth season during these months. In March, in Bwamba I have seen two females carrying young and collected a third one that was lactating. In Kampala, a pregnant female was collected in late June and a female with a clinging young was collected in early August. If a birth pattern emerges from these scattered records it is of two birth seasons or peaks, centring on February and July.

One young is usual but twins have been recorded. Dimorphism of the sexes is already evident at birth in the shape of the muzzle. The drawings opposite show a very young male together with its mother.

The male prerogative of a loud call suggests that it serves primarily as an advertisement directed at the females; if there were a clear-cut birth season, equally clear-cut periods of "singing" might be expected. While there is some evidence of fluctuations in the frequency of calling, this could be due to the mobility of the bats and cannot yet be assigned with certainty to a sexual season.

Long-tongued Bats, Nectar Bats

Macroglossinae

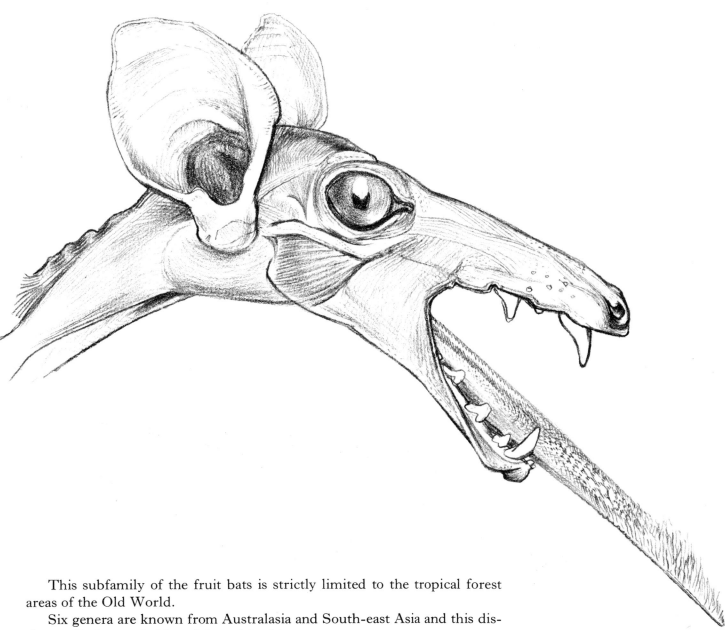

This subfamily of the fruit bats is strictly limited to the tropical forest areas of the Old World.

Six genera are known from Australasia and South-east Asia and this distinct radiation probably originated in this area.

The monospecific African genus presumably derives from a period of forest connection between Africa and the Oriental region, perhaps during the Oligocene.

All the nectar bats are very small.

**Nectar Bat,
Long-tongued
Fruit Bat
(Megaloglossus
woermanni)**

Family
Order

Pteropodidae
Chiroptera

**Measurements
head and body**
64—82 mm
forearm
40—46·5 mm
weight
12—20 g

Nectar Bat, Long-tongued Fruit Bat (Megaloglossus woermanni)

Race

Megaloglossus woermanni prigoginei

This little bat is quite unmistakable because of its small size, large brown eyes and long pointed muzzle.

The soft fur is pale beneath but appears to be "smoked" with brown or sepia on the tips of the hairs. Males have a pure white ruff of stiff hair.

Although the nectar bat does not differ very greatly in general conformation from other fruit bats, the structure of its head and skull betrays a high degree of specialization. The muzzle is greatly elongated and, except for the canines, the teeth are very poorly developed, the lower molars scarcely breaking the surface of the gums. The tongue is tapered and may be as long as 30 mm. Its tip is armed with a brush of papillae (see drawing, p. 175).

These developments are connected with a diet of nectar and they are shared by nectar-feeding animals from totally different groups. Indeed, there is a striking convergence in some American phyllostomid bats, and a comparison of the skull of one of these bats with that of *Megaloglossus* is interesting, particularly if we add the skull of a nectar-feeding marsupial, *Tarsipes*, to the comparison.

The nature of this convergence is well brought out if the skull of each nectar feeder is placed beside that of a fruit-eating relative. The great differences between fruit bats, phyllostomid bats and marsupials are more clearly evident in these less specialized fruit eaters. The physical modifications common to the flower probers have something in common with those of that other peculiar group, the ant-eaters, in that all these specialized animals have reduced teeth and tubular snouts.

Megaloglossus is a typical lowland forest animal in its range from Guinea to Uganda and through the Congo basin to northern Angola. The Uganda animals are rather larger than those from West Africa. It has seldom been collected while roosting, but two have been found together under a banana frond and other specimens have been brought in from banana gardens by cultivators. Rosevear records one roosting in a hut, and another in a house near a bunch of bananas. The natural roosts are probably in dense forest foliage.

The food of these bats is nectar and perhaps pollen, but it is not known what kinds of flowers are visited. Jaeger (1954) provides a recent summary of the literature of chiropterophily and Baker and Harris (1957—58—59) and Harris and Baker (1959) discuss pollination by bats in West Africa. Osmaston (1965) remarks

"many species of trees and lianas have been recorded as being frequented by bats for the purpose of eating nectar, pollen or the whole flower, but positive

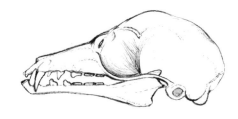

Pteropid bat (*Megaloglossus*).

Marsupial (*Tarsipes*).

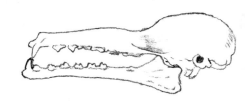

Phyllostomid bat (*Musonycteris*).

evidence of the bats' role in pollination is not always available and sometimes they may be only incidental or merely destructive. Among these plants are the following genera, of which representatives occur in East Africa: *Adansonia, Dombeya, Parkia, Bauhinia, Erythrina, Markhamia, Kigelia, Agave, Musa, Eucalyptus, Grevillea* and *Ceiba* (the last five being cultivated)''.

Kigelia and *Musa* are the only species known so far to be visited by *Megaloglossus*. Eisentraut (1956a) noticed a tropical American tree, *Crescentia cuyete*, in the botanical gardens at Victoria, Cameroons, which was bearing numerous fruit. As a phyllostomid bat, *Glossophaga soricina*, is known to be the principal pollinator of this tree in America, he wondered whether *Megaloglossus* might not be visiting the blossoms of the transplanted tree. A pollen analysis of the gut content has given nothing more conclusive than ''possibly a genus of Caesalpinaceae''.

The tongue of *Megaloglossus* is very extensible and, as in other nectar feeders, the lips and teeth are modified at the tip of the muzzle, so that the tongue can dart in and out of the mouth with minimal opening of the jaws.

The gut is only twice the length of the body (compared with $6\frac{1}{2}$ times in fruit bats), indicating how assimilable and nourishing nectar must be compared with fruit, of which greater quantities must be eaten.

Captive nectar bats seem to be quite silent and soon die. Brosset managed to keep them a few days on honeyed water, but they refused fruit and did not survive.

The social life of this bat is not known. In recent years numbers have sometimes been caught in mist nets stretched over water. Brosset (1966b) caught 86 in Gabon, all of which were males. The few catches of this species in Uganda have also sometimes netted only one sex, but the series have usually been small, and one series of seven bats from Bwamba netted during one November night was made up of both males and females.

The occasional appearance of this bat in localities where it has not been recorded previously, in spite of sustained netting on the same site, raises the possibility that it might migrate, or at least follow the local flowering cycles of its favourite trees. It may be of interest to note that a Mexican nectar bat, *Leptonycteris*, is known to migrate (Novick, 1969) but this is not a tropical forest species and may be subject to greater seasonal fluctuations in its food flowers.

The most easterly records are from Entebbe and Mawokota (near Kampala); all were females, netted in late February and September.

The breeding behaviour is not known. I collected a lactating female in Bwamba Forest in mid December and Okia has collected two lactating females with a free-flying juvenile near Entebbe in late February (personal communication). An adult female from north Kigezi in western Uganda, was not in breeding condition in June. A subadult animal was caught near Kampala in August.

There is no difference in size or weight between the sexes, but the male is distinguished by a brilliant white ruff. The minimal forearm measure for adults of *M. w. prigoginei* is probably not much below 40 mm and the average adult forearm is about 44 mm long. Brosset estimates a 42 mm average and Rosevear (1965) gives 37 mm as the lowest adult measure in *M. w. woermanni*. For *M. w. prigoginei* this size would still be subadult.

Megaloglossus

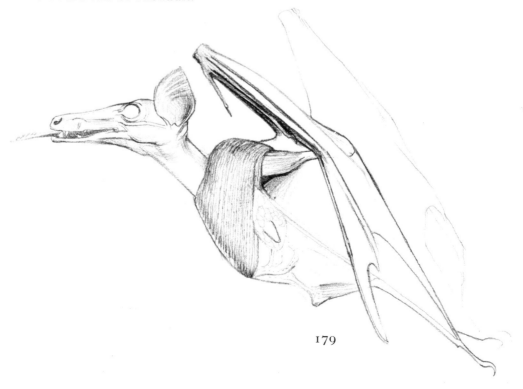

left:
Megadermatidae

right:
Nycteridae

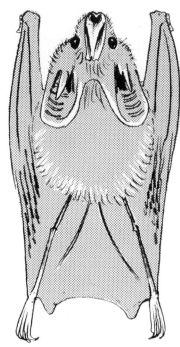
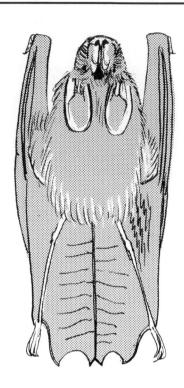

left:
Rhinopomatidae

right:
Emballonuridae

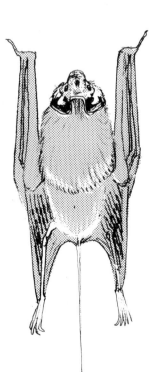
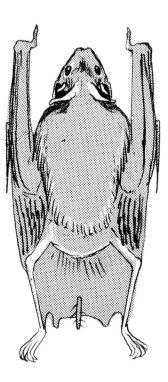

MICROCHIROPTERA

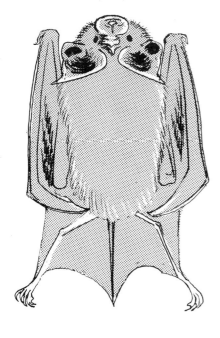

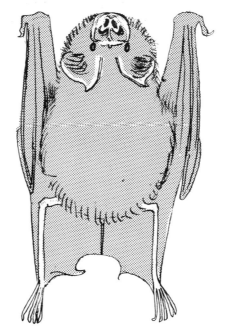

left:
Rhinolophidae

right:
Hipposideridae

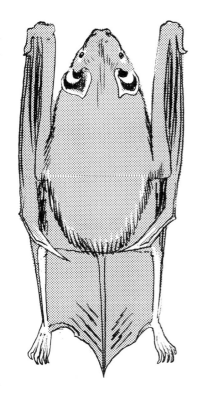

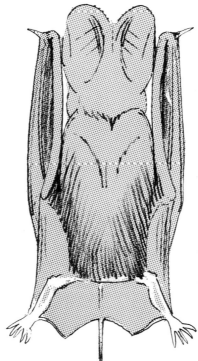

left:
Vespertilionidae

right:
Molossidae

Insect Bats

MICROCHIROPTERA

The most obvious features of the insect bats are their small eyes, the absence of a claw on the second digit and the more complex form of the ear. An invisible but important distinction is the use of high frequency echolocation by all insect bats—the relatively low frequency click of *Rousettus* bats is an independently evolved and probably much cruder system of echolocation.

The range of forms within the Microchiroptera is greater than in any other suborder of mammals and this diversity can be related to the undoubtedly great age of the group.

Eight families are represented in East Africa and their differences can be readily appreciated from the pictorial key (previous page). The most conspicuous distinguishing characteristics are the structure of the interfemoral membrane and tail and the form of the skull, teeth, ears and noseleaves.

Within each family the genera and species are most easily distinguished by tooth structure and the shape of the ear tragi and noseleaf. A most important criterion within some genera is that of size, as most bat species seem to exist within a fairly narrow range of dimensions. There are indeed some bats, such as *Miniopterus*, which contain several species that are only really separable on size, as structurally there is little or nothing to separate the species.

In spite of the name Microchiroptera, a few insect bats are relatively large, with wing spans of over 60 cm and weights exceeding 200 g. However, the majority are fairly small. The most convenient and conventional measure of relative size is the length of the forearm and the table below represents this measure of all East African species, by families.

RANGE OF FOREARM LENGTH IN MILLIMETRES	
Rhinopomatidae	52—59
Emballonuridae	52—97
Megadermatidae	50—64
Nycteridae	32—60
Rhinolophidae	37—65
Hipposideridae	30—121
Vespertilionidae	26—80
Molossidae	28—72

Colour is often variable and is an unreliable guide to species, pale, red or dark phases being known in many species and practically all families.

Peculiar niches or diets are reflected in various features of the body, particularly in the skull and teeth, which may be excellent examples of functional adaptation (see Storch, 1968).

The dental formulae are listed in the profiles and the following table details the teeth numbers of families together with their world status as measured by the number of forms.

FAMILY	NUMBER OF TEETH	GENERA	SPECIES
Rhinopomatidae	28	1	4
Emballonuridae	30—32	13	40
Megadermatidae	26	3	5
Nycteridae	32	1	10
Rhinolophidae	32	2	50
Hipposideridae	30	9	40
Vespertilionidae	30—38	38	275
Molossidae	26—30	12	80

Within a family (such as the Vespertilionidae) tooth reduction is thought to be some indication of relative advancement (see Tate, 1942). Within the Microchiroptera as a whole, however, this is not reliable. *Rhinopoma*, for instance, has reduced the number of teeth to 28 but also has the most primitive wing structure of all insect bats (Revilliod, 1916). The families instead seem to reflect distinct adaptive trends in which some functional parts of the bat's body may become extremely specialized while other parts may remain relatively primitive.

The insect bats' unique development of the vocal-auditory apparatus at the expense of other senses and the separation of function between fore and hind limb—which is common to all bats—allows these functional areas a remarkable degree of independent specialized development as self-contained units. A comparison of the nycterids and molossids illustrates this interplay. The nycterids have modified their legs into fragile supports for a very extended interfemoral membrane. Leg muscles have atrophied and the slender-boned legs have the limited function of manipulating the membrane in flight and hooking the animal up on to a support. The membrane may also assist in the capture of flying insects. A roosting nycterid will hang, all its mobility restricted to twists and turns of the body and head. Flight is slow and the wing structure is relatively primitive (see Revilliod, 1916).

By contrast the molossids have "primitive" mobile legs which allow them to run about with agility on the walls of their shelters or on the ground. Their wings are suited to a sustained and fast flight and, unlike the nycterids' wings, are phylogenetically advanced and mechanically improved.

Discussion cannot proceed further without considering the most fundamental characteristic of insect bats—their use of echo-location to navigate in the dark. While peculiar noseleaves and complex ear folds and lobes have been found to be structural adaptations to serve the echo-locating system, there is little else to betray this ability to raise the sound frequency of the voice within a range of 20 to 160 kHz beyond the range of the human ear.

However it is this technique for navigating and hunting in the dark to which the insect bats owe their peculiar evolutionary success since it has allowed them to make use of a vast food resource. There are about 700 species of Microchiroptera and they are certainly the most effective harvesters of nocturnal insects. Probably the only serious competitors for this resource are the nightjars, Caprimulgiformes, but they are by no means as successful—with less than 100 species.

The abundance of the insect resource and the skill of bats can be measured by the speed with which bats can fill their stomachs. On a good night a bat of 50 g may eat 10 g of insects within an hour of emerging at nightfall. A second feeding spell in the early morning is likely and a couple of hundred such bats would consume more than a metric ton of insects in a year. It is difficult to imagine any man-made device that could harvest tiny insects on such a scale.

Because the sounds used in hunting and manoeuvring are beyond the human ear, because their nocturnal activity is largely invisible and because their world does not intrude in any obvious way into our own, investigation of the natural history of bats is largely a story of neglect punctuated by the occasional observations of unusually curious individuals. In 1767 Gilbert White described keeping a tame bat and watching myriads of bats flying over the Thames between Richmond and Sunbury, he noted that they drink on the wing "by sipping the surface, as they play over pools and streams. They love to frequent waters, not only for the sake of drinking, but on account of insects, which are found over them in the greatest plenty". In 1771 he described a bat, probably a noctule "within the ear there was somewhat of a peculiar structure that I did not understand perfectly; but refer it to the observation of the curious anatomist". Spallanzani (1774) appears to have been just such a curious anatomist and it was he who first determined that bats navigated by hearing. He blocked the ears of some bats and blinded others and found only the former unable to avoid obstacles in flight. At this early date he posed the question: how could ears replace sight? The subsequent history is told in several books: Griffin (1958), Novick (1969), Russel Peterson (1964). The question was not tackled again until Hahn (1908) determined that the inner ear was crucial, after experiments involving cutting off the ears and tragi of insect bats. Hartridge (1920) suggested that it was the echo of ultrasonic vibrations that the bats perceived, but it was in 1938 that Pierce and Griffin finally proved the existence of echo-location, using electronic equipment to record and translate into audible signals the cries and bleeps of flying bats.

This very late discovery and investigation of echo-location betrays how long it took for bats to catch the scientific imagination and it is interesting to ask why. The most obvious answer must be that the electronic techniques suited to the investigation had not been developed but the answer must also lie partly in the value systems of scientific communities—while fundamental questions about the nature of bats' special abilities remained unanswered, a moratorium appeared to be tacitly declared on bats as a worthy field of scientific study, in spite of their accounting for a very high proportion of the world's mammal fauna, both in number of species and individuals and their being the most effective insecticide known. As Kuhn (1962) points out, scientists pay due regard to the tools that are available for research and concentrate on problems that they have reason to believe they can solve.

As far as the natural history of African bats is concerned, the richest source of information for many years has been Lang and Chapin's painstaking field notes (1917)—(at this date it was thought that the bats' membranes were mysteriously sensitive). Lang and Chapin's notes are an appendix to the catalogue of a collection, and the bulk of bat literature before 1940 consists of catalogues, taxonomic discussions and descriptive anatomies. Since 1938, the date of Griffin's first paper on bat sonar, scientific interest in bats has flourished and their vital ecological roles have also begun to be appreciated.

Bat sonar has been shown to be a system that is capable of great variety and modification. Families and other groupings of bats have different and highly characteristic ultrasonic patterns. Correlations between the habits of different bats and their various sonar patterns are often impossible to make, largely because so little is known of both their habits and their acoustic structures. However, there are indications that the particular types of ultrasound used by some species are well-adapted to the environment, flight patterns and hunting methods natural to the species.

The voice is produced by the vocal chords, which are put under tension by well-developed cricothyroid muscles contained within a bony laryngeal box. The sound may be beamed through the mouth and lips, which assume a long trumpet shape in many simple-nosed types, or it may be emitted through the nostrils in the noseleaf species. Resonating chambers are found in the latter's nasal passages and they have a neat valvular fit between the larynx and the nasopharynx, which allows sound to travel directly from the larynx to the nostrils without any opening into the mouth. Nasal signals have the advantage that feeding and drinking do not interfere with sonar.

An ecological limitation may be sufficient to explain the division in that delicate and prominent noseleaves may inhibit the use of a wide variety of roosting sites in cracks and crevices, and opportunities for densely packed societies may also be lost because the animals cannot squeeze or abrade their noseleaves. It is also possible that the nasal sonar system and its apparatus could be subject to certain spatial limitations, needing a relatively wider field than the more directly "shouted" signals of simple-nosed bats. In any event, most leaf-nosed bats need to hang from branches or the roof of some roomy shelter, a dependence betrayed by their legs, which in the majority of species, are adapted to passive hanging, whereas the legs of most simple-nosed bats are more adaptable.

The peculiar fleshy lobes mounted on the rostrum of these bats probably have more than one function. Mohres (1953b) showed that the rhinolophid horse-shoe and sella help to concentrate sound into a beam. The same structures may also shield the ears from the full intensity of emitted signals and at the same time they might also deflect a controlled amount of sound back into the base of the ear. The cellular posterior process in rhinolophids and hipposiderids may cast an echo shadow on the pinnae.

It has already been remarked that the external pinnae of bats vary greatly and the key shows the arrangement in several families of microchiroptera. The hearing mechanism itself is not fundamentally different from that of any other mammal in that vibrations from the ear-drum or tympanic membrane are transmitted through the ossicles of the middle ear to the cochlea and sense receptors of the inner ear. The mechanism for damping sound transmission by altering the position of the ossicles is common to all mammals and functions to protect the inner ear from unnecessary stimulation. The two muscles operating this mechanism are the strap-shaped *tensor tympani* moving the malleus and the conical *stapedius* altering the angle of the stapes; they are quite exceptionally large and well developed in bats. The cochlea is also very large and, in many bats, the whole otic capsule is insulated from the skull by

R. hildebrandti.

a layer of fatty tissue instead of being solidly fused into the skull. All these peculiarities are adaptations to limit sensitivity to echoes and avoid the very high noise level of self-emitted bat signals. Henson (1967) found the *stapedius* of *Tadarida brasiliensis* contracted before each squeak; the maximum damping of transmission was correlated with the loudest moment of the bat's pulse and sensitivity increased by muscle relaxation as (fainter) echoes returned after long delays.

Insect bats use two principal types of sonar. In one type the frequency is modulated (F.M. signals); in the other the major component of the voice is at a constant frequency (C.F.). Short broad·band F.M. signals give the bat accurate range information. The C.F. system, instead, is best adapted to indicate velocity (see Pye, 1968a). The closer an object or obstacle the quicker will echoes of the bat's signal return. A bat uttering a short pulse might therefore hear the echo in the interval between one pulse and another. A longer signal will return echoes from nearby objects before the pulse is finished, so that the emitted sound and its echo overlap at the bat's ear. All bats increase the rapidity of their signals as they close in on prey and as they approach a roost to land. The pulse rate may increase by as much as ten times with changes in the structure of each pulse. Ultrasonic patterns may therefore vary with their context.

The anatomical comparison made earlier between two families of insect bats may be made again in relation to sonar. Molossids have F.M. signals that are slow, shallow sweeps at relatively low frequencies—below 60 kHz.

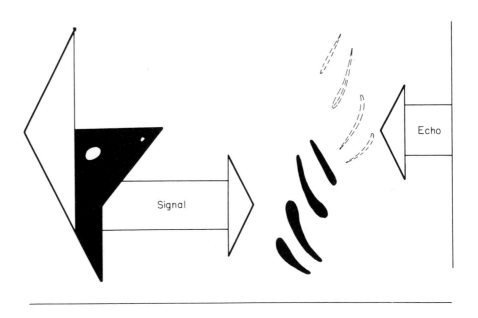

Sweeps run from 60 kHz down to 30 kHz with pulses lasting 10 millisecs. Nycterids and megadermatids instead have pulses of very short duration, 0·25

to 1 millisec. The low intensity of their ultrasound requires very sophisticated equipment to record it. Pye has found that these two families make short frequency modulated signals with several harmonics. The use of lower frequencies and relatively long pulses in molossids may be related to the fast flight and generally open habitat of these bats while Novick (1969) suggests that, by reducing output intensity, slow-flying bats like *Nycteris* can concentrate their attention on the numerous obstacles and difficult prey of their tangled environment.

The table below lists some acoustic data for microchiropteran families. (Data from Pye, 1968a, personal communication and monograph in preparation.)

Family	Frequency pattern	Band (kHz)	Pulse duration (msec)	Repetition rate per sec	Harmonics
Rhinopomatidae	F.M.	110—25	1—4	(prob. 100 max.)	several (5—4) prominent
Emballonuridae	F.M.	75—20	1—6 (poss. to 30 ms.)	200 max.	several prominent
Megadermatidae	F.M.	100—80 ?	0·25—1	200 max.	several prominent
Nycteridae	F.M.	100—80 ?	0·25—1	200 max.	several prominent
Rhinolophidae	C.F.	120—40	10—60	5—80	"pure" 2nd harmonic
Hipposideridae	C.F.	160—50	0—60	5—80	relatively pure 2nd harmonic
Vespertilionidae	F.M. (mainly)	120—20	0·25—4	10—200	mainly fundamental some 2nd
Molossidae	F.M. (or almost C.F.)	60—20	8—10 (less in buzz)	20—40	mainly fundamental

The figures above are approximate and are typical although probably not fully representative. Pye has found that many bats can use C.F. signals even when this is not typical of them and he believes that their sonar systems are more flexible than was previously thought.

Since the bat's perception of objects surrounding it is by means of echoes, movement, whether by the object or by the bat, causes a frequency change known as the Doppler effect. That some bats have the means of measuring the Doppler shift has been shown by Schnitzler (1968), who swung a pendulum near *Rhinolophus* of two species. When at rest these bats have specifically constant frequencies—i.e. 83 kHz for *R. ferrumequinum* and 104 kHz for *R. euryale*. Once their attention was excited by the movement of Schnitzler's pendulum, it was found that the bats regulated the frequency of signal *emission* in such a way that the echo *returning* from the pendulum always maintained the species' specific frequency. This discovery shows that these bats can discern an object clearly and can measure the rate of movement, a capability that must be of critical importance in hunting when a single insect is followed at a time. The dynamic control of the Doppler shift in order to receive signals within a narrow bandwidth of constant frequency would have the added advantage that hunting neighbours would not disturb or jam one another as their emitted frequencies would be constantly changing while hunting. Many leaf-nosed bats hunt like flycatchers, hanging from a perch and flying out to seize prey. Pye (personal communication) has remarked that movement against a background would be easily detected.

Schnitzler's discovery may be related to the well documented observation that signal emission and ear movements are correlated (Pye, 1960; Pye *et al.*, 1962; Pye and Roberts, 1970; Griffin *et al.*, 1962; Schneider and Mohres, 1960). The last authors cut the ear muscles of *Rhinolophus* and found the bats unable to orientate until they had "learned" to nod their heads. Pye and Roberts (1970) illustrated evidence of exact numerical correlation between pulses and ear movements at rates from 6 to 80 per second. The upper rates are remarkable and display the extraordinary refinement of sonar architecture in these bats.

Broad scanning movements probably made to seek the direction of prey or obstacles are of a different and grosser nature to these small vibrations, which might measure echo amplitude or perhaps produce "local Doppler shifts to increase the effective directionality of the pinnae". (Pye and Roberts, 1970.)

In horseshoe bats the zone of most intense sound is a beam of interference waves radiating from the nostrils. Mohres (1953b) has shown that the C.F. wavelengths of these bats measure twice the distance between the nostrils. It is possible that similar acoustic conformities may be found in laryngeal boxes or vocal chords of different size, so that simple-nosed bats may also have size-determined frequency patterns.

It has already been remarked (p. 182) that many microchiropteran genera contain pairs or even series of species that are distinguishable by size alone and each exists within a narrow range of measurements. The ultrasonic patterns of bats might conceivably turn out to be useful taxonomic criteria, for it seems possible that different-sized acoustic structures making qualitatively different signals might raise barriers between closely related forms (see Pye, 1972). The fragmentation of size clines or the geographic isolation of bigger or smaller races might initiate speciation of this sort. Metric differences within well-defined bat groups or complexes are a largely unexplored and interesting field of investigation. The table opposite lists some African bat genera containing "metric species", i.e. species that are separable principally on metric differences alone.

The extraordinary variety of bats—and presumably also their ultrasonic characteristics—became established at a very early date, as Eocene bats (the earliest known) had already diversified and some were recognizable as representatives of modern groups. The list of fossil bats (p. 190) not only reveals that the representatives of several families existed in the Eocene but that many modern genera were probably well established by the Miocene.

Their remarkable stability is betrayed by the very wide distribution of specialized types within families or genera. For instance, *Hipposideros* is an

FOREARM MEASUREMENTS

20 30 40 50 60 70 80 mm	Species	Species group
36—45	N. hispida	Nycteris hispida
57—66	N. grandis	group
32—36	N. nana	Nycteris javanica
36—46	N. arge	group
47—50	N. major	
50—60	R. fumigatus	Rhinolophus
62—67	R. hildebrandti	fumigatus group
45—50	R. darlingi	Rhinolophus
50—57	R. clivosus	clivosus group
40—48	R. landeri	Rhinolophus
49—59	R. alcyone	landeri group
59———71	H. cyclops	Hipposideros
72—77	H. camerunensis	cyclops group
41—44	H. beatus	Hipposideros
45—48	H. caffer	caffer group
48—58	H. ruber	
26—29	E. pusillus	Eptesicus
29—35	E. capensis	capensis group
43—50	S. leucogaster	
50—65	S. nigrita	Scotophilus
75—80	S. gigas	
27—31	K. muscilla	Kerivoula
30—32	K. harrisoni	argentata group
34—39	K. argentata	
35—39	M. minor	Miniopterus
42—47	M. schriebersi	
45—50	M. inflatus	

Old World genus containing several interesting species groups. Of the seven divisions listed by Hill (1963) several seem to represent groups that have adapted to distinct niches, common to the Old World tropics. Three groups are limited to Asia or Africa alone, but four groups have both African and Australo-Asiatic representatives. Convergence alone could hardly explain the situation; for the shared resemblances are very numerous. Tate (1941) described the noseleaves of the Afro-Australasian *Cyclops* group as "so peculiar, specialized and seemingly functionless that they are unlikely to have arisen independently". In his revision of the genus, Hill (1963) also remarks on this group's structural resemblances: "despite the curious distri-

CHART OF FOSSIL MICROCHIROPTERAN FAMILIES AND THEIR OCCURRENCE (TIME AND PLACE)

Family	Fossil form or ancestral type	Continental deposits	Age
Rhinopomatidae	Not known, but Eocene *Palaeochiropteryx* had similar humerus		
Emballonuridae	*Vespertiliavus*	Europe	Eocene—Oligocene
	Taphozous	Europe	Eocene? Oligocene
		Africa	Miocene
Nycteridae	Not known	—	—
Megadermatidae	*Necromantis*	Europe	Eocene—Oligocene
	Megaderma	Europe	Miocene?
		Africa	Miocene
Rhinolophidae	*Rhinolophus*	Europe	Eocene—Recent
	Palaeonycteris	Europe	Oligocene
Hipposideridae	*Hipposideros*	Africa	Miocene
	Palaeophyllophora	Europe	Eocene—Oligocene
	Paraphyllophora	Europe	Eocene—Miocene
	Pseudorhinolophus	Europe	Eocene—Miocene
	Asellia	Western Asia	Pleistocene
Vespertilionidae	*Nycterobius*	Europe	Eocene—Oligocene
	Myotis	Europe	Oligocene—Recent
	Pipistrellus	America	Pliocene—Recent
	Eptesicus	N. America	Pleistocene
	Miniopterus	Europe	Pleistocene
Molossidae	*Tadarida*	Europe	Miocene—Recent

Data from Romer (1945), Walker (1964), Grasse (1955), Piveteau (1958).

butional pattern of its members" (Papua, northern Australia and forested Africa) "they must be considered to share a common if remote origin". Of the major divisions of the genus he concludes: "It is evident that their separation occurred at a very early stage in the evolution of the genus".

The *cyclops* group have evidently specialized their echo-locatory equipment for they have enormous cochleae and ear pinnae, with a highly developed noseleaf. Perhaps, for all their specialization, they are unable to compete with other bats outside those limited forest areas or habitats in which they have acquired a stable niche.

The last forest connections between Africa and the Far East were broken at an early date and it is possible that *Hipposideros* might have dispersed through the tropics of the Old World at much the same time as the chevro-

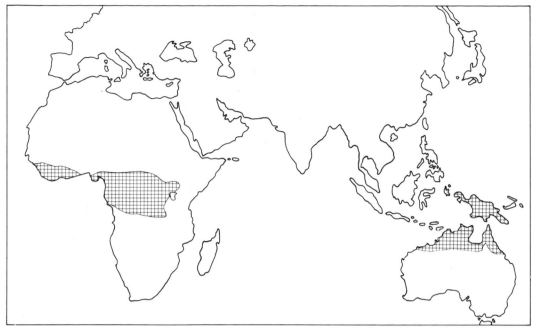

Distribution of *Hipposideros cyclops* species group.

tains, pangolins, lorises and apes, perhaps in the Oligocene and no later than the earlier Miocene. Whatever the period of dispersal, the genus had at this time already stabilized into several well-adapted forms and differences between contemporary species contained within these species groups are probably due more to long-term isolation than to major environmental considerations. Similar examples might be drawn from *Rhinolophus*. As in some birds there are a number of widely distributed genera that have been emancipated from regional limitations through their ability to fly across geographic barriers. *Miniopterus schreibersi* has a very extensive distribution in the Old World and Australia, while *Myotis*, *Pipistrellus*, *Eptesicus* and some other vespertilionid and molossid genera range as far as America. Attempts at understanding bat distribution patterns are obscured by many areas of ignorance: 1) the lack of adequate records, 2) the probability of migration in many more forms than we know of, 3) the great age of many bat species relative to other mammals, 4) lack of knowledge of their physiological, dietary and ecological needs, 5) ignorance of the role of competition between species, and so on.

Notwithstanding these limitations there is a surprising degree of correspondence between some bat distributions and those of other mammals within the continent. There are, for instance, relic groups living in the same refuge areas as other ancient forms. The most primitive species of *Rhinolophus*, *R. ruwenzorii*, is only found on Ruwenzori but has a related form, *R. maclaudi*, in Guinea. The very ancient insectivores of the genus *Micropota mogale* are likewise limited to the Ruwenzori and Guinea Refuges. There are forest species strictly limited to the main forest belt and others that extend into the outlying forests of East Africa. As with other mammals, western Uganda seems to harbour an astonishing range of species. Savanna-adapted

Hipposideros commersoni.

forms such as *Lavia frons* range all over the tropical African savannas. Very local species are not unusual and these distribution patterns are discussed in the text.

There are a number of desert or Middle Eastern forms that extend into the more arid parts of Kenya or along the coast, e.g. *Rhinopoma hardwickei*, *Asellia tridens* and *Triaenops persicus*. There is the interesting case of *Rhinolophus blasii*, a bat of the Mediterranean and South-west Asia which is also found in southern and central Africa but appears to be absent from the tropical belt; presumably there are competitors amongst the numerous tropical species.

The most widespread of all bat species is *Miniopterus schreibersi*, the dominant species in a subfamily containing only one genus. Although there are several more local "metric species", the extensive range of the migratory *M. schreibersi* seems to be due to its having effectively filled some niche that is not challenged by any other bat group; what exactly this niche is awaits definition, and the species is a strong candidate as a subject for further research (see p. 308).

Some species seem not to range much above sea level, e.g. *Triaenops persicus* and *Miniopterus minor*. Some species are severely limited by ecological factors. For instance, in some habitats, *Lavia frons* has to change its roosting sites during the dry season, because of drought and fires removing all the leaves and large tracts of land are not fully exploited by bats because of the absence of suitable shelters. Where species have adapted to peculiar roosts, like *Platymops* under loose stones and rocks, or *Mimetillus* and *Laephotis* in bark crevices, these bats are probably at a special advantage when alternative shelters are scarce.

The roofs of modern houses are well suited to shelter bats and some species seem to colonize them more readily than others. The bat fauna of Africa has probably undergone very considerable changes in recent years as a result of development. More houses being built favours species like *Tadarida pumila* and *Tadarida condylura*, *Scotophilus nigrita* and *Hipposideros caffer*, while the continuous felling of trees and clearing of bush and forest is probably making other species very much more local. Modern timber management, which aims to maintain the seral stage of the forest cycle, will ultimately deprive many bats of the shelter of hollow tree trunks. These bats, like orchids, anomalures and many other life forms dependent on old, decaying trees can be expected to become rare or extinct in the managed forests. The digging of mines may also lead to the creation of new shelters, so that many thousands of bats may move in and breed in areas where their numbers were formerly restricted. According to local informants, such a situation appears to have taken place in the West Nile district since 1935, when mines near the banks of the Albert Nile were abandoned.

The feeding habits of bats are closely tied to their ecological niches and there are specialists in moths, in mosquitoes, in beetles and flying ants, etc. However, opportunism has been the order of the night when I have netted on occasions when termites were swarming—every bat species caught had its belly distended with winged termites.

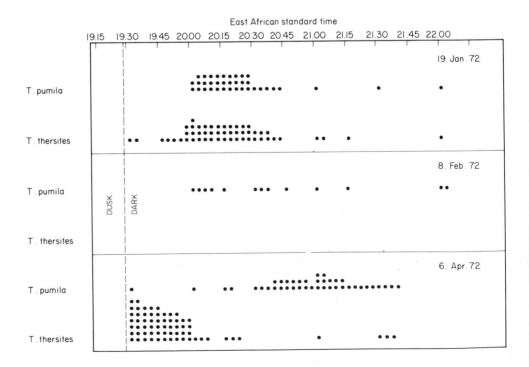

The pattern of feeding and drinking varies a great deal from night to night and from season to season as is revealed in the following records of catches made over a pool near Kampala. The molossids, in particular, seem to come to drink after feeding and the timing of their drinking seems to vary with the success of the evening hunt. The first graph shows the times of coming to drink on three different evenings, pointing out the variation in two molossid species, *Tadarida pumila* and *Tadarida thersites*. January 19 was an evening after rain when both species had been feeding on flying termites and all bats had full stomachs (mean stomach weight about 2 g). February 8 was a dry night and the few bats caught had eaten very little (mean stomach weight 0·5 g). April 6 was another night when termites were flying. The great majority of *T. thersites* were females that had recently given birth and were lactating, which probably explains the difference in pattern between the two species,

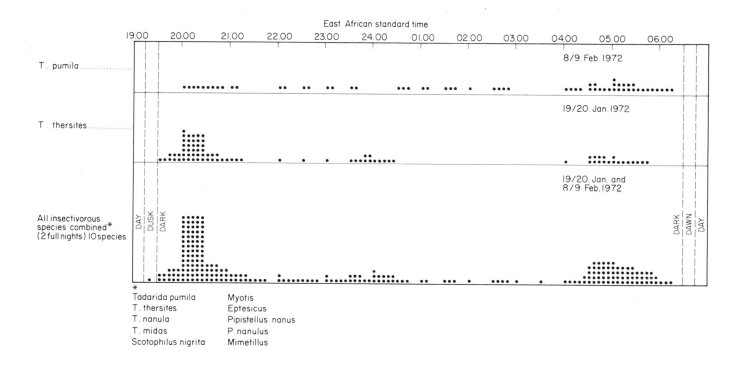

*
Tadarida pumila   Myotis
T. thersites   Eptesicus
T. nanula   Pipistellus nanus
T. midas   P. nanulus
Scotophilus nigrita   Mimetillus

for *T. pumila* is not a seasonal breeder. The second graph shows the general pattern of drinking early in the night and then again before dawn. February 8 (shown for *T. pumila*) was a night when food was scarce. January 19 (shown for *T. thersites*) was a night with abundant food. The third figure totals both nights' netting for all insectivorous bats coming to drink during the night. There is some evidence of a minor drinking visit at about midnight.

The use of sonar in feeding has been touched upon in the discussion of sonar; some catch fast insects in flight, others slow ones as they flutter about, others pick insects off leaves or off the ground. The prey is often seized directly by the open mouth in flight. Other bats have been observed to use their wing or tail membranes to "field" insects. Some bats temporarily hold insects in the cheeks, presumably so that they can keep on hunting and echo-locating. A *Mimetillus* that I caught had the head parts of a large-jawed species of

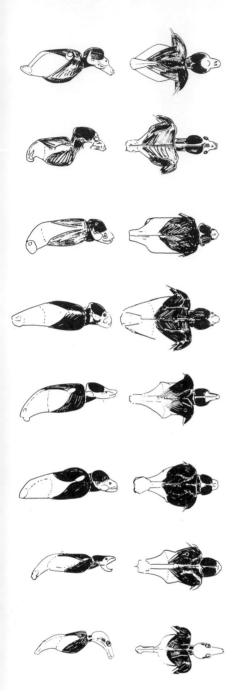

winged ant attached to the corner of its mouth, which had been severely mangled.

Most microchiroptera have glands which fluctuate in size and appear, in most instances, to be connected with sexual behaviour. Emballonurids have glandular secretions with which they smear their favourite roosting perch, so that a sexual function for scent may be augmented by a territorial one.

I have watched male *Taphozous peli* scratch their gular pouch with the hindclaws and in subsequent grooming transfer the gular secretion to other parts of the body. Likewise, male *Hipposideros commersoni* scratch the peculiar glandular pocket behind the noseleaf.

Most vespertilionids have glands on the muzzle or face which swell immensely during the breeding season. The orange secretion of the male of *Eptesicus tenuipinnis* is so copious at this time that it stains the entire bat a pale pink and the face tends to lose its fur—perhaps from too much rubbing and scratching. Grooming with the feet and teeth occupies long hours of the daily lives of most species.

The drawings of bats in this volume can give only a limited idea of the character of bats, but in the field it is soon apparent that each species is highly distinctive. Each form hunts differently, hangs or crawls differently, shouts differently, flies its own way at its own speed and so on. The silhouettes of bats' wing shapes on p. 110 give some idea of the variety of their aerodynamics. Even more impressive is a comparison of the body shape of some Microchiroptera. Slender streamlined forms, blunt tubby ones, large-headed, broad-shouldered, small-headed flat forms; indeed, the fascinating variety of types dealt with in this volume invites an exploration of form and function.

It is beyond the scope of this book to embark on a comparative study of bat forms, correlating shape with information on flight speeds, hunting methods, roosting posture and other behavioural and ecological data, even if these were known in any detail. However, information is accumulating and the profiles may help to encourage such study. The schematic outlines (opposite) emphasize the diversity of body form in bats.

Bats seem to be peculiarly long lived. A *Miniopterus* ringed in France is known to have lived 15 years, while a German *Rhinolophus* has been recorded still living 24 years after it was first ringed.

The societies of bats comprehend every type of aggregation, stable male-female pairs, solitary mothers with offspring, nursery groups, all-male or all-female groups and mixed crowds of many thousands. Some species space themselves (emballonurids), others hang in clusters while others pack in tight wadges of close bodies. The latter are often restricted to a few favourite caves where they may live in thousands. The very few known roosts of *Otomops*, for instance, are tenanted by vast numbers and it seems unlikely that small well hidden groups are scattered about but, rather, that major roosts of these fast, long-distance flyers serve a very large area, perhaps comprising many thousands of square miles.

Bat migrations are a well studied phenomenon in Europe and America and there is little doubt that they occur among many African bats too. At present this is suggested only by the appearance of species during particular periods of the year and in localities where they have not been caught at other times.

Nycteris (Nycteridae).

Cardioderma (Megadermidae).

Tadarida (Molossidae).

Taphozous (Emballonuridae).

Myotis (Vespertilionidae).

Glauconycteris (Vespertilionidae).

Mimetillus (Vespertilionidae).

Megaloglossus (Macroglossinae).
        (Megachiroptera).

196

The most important predator of tropical bats is probably the bat hawk, *Machaerhamphus*. This bird is rather like a falcon in shape and its aerobatics when hunting bats are most impressive. It seems to prefer coming at its prey from below and makes tight twists and turns as it strikes at flying bats. I have on three occasions netted bats with serious wounds that had probably been inflicted by this bird. Other birds that have been recorded killing bats are kestrels, *Tinunculus*, harriers, *Circus*, and pied hornbills, *Tockus fasciatus*. Britton (1972) saw an early-flying bat buffeted and attacked by swallows as it flew over a pond in Kenya. It was finally forced to hide in some reeds.

As bats often live in houses or caves, they sometimes come into relatively intimate contact with people. One of the most widespread human parasites,

Hipposideros commersoni.

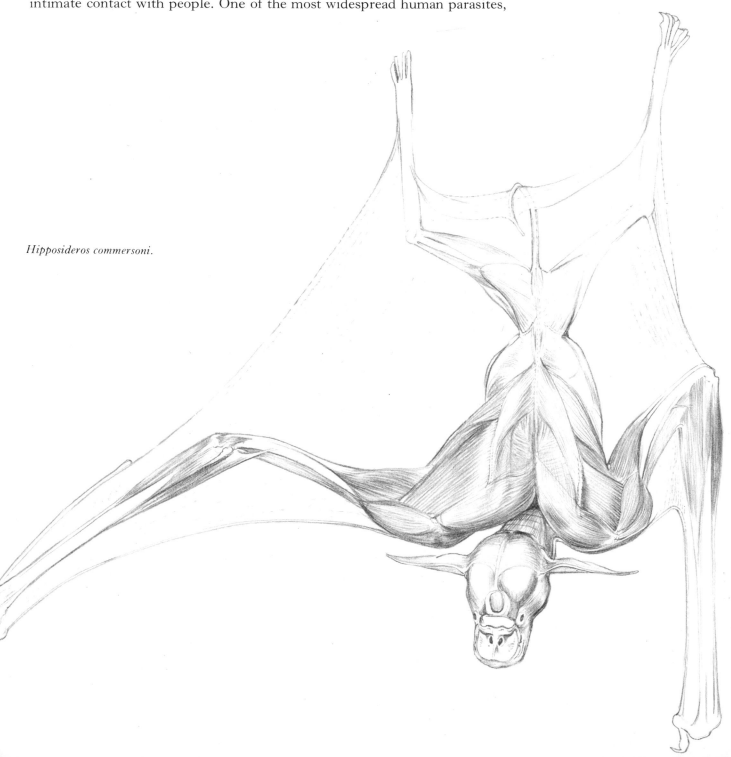

the bed-bug, *Cimex*, has probably taken to man as a secondary host. The majority of *Cimex* species live on bats, and Usinger (1966) considers it likely that bed-bugs adapted to live on man in shelters that were shared by man and bats.

In the Americas, rabies has been transmitted to man through bites from bats. This is not known for Africa in spite of extensive investigation. Bats, however, have been used as experimental animals in virus research, notably at the East African Virus Research Institute at Entebbe. Viruses that have been isolated in East African bats are listed below.

EVIDENCE OF NATURAL INFECTION IN BATS

Virus	Reference *(See Wimsatt, 1970)
Chikungunya	Shepherd and Williams, 1964
Semliki forest	Shepherd and Williams, 1964
West Nile	Taylor *et al.*, 1956
Usutu	Simpson *et al.*, 1968
Mt Suswa bat	Henderson *et al.*, 1968
Dakar bat	Williams *et al.*, 1964
SAH 336	Shepherd and Williams, 1964
Uganda S.	*Andral, 1968
Yellow fever	Shepherd and Williams, 1964
Zika	Shepherd and Williams, 1964
Entebbe	Lumsden *et al.*, 1961
Ntaya	Simpson *et al.*, 1968
Bukalasa	Williams *et al.*, 1964
Bunyamwera	Shepherd and Williams, 1964
Lagos bat	*Boulger and Portfield, 1958
Mt Elgon bat	*Metselaar *et al.*, 1969

Experimental infection of African bats with human pathogens has also been reported, notably plague, *Pasturella pestis*, by Leger and Baury (1932). Buruli ulcer, *Mycobacterium buruli*, has been reported by Church and Griffin (1968), and sleeping sickness, *Trypanosoma*, by Stiles and Nolan (1931). In no instance were bats thought to be natural hosts of these diseases. A type of relapsing fever, *Barrelia*, has been found in bats. Rousselot (1958) claimed that bats in the roof of a school in West Africa caused severe headaches and illness. Spores of the fungi, *Histoplasma*, *Cladosporium* and *Microsporium* have been found in bat guano in America and these can cause illness when inhaled.

However, the greatest importance of bats to man in the tropics lies less in their role as experimental animals or as vectors of rare diseases than in their role as non-toxic insecticides. Where important crops are harassed by flying insect pests, it is not too far-fetched to suggest research into the possibility of providing bat shelters.

Much remains to be learned about the breeding of tropical bats. They have rather lengthy gestation periods of three or four months, but post-natal growth is generally very rapid indeed. There is sometimes sexual dimorphism; the female is larger in *Lavia frons*, *Nycteris hispida* and *Coleura afra*, the male in *Rhinopoma*. The male may have special glands or glandular tufts as in many species of *Rhinolophus* and *Hipposideros*.

Young weigh about 20% of the mother's weight at birth and most species bear a single young, although twins are common in some vespertilionids. The milk is so rich in fats that it is sometimes semi-solid. The young are born with large heads and feet and short forearms, proportions that change rapidly with growth.

African sheath-tailed bat
Coleura afra

Naked-rumped tomb bat
Taphozous (Liponycteris)
nudiventris

Black hawk bat
Taphozous
(Saccolaimus)
peli

GENERA AND SPECIES

**Mauritian sheath-tailed bat
Taphozous mauritianus**

**Hildegards sheath-tailed bat
Taphozous hildegardeae**

**Perforated sheath-tailed bat
Taphozous perforatus**

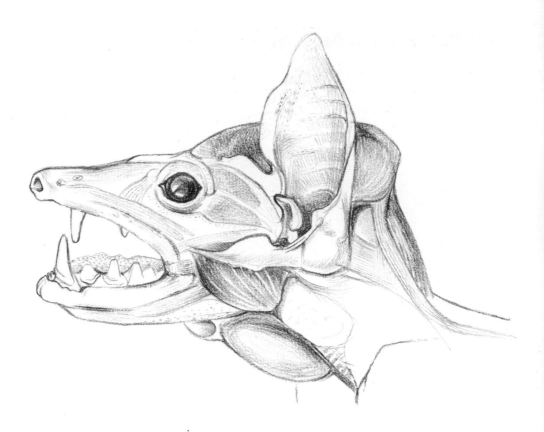

Sheath-tailed Bats

Emballonuridae

This family is among the more primitive of insectivorous bats but is, nonetheless, a successful and widely distributed group, ranging from Oceania and Australia through the tropics and sub-tropics of both the Old World and the Americas.

Fossils of this family are known from as early as the Eocene in Eurasia and the Miocene in Kenya.

A great deal remains to be discovered about the emballonurids and their behaviour. Scents clearly play a great role in their life and almost all species possess a powerful glandular secretion, which is probably connected with their sexual and territorial behaviour.

The peculiar arrangement, whereby the tail is partly enclosed within the inter-femoral membrane and partly free, seems to be a device that allows the tail to be used both as a sensory probe when at rest and as a support and control for the membrane when in flight. Lang and Chapin (1917) described this mechanism in *Taphozous peli*: "By stretching their hindlimbs this membrane slipping quite easily over the tail vertebrae can be considerably lengthened. So by pulling in or moving out their legs these bats can set their sail, as one might say".

The genus *Taphozous* contains the largest number of species and is considered to be the most highly evolved within the family.

African Sheath-tailed Bat
(Coleura afra)

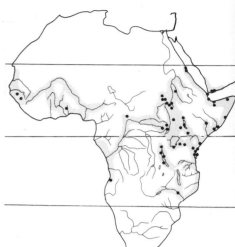

This small bat has a short, vertically held tail, sooty-brown colour and an exaggerated emballonurid posture. The tragus of the ear has a simple rectangular shape with a distinct pimple on its outer edge.

The lower jaw can be distinguished from that of *Taphozous* by the presence of six incisors instead of four. The nose overhangs the lower jaw, giving the impression of a miniature trunk.

This species ranges through the African woodlands and also into Somalia and Arabia. It has a marked proclivity for roosting in caves along the sea or lake shore but it is also found in rocky outcrops and in houses.

It feeds on small insects and moths. The flight of this bat as it emerges from the home cave has found scant mention in the literature and individuals away from their roosts appear to be inconspicuous. Although it may be netted in the immediate vicinity of the cave, experienced collectors have failed to catch *Coleura afra* in the open or over water with mist nests, a fact which may either reflect on the species' feeding and drinking habits, or on its capacity to detect and avoid nets. At any rate its hunting methods and flight patterns are not known.

In caves, the common posture of these bats is on all fours, belly to the wall and the head raised so that the nose points prominently upwards. More rarely they are found hanging freely from the roof of the cave by their feet. They dislike flying in daylight and are reluctant to leave their roosts, which often harbour very large numbers of individuals; colonies of thousands are known. The adult bats are generally well spaced, but younger animals sometimes cluster in large groups and clamber over each other. Sexes are thought to be evenly matched and a one to one sex ratio is apparent at birth. Kock (1969) found that females were on average slightly larger than males.

Sheath-tailed bats often share their caves with other species of bats but seldom intermingle, except with *Taphozous mauritianus* in some coastal caves. They often occupy the lighter areas nearer to the cave's entrance or exit, while other species occupy deeper recesses. *Hipposideros caffer*, *Taphozous*

African Sheath-tailed Bat (Coleura afra)

Family	Emballonuridae	Measurements
Order	Chiroptera	**head and body**
		52—66 mm
		tail
		10—19 mm
		forearm
		45—55 mm

spp., *Triaenops*, *Rhinopoma* and *Asellia* have all been recorded from the same caves as *Coleura afra*.

Populations seem to vary in their seasonal stability. Fluctuations in numbers have been noticed at Lake Baringo, Kenya, and also at Suakin on the Red Sea coast, and it is possible that, in those rather dry parts of their range, a seasonal shortage of insects forces them to migrate. However, numbers are thought to remain stable in some other caves.

This species is apparently very sensitive and Kock (1969) had a large proportion of his adult captives die within half an hour of being caught. Juveniles appeared to be more resistant.

Breeding has been found to be markedly seasonal in the Sudan, where the young are born in October, at the end of the rains. Pregnant females have been recorded from the Tanzanian coast at the end of December. Births may therefore, occur in the dry spell between the short and the long rains. A high proportion of the females have young each year, and Kock estimated a 75% reproduction rate. Only one young is born at a time and it is carried continuously by the mother, which will not accept any other young. The young are able to fly on their own when the length of the forearm has attained 38 to 39 mm.

Tomb Bats (Taphozous)

Taphozous: typical emballonurid bats with a pocket on the wing (radio-metacarpal pouch), skull with deep frontal depression.

Species

Taphozous mauritianus
Taphozous perforatus
Taphozous hildegardeae

T. mauritianus: common, widespread bat found in all moist open habitats. Grizzled grey with white belly. Male has large gular sac.

T. perforatus: drab coloured bat without gular sac in either sex. Found in semi-arid habitats.

T. hildegardeae: endemic to East Africa, mainly a coastal species. Without gular sac, but with conspicuous dark patch on throat, pale, whitish underside.

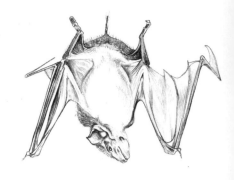

T. mauritianus.

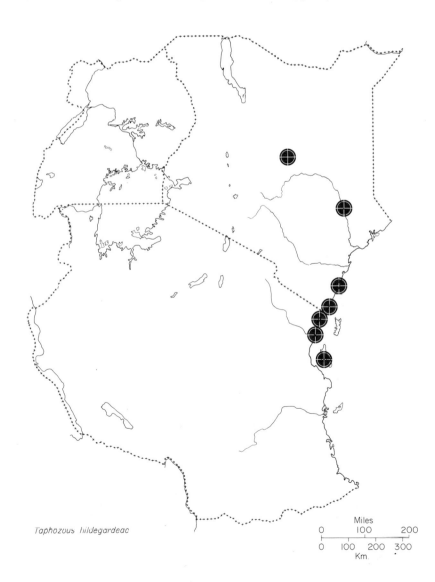

Taphozous hildegardeae

Miles
0   100   200
0   100   200   300
Km

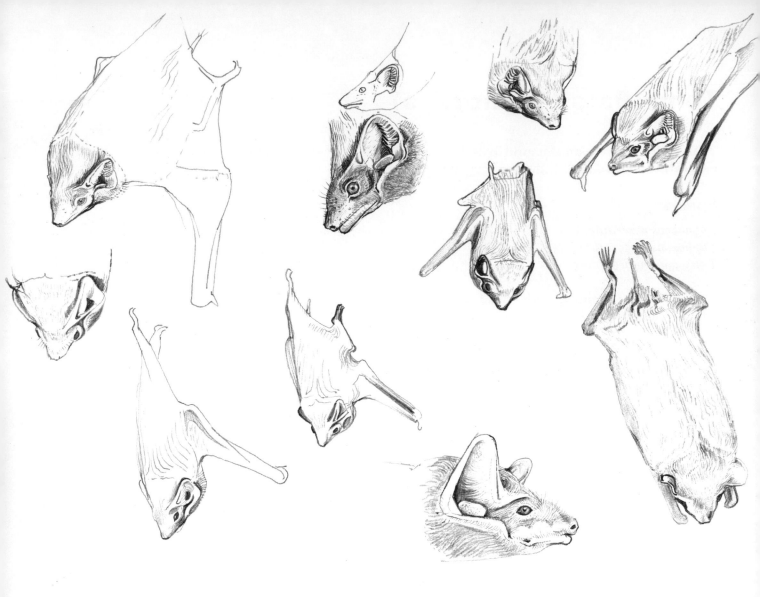

Tomb Bats
(Taphozous)

Family	Emballonuridae
Order	Chiroptera

Measurements
head and body
75—93 mm (*T. mauritianus*)
tail 22—28 mm
forearm 58·5—64·5 mm
weight 20—25 g

head and body
71—85 mm (*T. perforatus*)
tail 25—31 mm
forearm 60—67 mm
weight 22—25 g

head and body
76—87 mm (*T. hildegardeae*)
tail 23—33 mm
forearm 65—70 mm
weight 20—27 g

T. mauritianus is perhaps the most familiar insectivorous bat in East Africa, due to its conspicuous resting places on the outer walls of buildings and on the trunks of large trees, such as mango trees, palms etc. The habit of scurrying along walls sideways with its head down when people approach too closely, draws attention to its alert little face and bright eyes. In choosing its roosts it disregards light, although it dislikes the direct rays of the sun. Once a roost has been taken up by a pair or more of these bats they become permanently attached to that place and will return to identical stations after every disturbance. These stations become marked and stained either with urine or with the secretion of the gular sac. Individuals are generally spaced somewhat apart from one another; nonetheless, they form a recognizable group numbering from 2 to 12 animals (see drawings opposite).

T. mauritianus.

This species feeds primarily on moths and generally hunts in the vicinity of its day roosts at fairly low levels. It emits ultrasound through the mouth and flies very rapidly. Several observers have noticed that they follow very definite flight-ways, and may sometimes be seen chasing one another along these aerial highways, at which time a whirring can be heard (presumably made by their wing-pockets). Although this species is apparently awake and alert throughout the day, it often does not start hunting until it is really dark. Occasionally, however, it may fly before sundown, taking off from the vertical surface of its roost in a low dropping swoop. Indeed, roosts seem to be chosen to allow this.

In addition to the day roosts there are also favourite resting spots used only at night. After about three hours of intense activity after sunset, there are long rests interspersed with short flights.

When these bats are disturbed they make audible chirrups and afterwards they may squabble noisily while jostling to regain their position. Lang and Chapin (1917) reported that they fought both at the roost and in the air. These observations, the intervals between resting individuals and the marking of resting stations suggest that this species is highly territorial. The extraordinary tenacity to one spot is unusual in bats. Furthermore, when individuals or even entire groups are killed, the roost is often recolonized within a year. It is difficult to know whether the well-established smell of these spots is the principal attraction, or whether such sites are specially attractive because of ideal physical conditions.

Verschuren (1957) found evidence of an April—May birth season in northeastern Congo (Zaire). Lang and Chapin also recorded young in November and December. The single young is carried on the mother until it is able to fly on its own.

Taphozous perforatus is a bat that is found in North-west India, Arabia and right across the Sudanic woodland belt in Africa. It also occurs in eastern and central Africa, with isolated records from the Limpopo and Botswana. It will probably turn out to be more widely distributed than present records suggest, possibly throughout the southern woodlands as well. It is numerous along the Nile and the tomb bats acquired their name from this species, which roosts in most of the ancient monuments of Egypt. *T. perforatus* often hide within the narrow cracks found in masonry or rocks, but where they occur in large colonies, dense associations may be scattered over walls or rocks. A large colony in a deserted ruin on Lamu Island occupies the darker recesses

T. perforatus.

of an ambulatory round an enclosed courtyard. They seem reluctant to fly into the open in daylight. This species may be found in the same shelters as *Rhinopoma* and *Nycteris thebaica*, but the species remain separate.

They feed on moths and beetles and show a seasonal fluctuation in the accumulation of fat. Harrison (1958) found this species breeding twice a year in northern Nigeria.

T. hildegardeae is found in coral caves along the coast and, inland, as far as Chandler's Falls in central Kenya; it is very local. This species is apparently more closely related to an Indian *Taphozous* than to the other African species and might therefore be an isolate of a predominantly Oriental stock, or alternatively, a relic species. A comparative study of *T. hildegardeae* and *T. perforatus* in the area where they are sympatric might reveal what features limit the former or contribute to the success of the latter.

T. hildegardeae live in large colonies and each individual lies pressed close to the rock's surface. They occupy the lighter, more open parts of caves, associating closely with *Coleura afra*. *Triaenops* and *Hipposideros caffer* are found in the deeper recesses of the same caves.

They have a rapid flight and presumably feed on insects similar to those that form the diet of other *Taphozous* species.

Naked-rumped Tomb Bat (Taphozous (Liponycteris) nudiventris)

| **Family** | Emballonuridae |
| **Order** | Chiroptera |

Measurements
head and body
88 (83—93) mm
tail
27—34 mm
forearm
73 (69—81) mm
weight
approx. 60 g

Naked-rumped Tomb Bat (Taphozous (Liponycteris) nudiventris)

This greyish bat resembles other Taphozous species but can be distinguished by its naked rump and angular ears with large hatchet-shaped tragi.

Probably in correlation with its larger size and stouter teeth, the skull of *nudiventris* is more streamlined than that of the smaller *Taphozous* species; occipital and sagittal crests reduce the prominence of the cranium and thus give a look of the large *Taphozous peli* to the bat's head.

The distribution of this species follows a broad sub-Saharan belt. This bat occurs widely in the Middle East, India and the semi-arid areas of East Africa and Somalia.

The unique naked rump is possibly related to the habit this species has of carrying the young on the back rather than on the belly as other bats do (Brosset, 1962). Only the female carries the young but there must be some special stimulus for the behaviour, of which the naked rump is a manifestation.

T. nudiventris shelters in crevices in caves and dark buildings. In Iraq, Al Robaae (1968) found that they change quarters to escape spending the winter in cold buildings or caves and enter old buildings with wooden or rush ceilings. In East Africa as well as in India and Iraq there is a seasonal accumulation of fat in this species. In Iraq this precedes complete dormancy during the winter, but in India the winter is mild and the fat deposits merely coincide with a period of reduced flight activity and the bats do not hibernate.

Colonies range between about 200 and 1,000 individuals and their composition changes during the breeding season, as the males leave—or are perhaps driven off—by the females shortly before these give birth. The males rejoin the females only after the young are able to fly and, during the late pregnancy and lactation periods, all-female groups are the rule, although all-male groups may roost nearby.

Al Robaae has found delayed fertilization in Iraq bats, the animals mating shortly before going to their hibernation roosts. For the whole winter the sperm is stored in the uterus but, within a few days of the animal's emerging and flying off to summer roosts, the ova mature and descend to become fertilized. Gestation lasts nine weeks and the young are nearly all born within a period of ten days. This remarkable synchronization must depend upon the closely related stimuli that cause the bats to emerge from their winter torpor and fly out to their summer roosts. It would be interesting to know if there is a similar breeding pattern in tropical Africa.

The single young is born blind and naked with head and body measurements of about 45—50 mm and forearm measurements of 25—30 mm. The young cling to the mother's back for the first few weeks or so but leave her briefly to roost on the rocks beside her at the age of two weeks. From this time on the mother hunts alone, leaving the young hanging by themselves but close to one another. If disturbed they will bunch together. Growth is rapid and the eyes open before the end of the first week. At the age of one month the forearm measures 39—47 mm and reaches adult size before the animal is six months old. Hair starts to cover the body at one month and by the seventh week Al Robaae found that their stomachs contained a few half-digested insects.

A comparison of the habits of tropical populations with those of Iraq would provide interesting information and reveal which elements of behaviour are determined by the environment and which are specific features.

**Giant Pouched Bat,
Black-hawk Bat**
(Taphozous
(Saccolaimus) peli)

**Family
Order**

Emballonuridae
Chiroptera

**Measurements
head and body**
110—157 mm
tail
27—36 mm
forearm
84—97 mm
weight
92—105 g

Giant Pouched Bat, Black-hawk Bat (Taphozous (Saccolaimus) peli)

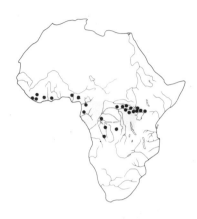

This very large black bat has a broad flat head and shoulders, large eyes, small ears and short greasy fur. Both sexes possess a gular sac. It has pointed wings, and its falcon-like flight has impressed many observers. It is a forest species, recorded from Liberia in the west, across to the forests of the Congo basin, to Uganda and to Kaimosi in western Kenya. It is restricted to the dense forest zones, wherein it shelters by day; in the evening it comes out to hunt along the margins of the forest, in clearings or along river valleys within the forest. It can generally be seen at dusk, hunting for moths and beetles. As it first appears it flies above the forest canopy crying loudly. Its call, which can be heard clearly even at a distance of a few hundred metres, has been rendered as "chuwee, chuwee". As darkness deepens, these distinctive cries, which have a timbre similar to those uttered by the fruit-bat *Eidolon*, betray that *Taphozous* is descending to lower levels and, on moonlit nights, it can be watched hunting in forest clearings until about 8.30 p.m., after which time it can no longer be heard or seen. In good weather this bat can feed to capacity in less than an hour. That such large bats can fill themselves with quantities of small insects in so short a time is some measure of the vast resources available to bats in tropical forests and of the effectiveness of this species in harvesting them. I have found the remains of small black and green beetles in their stomachs.

Falcon-like swoops are characteristic of this species, and these bats are also able to glide briefly in between their precise, rapid and clearly visible wing strokes; their aerobatics are sometimes reminiscent of arctic skuas when they make sudden snatches at flying insects. Brosset (1966b) suggests that they are opportunistic hunters, without fixed hunting grounds or flyways, as he found them disappearing and re-appearing sporadically in Gabon. In West Uganda, I have seen large numbers on clear nights and none on wet ones, and I have also noticed a greater abundance at some times of the year than at others, but no seasonal pattern has emerged. On a clear March evening I once saw some 25 or 30 of these bats all at once, dispersed over a wide area along the forest edge. Their calling was most noticeable as individuals converged on one another, and there seemed to be a distinct tendency for them to associate in pairs. Lang and Chapin found them roosting in hollow trees and noted that they remained well spaced and did not huddle together. It would be interesting to know if these roosting sites are marked with smears in the way that those of *T. mauritianus* are, and to learn more about the role of the gular sac.

The possession of a gular sac by both sexes suggests that its function in this species may differ from that in other members of the genus, where only males have the sac. The gland itself is very large. The typical posture of a resting bat is hanging with the belly flat against a surface with its head raised; this position puts the opening of the pouch in a most prominent place *vis-à-vis* any other approaching bat.

Captive animals frequently scratch their gular pouches with their hindfeet and I noticed that the yellow secretion was more copious in a male with enlarged testes. The scratching sometimes opens out the gland; the subsequent grooming of other parts of the body must help to spread the scent all over the animal.

Emballonurids have the reputation, unusual among bats, of being gentle creatures, but this is by no means true of this species; when these bats are caught they menace and bite ferociously. They do not, however, bite one another or harm other species put in the same cage with them. I have kept several wounded animals for varying periods of time. One male lived in captivity for nearly three weeks on scarcely any food or drink. It refused to eat or drink by itself and resisted being force-fed. This capacity to survive without food and suffering considerable dehydration is most unusual among bats that generally die rapidly when subjected to dehydration or stress.

The few available records give little idea of the species' reproductive pattern. Lang and Chapin recorded pregnancies in June and December for the eastern Congo. In Uganda, I collected four females in March and found that one had enlarged mammae and traces of milk glands, one had apparently just been fertilized, a thin animal was neither lactating nor pregnant and one had the uterus distended with semen. The latter record is of special interest, because *Taphozous nudiventris* is known to have delayed fertilization (see p. 209).

T. peli scratching pouch.

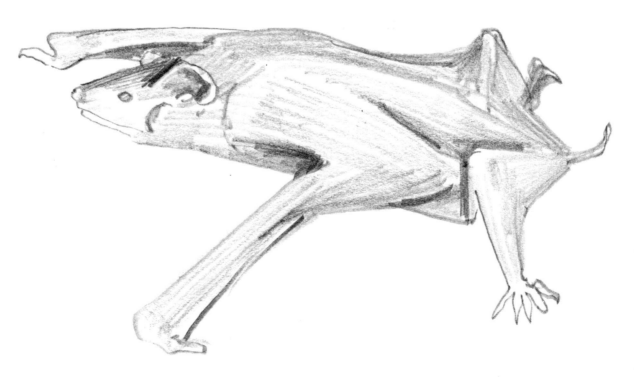

Leaf-nosed Bats

Rhinopomatidae (rudimentary noseleaf)
Nycteridae
Megadermatidae
Rhinolophidae
Hipposideridae

The leaf-nosed bats are distinguished by peculiar growths on their noses. These structures are related to the emission of signals through the nostrils rather than the mouth. They probably act as megaphones and acoustic beam focus mechanisms as well as sound baffles. Some lobes and flanges may deflect nasal sound from the ear pinnae, while others may actually move and thus open or close sound channels leading from the nose to the ear. The noseleaves of rhinolophids and nycterids have two channels emerging from the nostrils, one bearing out and forward, the other leading back to the conch of the ear. It is tempting to guess that this allows the bat to compare emission and reception in some way, but too little is known at present about the workings of bat ultrasonics.

It is known that these bats are immensely old; the Rhinolophidae, for example, are recognized in the Eocene. The noseleaves are therefore the product of over 60 million years evolution, and it would be surprising to find evidence of how such complex structures arose. However, the various living families exist at quite different evolutionary levels. *Rhinopoma* is a very primitive form and its noseleaf is only one of several anatomical features that betray a primitive status. It must be remembered, however, that all surviving families are specialized in some way or other, and individual genera and species are a mosaic of primitive and specialized features. Notwithstanding this, it is possible to suggest how the noseleaves of some forms became elaborated from simpler forms. The diagrams opposite represent in a schematic form how the noseleaves of a nycterid bat might have evolved from a simple-nosed bat. In order to show that the exercise is not altogether fanciful, and also to allow meaningful comparisons, I have drawn portraits from four living families.

Although the diagrams relate only to elaboration in the nycterids and the other families illustrated are un-related to this development, the comparisons give some idea of how a startling variety of noseleaves could derive from much simpler forms; they also show that behind the variety there are some common structural elements. More detailed illustrations and discussion will be found in the nycterid profile.

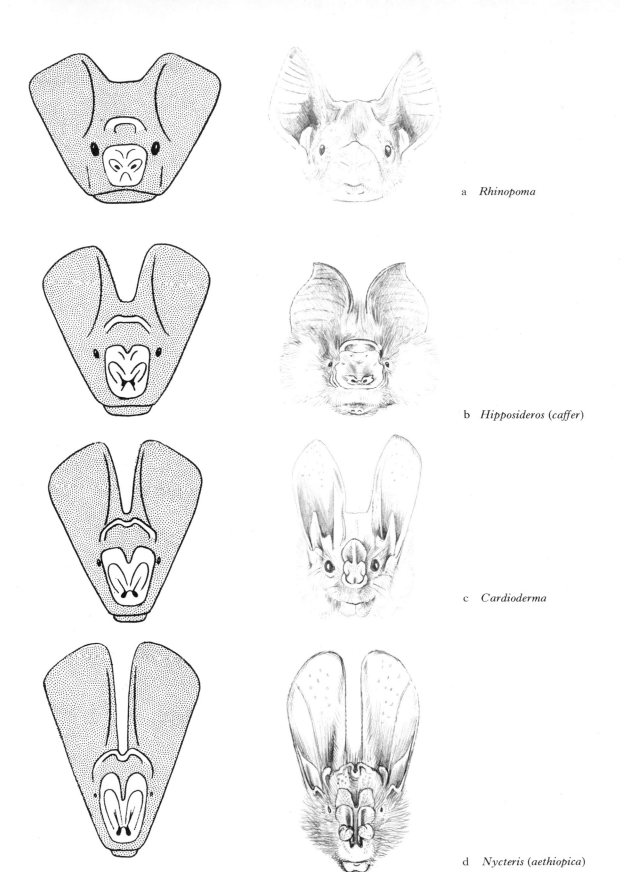

a   *Rhinopoma*

b   *Hipposideros* (*caffer*)

c   *Cardioderma*

d   *Nycteris* (*aethiopica*)

Mouse-tailed Bats

Rhinopomatidae

This family is regarded as the most primitive of the Microchiroptera. A very interesting feature of this family is the presence of a rudimentary nose-leaf and curious valvular nostrils. Because it emits signals through the mouth *Rhinopoma* is probably not in a strict sense a noseleaved bat.

The four species of this family and genus—*Rhinopoma*—range from South-east Asia through India and Arabia to North-west Africa. They are common bats in India and Egypt and they seem to have successfully specialized for life in the more arid areas. Northern Kenya marks the southern limit of their range.

| **Mouse-tailed Bat** (*Rhinopoma hardwickei*) | **Family** **Order** | Rhinopomatidae Chiroptera | **Measurements** **head and body** 52—63 mm **tail** 48—68·5 mm **forearm** 52—59 mm **weight** 10—12 g |

216

Mouse-tailed Bat
(Rhinopoma hardwickei)

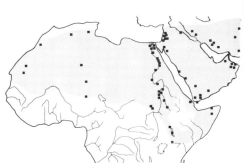

This sandy-coloured bat is distinguished by a long wispy tail and valvular nostrils set in a small pad-like noseleaf, which the scientific name describes as "nose lid". Immediately behind the nose is a depression and behind this and above the eyes is the membrane joining the large rhomboid ears. The eyes are large and rather prominent. This bat acquires fat deposits, which can grow to be very conspicuous around the base of the tail and hindlegs. When at rest, the forearm is strikingly long and thin.

The structure of the wing has been discussed by Revilliod (1916) who points out that the humerus resembles that of the primitive fruit bat type with small and underdeveloped tubercles. The outermost third finger is shorter in relation to the forearm than in any other bat, and Revilliod remarks that the long radius is a meagre compensation for the inadequate wing area. It looks as though *Rhinopoma* has a less highly evolved flight capacity than other insectivorous bats but that this is offset by physiological adaptations which allow it to live successfully in arid habitats.

This species ranges from Egypt, Sudan and Somalia across the Sahara to Morocco and northern Nigeria. Northern Kenya is the species' extreme southern limit. It is common on Gibraltar Island in Lake Baringo. *Rhinopoma* is found in fairly wide rock crevices and in caves. The peculiar length of the radius and the primitive humerus may have a connection with this bat's need for broader flight passages than other bat species. For instance, Kock (1969) found this species could not negotiate small spaces that were flown down by *Rhinolophus landeri*, a bat of similar body measurements and total wing length but with a shorter radius.

In the Saqquara ruins, in Egypt, Kock found this species sheltering in relatively unprotected but dark recesses. He suggested that the species may be able to shelter in drier situations than most other bat species. Such a tolerance for low humidity together with the capacity to store fat may give the rhinopomids an advantage that has allowed the group to survive and flourish in dry areas.

Rhinopoma feed on small insects; they have been found eating small beetles in Kenya. They hunt at a height of about five to ten metres, often flying in undulating swoops with long glides. They have sometimes been seen to flutter in an approximation of hovering. They use short frequency-modulated pulses with prominent harmonics; the signals are emitted through the mouth and it is possible that the operation of the muscles in the valvular nostrils could be linked with the emission of signals.

The deposits of fat on the hindquarters of adult *Rhinopoma* have caught the attention of all naturalists acquainted with this bat. The fat reserves which are white, soft and almost fluid have been noticed to fluctuate seasonally. Little is known about the timing and the duration of these fat deposits and observations are needed. Kenya specimens are fat in January, after the short rains, and Kock found fat bats in the Central Sudan between December and

March, and thin ones in March and April. It is probable that the mechanism is an aid to survival during periods when insects are scarce. Early authors thought it might be connected with aestivation, but observations have failed to reveal seasonal changes in activity and Kulzer (1965a) found that *Rhinopoma* maintained a constant temperature.

Rhinopoma clamber over vertical surfaces with some agility, they can also hang by the feet or, more rarely, by their thumbs. Both thumbs and feet are very slender, small and delicate. They have been seen to enter crevices backwards, at which time the tail appears to have a function similar to that of a cockroach's antennae, as it weaves about in all directions. The tip of the tail is armed with very fine hairs; these and probably the tail itself have a sensory role. It is commonly held at a stiff right-angle to the body while the animal scuttles over the walls of a cave. They can often be heard squeaking in their diurnal shelters and if disturbed may fly about the entrance in daylight. They fly out soon after dark.

These bats live in small scattered groups which generally contain both sexes. However, Panouse (1951) found all-male and all-female groups in Morocco. Colonies may be associated with other bat species and even with mice, *Acomys* species, and with hyraxes, *Procavia*. Kock lists *Taphozous perforatus*, *Nycteris thebaica*, *Rhinolophus landeri* and *Asellia tridens* as commonly associated bat species.

In some areas *Rhinopoma* may migrate and Panouse found that Moroccan colonies were subject to massive changes in numbers. Hoogstraal (1962) found that the numbers of this species diminished during the Egyptian winter, although some remained present throughout the year.

In the Sudan there appears to be a birth season in July. Kock found embryos in March, young during October and no pregnant females in December or January. Two females from Kenya were not pregnant in January. Males are on average slightly larger than females.

False nipples or "fastening teats" are a conspicuous feature of females and are situated on the lower belly, just above the os pubis and the associated fat deposits (see drawing above).

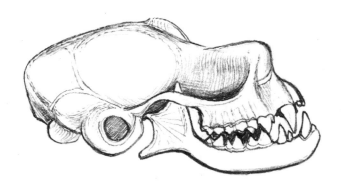

Slit-faced Bats

Nycteridae

N. nana.

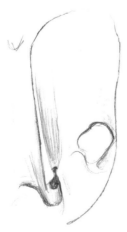

N. hispida.

This family is immediately recognizable by an elongated and foliated trench which runs forward from the forehead to include the nostrils; long ears appear to be concomitant with the structure. The bifurcation at the tip of a long tail—forming a Y shape—is also a typical feature of these bats, and all species have long silky fur.

The family contains 12 species within a single genus and, excepting one species in South-east Asia and one in Madagascar, it is restricted to the Ethiopian region.

The deep trench down the forehead, from which this group gets its English name, is surrounded by a unique structure of lobes and flanges, which is probably connected with the transmission and control of high-frequency "whispers" uttered by the Nycteridae. As they hang in their roosts they appear to take their bearings in a similar fashion to rhinolophid bats, twitching their noses and ears. It is thought that interference from the environment may be reduced by lowering the intensity of sound pulses. Also the close range of obstacles and prey may necessitate the very short pulses made by Nycteridae. Novick (1969) recorded members of this family and Megadermidae using pulses of 1 millisec to less than 0·5 of a millisec duration. By contrast, 50 to 60 millisec pulses were recorded from Hipposiderids.

Although the most obvious distinctions between species involve differing proportions, a closer look at the hairy noseleaf reveals that this structure, too, differs a great deal from species to species. Furthermore, some species appear to have elaborated the structure further than others. A pattern for the evolution of the nycterid noseleaf was illustrated earlier (p. 215). This very simplified scheme suggests that originally a channelled leaf developed round the nostril, presumably allowing some ultrasound to reach the ear directly. The nostril itself is in a steeply inclined groove facing forward; this must be the pathway for the bats' ultrasonic beam. Immediately above the nostril hole, the edges of the leaf fold forward to meet one another and serve to separate the two channels. Above these very small structures (see margin drawing opposite page) there is a special elaboration at the anterior end of the first leaf, which seems to reinforce the separation. These rounded, thickened lobes lie immediately above the nostrils. In the *aethiopica* group these stool-like lobes are free and constitute the topmost in a series of 3 layers of leaves. In the *thebaica* group, they still show some connection with the first leaf, suggesting that the *aethiopica* group derived from a *thebaica*-like stock. The *hispida* group show a more primitive condition with the lobe still an integral part of the first noseleaf. On the basis of their teeth and noseleaves, *N. arge* and *N. nana* are the most primitive species; both are relatively rare forest forms.

Nycterid bats have adapted to relatively slow foraging rather than fast hunting. They hover or flap about, rather like large moths, picking up stationary insects from a wide range of situations and habitats, or alternatively, they catch the slower species in flight, particularly when these form natural concentrations such as flying termites or ant swarms. These bats are com-

N. aethiopica group. .

220

monly seen feeding on insects attracted to lights. They have also become the principal type of insectivorous bat to exploit the rich, but rather dispersed, food resources of thickets, forest undergrowth and reedbeds. Verschuren (1957) found a high proportion of diurnal insects taken by nycterid species in the eastern Congo.

Slow flight may be an archaic characteristic as the humerus has a primitive form with small tubercles and the sternum is also undeveloped. A primitive status is also betrayed by the retention of premaxillaries and two incisors. On the other hand, the form of the nasal structure is peculiarly specialized. This combination of characteristics suggests that the ecological advantages of slow flight were exploited by the unspecialized ancestral nycterids. To exploit this niche to the utmost, the group has developed a peculiar sonar system to detect prey more efficiently and to navigate their environment.

Their low aspect ratio wings and low intensity ultrasound are both adaptations to the slow, close-range foraging in environments that tend to obstruct free flight and are also full of interference for the bats' echo-based sonar.

When resting they hang from their long, delicate legs and small feet which lack the *perineus* muscle. They are alert and sensitive to sound and movement. Their roosts are always well-shaded and very often in the cool, dark interior of hollow trees, caves, holes or buildings, although for a few species the foliage of trees and thickets may provide adequate shelter.

The slit-faced bats are physiologically delicate and very difficult to keep alive in captivity, apparently dying of dehydration soon after capture.

Social structure varies from largely solitary species—*N. aethiopica*—to highly gregarious species—*N. thebaica.*

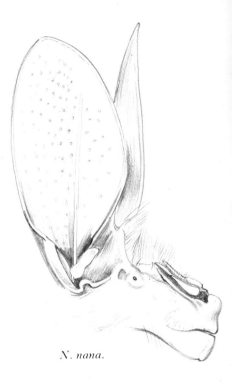

N. nana.

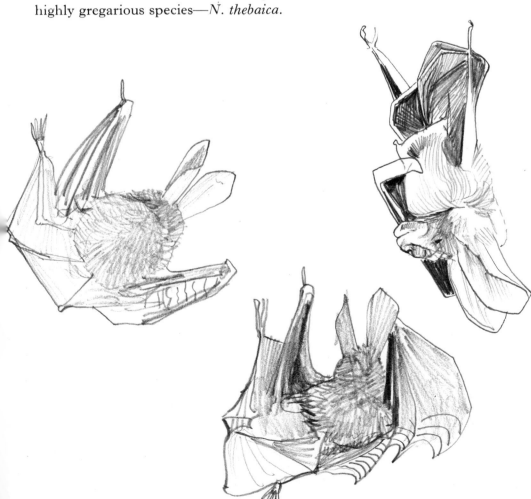

PICTORIAL KEY TO

hispida group

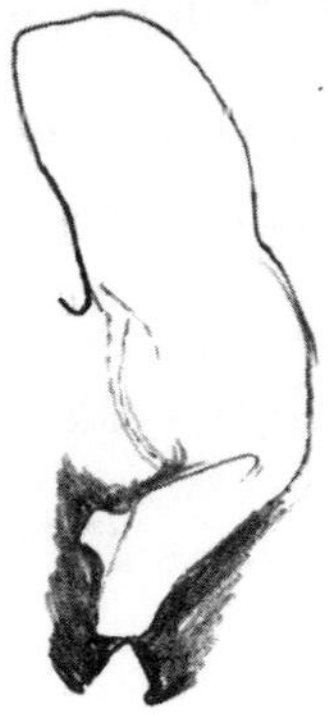

N. hispida

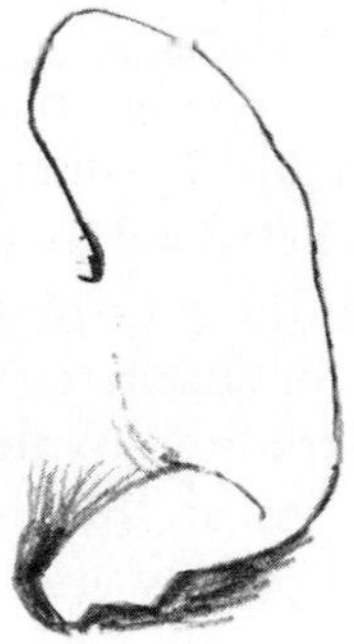

N. grandis

aethiopica group

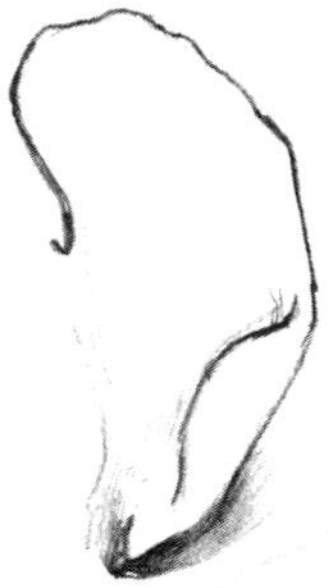

N. aethiopica

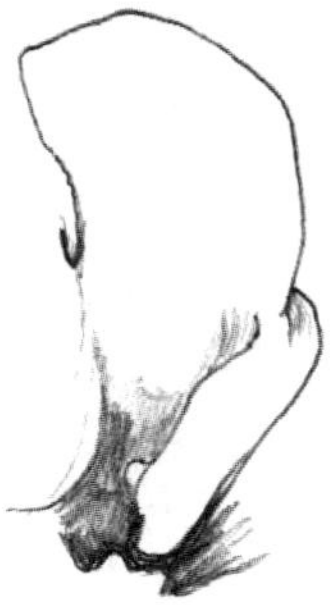

N. macrotis

thebaica group

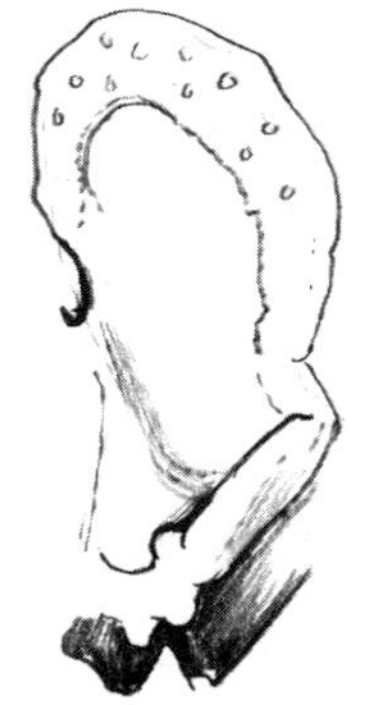

N. thebaica

arge group

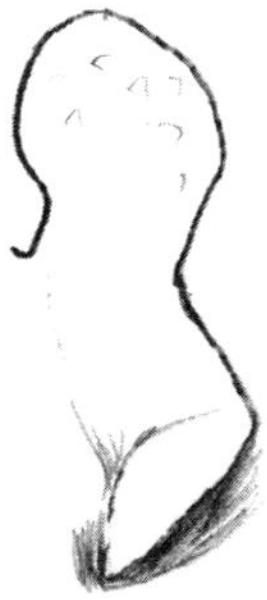

N. arge

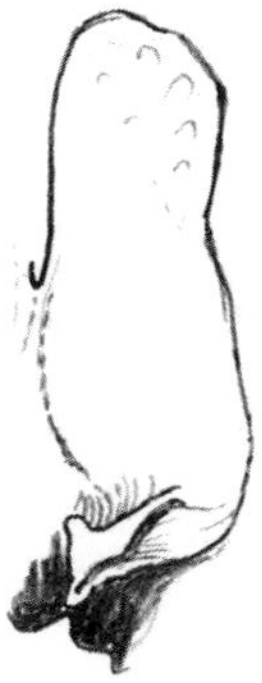

N. nana

THE GENUS NYCTERIS

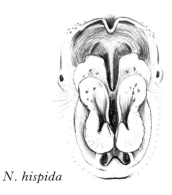

N. hispida

N. hispida
H. & B. 32—50 mm
Forearm 36—45 mm
Ear   17—25 mm
Wt.   6—10 g

N. grandis
H. & B. 63—93 mm
Forearm 57—66 mm
Ear   26—35 mm
Wt.   25—36

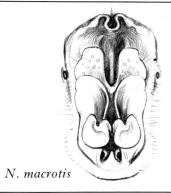

N. macrotis

N. aethiopica
H. & B. 54—60 mm
Forearm 45—52 mm
Ear   22—38 mm
Wt.   10—15 g

N. macrotis
H. & B. 52—63 mm
Forearm 45—50 mm
Ear   26—34 mm
Wt.   9—14 g

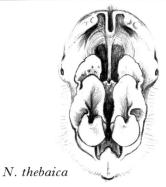

N. thebaica

N. thebaica
H. & B. 44—77 mm
Forearm 42—52 mm
Ear   28—37 mm
Wt.   13—28 g

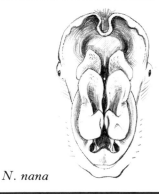

N. nana

N. arge
H. & B. 48—58 mm
Forearm 36—47 mm
Ear   25—33 mm
Wt.   8—10 g

N. nana
H. & B. 39—45 mm
Forearm 32—36 mm
Ear   20—22 mm
Wt.   5—8 g

**Slit-faced Bats
(Nycteris)**

Family	Nycteridae
Order	Chiroptera
Local names	
Ndema(Kikami), Tori (Kinyaturu)	

Slit-faced Bats (Nycteris)

Species

Nycteris hispida
Nycteris grandis
Nycteris aethiopica
Nycteris macrotis
Nycteris thebaica
Nycteris arge
Nycteris nana

Nycteris probably evolved in the first place as a forest bat. It is interesting, therefore, that many of the better known forms come from very dry areas. However, this secondary radiation has not led to the group's relinquishing their special ability to detect and catch insects in enclosed areas or in difficult terrain, and *Nycteris* is at a special advantage wherever there are reed beds, thickets or broken ground and in places where insects may be able to shelter from other bats. The pictorial key illustrates the diagnostic form of the tragus and noseleaf (pp. 222 and 223 respectively).

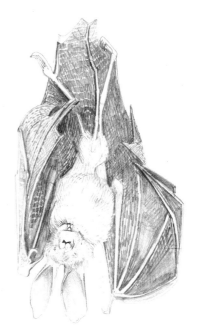

N. hispida is very common and widely distributed. It is frequently seen in and around houses and settlements, both flying at night or sheltering in roofs or deserted rooms during the day. It may also rest up in well-shaded bushes, aardvark holes, termitaries, hollow trees and papyrus crowns. This wide variety of roosting places and its widespread, conspicuous abundance contrasts with some other *Nycteris* species that are more exacting in their roosting requirements. They tolerate a wide range of climatic and ecological conditions, being found in dry open country as well as in moist vegetation types such as woodland and marsh. *N. hispida* frequents papyrus swamps, where it negotiates the channels and spaces between the clumps with great skill.

Like other Nycterids they are slow but agile flyers and easily avoid the efforts of people attempting to catch them with butterfly nets or racquets, even when in small confined rooms. The efforts of Lang and Chapin (1917) must have been repeated by many keen naturalists since.

"Sometimes we would close all the windows and try to catch them with butterfly nets—a warm but exhilarating sport. Now thoroughly frightened, they flew round and round the room, close to the ceiling, dodging our strokes with incredible agility, and hanging up from time to time in distant corners. Or down they fluttered repeatedly to the doors or windows by which they had entered, seeking a way out. Seldom could we net them until they were completely tired out, and not infrequently they eluded us entirely".

They feed on small moths and other insects; these may be picked off vegetation, from brightly lit walls or off the ground. Verschuren (1957) thought that this species fed during the early part of the night and it must have a very rapid digestion, as he found bellies and intestines were empty by the following day. They are nocturnal but can be alert during the day and may even start flying about within their shelters some two hours or so before dusk.

Groups of *N. hispida* are generally small, ranging from pairs up to about 20 individuals. Solitary bats are not uncommon. They are very attached to a favourite roost and after disturbance will return to the exact spot from which they were frightened away in the first place.

They are sometimes killed by predatory birds, owls and a marsh harrier have been recorded eating them. Kock (1969) had a captive sharing its cage with *Scotophilus nigrita* killed and eaten by it, but whether they would be attacked by other bats in the wild seems unlikely.

When netted, *N. hispida*, like other nycterids, lies still without tangling itself or struggling and it often flies out free again before being secured.

In the Garamba Park in northeastern Zaire, Verschuren found a biannual breeding season with most births in March and April and another minor season at the end of September. The new born has a forearm length of about

N. hispida.

N. grandis.

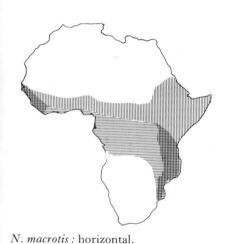

N. macrotis : horizontal.

N. aethiopica : vertical.

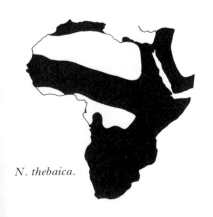

N. thebaica.

15 mm and may suckle while still attached by the umbilical. Adult size is reached in less than two months and the mother can fly carrying young of about her own size. Females may give birth in the company of males. Pairing seems to be common in this species. I once caught a female flying with her half-grown young attached. When the young was separated from her, its calls attracted another bat out of its dark roost into the daylight. This adult came and hovered close to it every time it called.

N. grandis is a large bat generally found in forests or forest relics across tropical Africa. It shelters in hollow trees and holes in the ground within the forest. Verschuren records three instances of finding them in hollows of *Mitragyna stipulosa*, a typical swamp forest tree.

Accumulations of large moth wings beneath their feeding roosts betray their food preference. They hunt mostly at low levels, fluttering close to the ground and seizing moths in flight or while settled. They do not eat while flying but return to the roost to eat. The unmanageable nature of their prey probably influences this habit.

This species has generally been recorded in pairs or as solitary individuals, but small groups have also been noted. Brosset (1966b) records foetuses or young in Gabon from April, August and November.

N. aethiopica is a relatively rare species but is known from a wide range across the northern savannas of Africa and extending as far south as southern Tanzania. Only solitary individuals have been found. Loveridge (1937) found them roosting in hollow baobabs in central Tanzania, where several females were nursing naked young in November.

N. macrotis may be an ecological race of *N. aethiopica*; it is primarily an equatorial form living in the forest and woodland belts of Africa. Occasionally it is found in open or even arid country but generally in moist situations. Its natural shelters are caves, rock holes and earth burrows and it is also found in houses, road culverts and in water reservoirs. They are often solitary but over a dozen have been found together, one group containing 15 males and 2 females. A female from Bagamoyo was pregnant in late December.

N. thebaica is generally pale in colour; it sometimes has an orange phase. Its ears are the largest in the family. It is common in open country across most of Africa south of the Sahara up to quite high altitudes and shelters in caves, rock holes, ruins and the roofs of houses and huts. It has been found deep in an aardvark hole. This species feeds on a variety of insect foods and in some areas feeds regularly on scorpions. Felten (1956) described this habit, which he observed in South Africa. *Antrozous pallidus*, an American bat of a different family but with broad wings and long ears like *Nycteris* has also acquired a similar facility in dealing with scorpions. (Photographs in Leen and Novick, 1969.) *N. thebaica* generally hunts in the vicinity of the day shelter and carries its prey back to the roost to eat it. Felten found that this species started hunting at about 9 p.m. Timing may not only be related to food abundance but, in very dry climates, it is possible that these bats, which have been found to die very easily of dehydration, may find the night climate at a later hour more suited to their delicately balanced physiology. They are also reported to avoid flying on windy nights.

This species is sometimes highly gregarious and Felten (1956) describes thousands in a cave in South Africa. Colonies of this size have not been reported for East Africa but they have been found in scores. Occasionally single individuals or smaller groups are also encountered. They associate with other species of bats and have been found in the same caves as *Rhinopoma*, *Rousettus aegyptiacus* and *Taphozous perforatus*. Loveridge (1922) found *N. thebaica* in Tanzania heavily parasitized by a small red acarine which congregated on the margins of the ears and wings and apparently caused the bat to damage its membranes by scratching at the irritation.

This species may be polyoestrous, as Harrison Matthews (in Harrison, 1958) found nine females from Tanzania both pregnant and lactating at the same time.

N. arge.

N. arge is a forest species found from Sierra Leone to western Kenya. It lives in small groups or pairs, sheltering in hollow trees, which are generally characterized by a low escape hole, so that the animal must drop down from its perch to leave the shelter. It is also found in houses and is attracted to small insects round lights. Verschuren (1957) found that this species, like *N. hispida*, fed during the early part of the night and had probably finished hunting by 10 p.m. It is commonly seen in village clearings in the forest. It roosts in hollow trees that are often shared with *Hipposideros cyclops*, *Rhinolophus landeri* or *Idiurus zenkeri*. Male bats have been collected, with either active or quiescent testes, at the same time of the year. The females are reported to nurse their young for about two months.

N. nana is a species very closely related to *N. arge* but distinguishable by its smaller ears and body size. It is a forest species only known from western Uganda. On several occasions, over a period of eight years, I have found a pair in the same hole—one dug by a giant pangolin. It is not known whether they were the same pair. The hole was deep in high-canopy secondary forest and when the bats were disturbed by thumping or by cigarette smoke they flew up into the surrounding undergrowth. Their flight is very dexterous. If they meet an obstruction they instantly perch, if only for a second, before continuing to dance their way through the undergrowth.

N. nana.

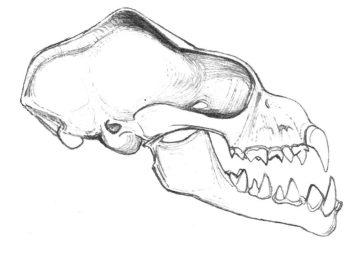

Nycteris grandis.

Large-winged Bats

Megadermatidae

This family is restricted to the Old World Tropics and Australasia. Members of this family, which are often referred to as the false vampires, resemble the nycterids in being relatively low flying and in having even broader wings. They also possess a mixture of primitive and evolved characteristics. They have a long erect leaf on the nose, but this is carried on a relatively flat nasal area, instead of being arranged over a pit—as in *Nycteris*—or mounted upon a bony swelling, as in *Rhinolophus*. They lack a tail.

All members of the family have a reputation for ferocity and in captivity they readily kill and eat other bats. The East African species feed mainly on insects but it is possible that they may also kill small vertebrates like their Oriental and Australian relatives.

Dougras (1967) has noted that *Macroderma gigas* will, on occasion, also eat fruit; a very interesting observation and a reminder that the diets of the two suborders of bats do not set an impassable barrier between them.

Heart-nosed Bat
(Cardioderma cor)

Family	Megadermatidae
Order	Chiroptera

Measurements
head and body
70—77 mm
forearm
54—59 mm
ear
35—40 mm
weight
21—35 g

Heart-nosed Bat
(Cardioderma cor)

    This relatively large bat is easily recognized by its big eyes and joined ears with sharp, bifid tragi. A robust muzzle is surrounded by a heart-shaped noseleaf. There is no visible tail.

    This species has a rather restricted distribution, down the eastern side of Africa from Eritrea to northern Zambia. Dry acacia bush and the coastal strip are its favourite habitats.

    Bats of this species are occasionally found in deserted houses or ruins.

229

Kulzer (1962a) found a colony at Tiwi, in a dry cave some 10 metres from the beach. The bats entered and left this cave through a perpendicular chimney. Light reached their resting places and they were apparently alert and awake throughout the day. They seldom share shelters with other bats and J. Williams has found the remains of other bat species under their roosts. This suggests that *Cardioderma* may have killed and eaten them (D. Pye, personal

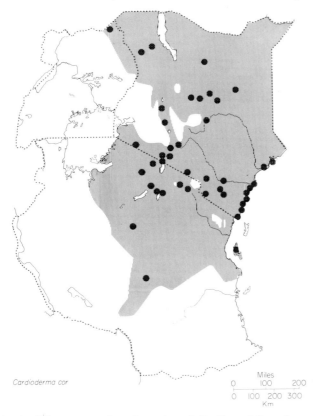

Cardioderma cor

Miles
0   100   200

0   100 200 300
Km

communication). The very closely related Indian *Megaderma* is known to kill mice.

They start hunting before it is really dark and these bats can be seen hanging from the lower branches of bushes scanning the surroundings for insects. They do much of their hunting from vantage points. They fly out to seize their prey and then return to their perch to eat it. Much of their food is taken directly off the ground and flightless insects probably make up a large part of their diet. The stomachs and guts of bats collected from a cave in August contained nothing at all, suggesting either that the bats were not getting much at this time or that digestion is very rapid. Dougras (1967) found that the Australian *Macroderma gigas* can take as long as six days to excrete its meal.

Cardioderma form small colonies or groups; they roost hanging by their feet with rather wide spacings between individuals.

Members of this species have been found to have a gestation of about 3 months. The young are born blind and hairless and are said to be carried by the mother for a period approaching two months. A new-born young has been collected in Kenya in August and 3 pregnancies, 3 lactating females and 2 juveniles for the same month suggest that August is in the middle of an extended breeding season.

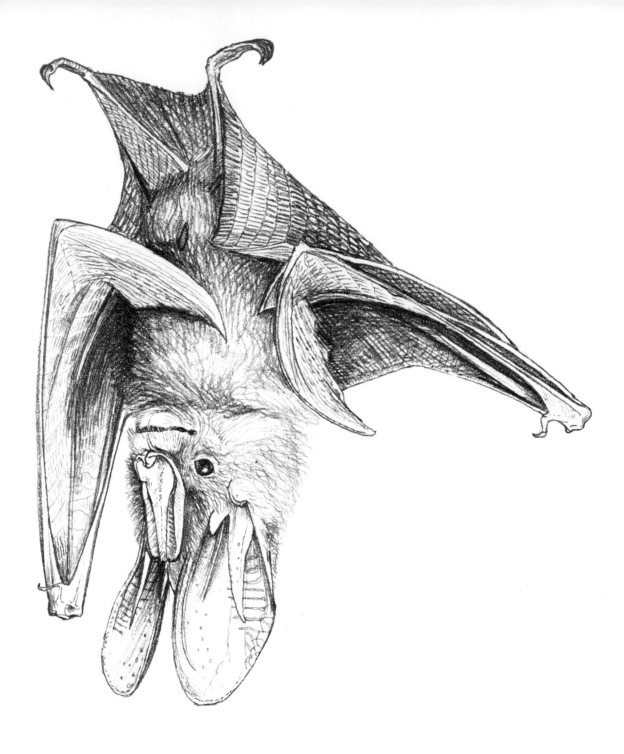

Yellow-winged Bat
(Lavia frons)

Family Megadermatidae
Order Chiroptera
Local names
Popo manjano (Kiswahili)
Lumenwa (Karamojong)

Measurements
head and body
63—83 mm
ear
3·5—4·7 mm
forearm
55—64·3 mm
weight
28—36 g

Yellow-winged Bat (Lavia frons)

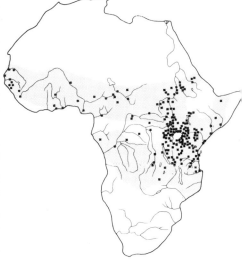

This bat is one of the most extraordinary and spectacular of all bats. It has the largest eyes of any of the insectivorous species and ones that in the light of a torch return a reflection. It has bright yellow or orange wings and ears which contrast with bluish fur. The ears are immense, with pointed, forked tragi. The nose carries a long spear-shaped leaf which resembles a third pointed ear. The males have a gland on the lower back which, when active exudes a yellowish secretion that discolours the fur of that area. In other details this bat is not dissimilar to its relative, *Cardioderma cor*.

Lavia is found over much of the low-lying savannas and open woodlands of tropical Africa, but it is generally limited to vegetation in the vicinity of water, or else within relatively moist climatic belts. It is common in riverine *Acacia* and in pseudo savanna. In open rangelands, termitary thickets or *Euphorbia* trees are favourite roosts. They seem unable to tolerate cold and temperature must limit their altitudinal and latitudinal distribution, for they are not found much above 2,000 m nor much south of 13° latitude. Verschuren (1957) has described *Lavia* as a gallery forest species as he found this bat in the Garamba Park associated with *Irvingia* trees in narrow forest galleries.

They are commonest, perhaps, in the bush country surrounding lakes, marshes, along rivers and near the sea and, although they have not been reported to drink, it is possible that this may have been overlooked.

Although they will change their perch sometimes when the sun shines directly onto them, they are unusually tolerant of conditions that would desiccate other species of bats within a very short time. Loveridge (1922) even describes them deliberately sunning themselves in the last rays of the evening.

They have a wide range of foods. They eat both hard and soft-bodied insects, ranging in size from very small to relatively large; grasshoppers, beetles, butterflies, moths, mosquitoes and flies have all been recorded. Prey composition probably varies with the locality and the season.

There are numerous reports of this species hunting before dark, and most local naturalists are acquainted with the sight of their yellow-winged forms flitting restlessly from one tree or thicket to another in broad daylight. It is misleading, however, to regard them as diurnal bats since, throughout their range, their principal feeding period is definitely after sundown. The use of ultrasound in hunting seems to be minimal in this species and the use of sonar is probably restricted mainly to orientation. Their method of hunting appears to involve both sight and hearing and unlike that of nycterids it is conducted at a relatively long range. The animal hangs from its perch on an exposed branch, swivelling itself from side to side, constantly adjusting its long ears and peering down alertly. When an insect is located the bat launches itself upon it; the prey may be on the ground, in the air, or even among stems of grass (*Lavia* has actually been flushed out of grass where it was probably hunting). Lang and Chapin (1917) describe a not uncommon scene:

"Late in January, not far from their roosting trees the grass was set afire just about noon. As usual, insects, to escape the advancing flames, tried to hide in great numbers in the grass. These bats would then dive after some of the insects

233

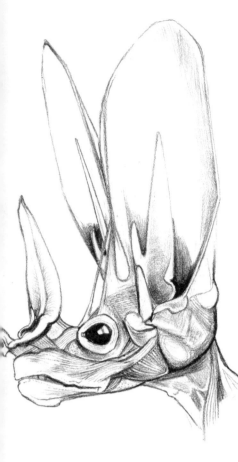

that passed nearest, just as eagerly and skilfully as the birds that snatched them far above the flames.''

The bats always perch again after catching prey and eat unhurriedly. Sometimes they are attracted to insects around house lights, but this is probably because their habitual roost is already in the vicinity.

The wings of *Lavia* are amongst the broadest to be found in bats and are adapted to allow maximum dexterity. This bat is indeed adept at negotiating thorny thickets and other dense vegetation.

Yellow-winged bats tend to hang in pairs but sometimes four or five individuals may share a tree. These pairs or groups generally contain both sexes. When disturbed they will often return to their first perch after a short while. They make a curiously bird-like contact call. When resting they cover their faces with their wings and are often very difficult to see, as they resemble dead leaves. Their cryptic form probably protects them from predators. On the other hand, when they spread out their wings to fly off in alarm, the sudden flash of vivid colour has a startling effect. Their colour, therefore, may have something of the same survival value as has been demonstrated for moths with cryptic forewings and brilliantly coloured or patterned underwings.

A *Lavia* bat has been retrieved from the stomach of a mamba, *Dendroaspis*. However snakes are probably infrequent predators for *Lavia* individuals are scattered widely and are also possessed of acute senses, so that they are unlikely to be more than occasional victims. They have been seen chasing after other bats and there is some evidence that they actually attack them (Pye, personal communication).

The gestation period of *Lavia* approximates three months as it does for other members of the family. Records of pregnancies from the eastern Congo (Zaire), Uganda and the Sudan are more numerous in the months from January to April, although there are scattered records from other times of the year. During the height of the rains and when vegetation is thick, *Lavia* tend to be inconspicuous, so that there is likely to be a "collector's bias" in *Lavia* specimens available for study. However Verschuren took five pregnant females and a juvenile in March and April. The identical stage of development suggests that there is a clearly defined breeding season in Garamba, with births at the beginning of April.

Kock (1969) describes an 11·4 millimetre embryo. The ears of this embryo had not yet joined and the noseleaf was proportionately smaller than that of adult *Cardioderma cor*, while the eyes were very large. Therefore *Lavia's* prominent features appear to be extreme developments of a trend that is perceptible in other Megadermatidae. The wide range and obvious success of *Lavia* within certain ecological, climatic and geographic limits, contrasts with what looks like a relict status for *Cardioderma cor* (and perhaps for some other non-African megadermatids).

The peculiarities of *Lavia* deserve further study, as this species has departed widely from the typical conformation and habits of conventional bats.

234

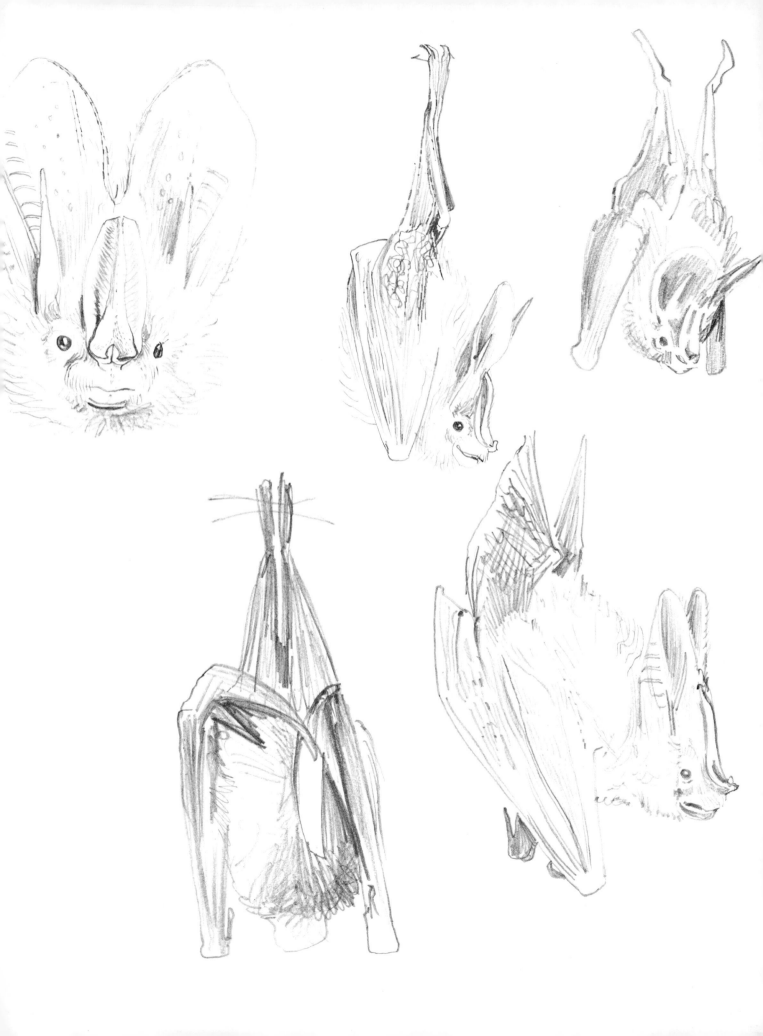

Horseshoe Bats

Rhinolophidae

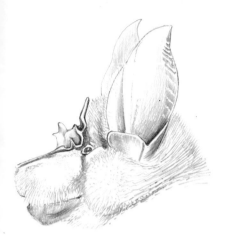

R. landeri.

This family is very widely distributed in both temperate and tropical regions of the Old World. Bats belonging to this family have evolved a highly effective type of sonar, involving the development of elaborate noseleaves and the emission of sound through the nostrils. They catch insects in flight with great agility, in spite of being relatively slow flyers with broad wings.

The great development of the noseleaves is also reflected in the skull, in the form of a swelling or podium behind the nasal openings, on which the vertical elements of the noseleaf are mounted.

The form of the noseleaf is an important aid to the identification of species. The muzzle in front of the nostrils varies very little, all species having a horizontal plate-like leaf or "horseshoe" with mobile edges allowing it to form a bowl-like shape. From the centre line between the nostrils two delicate lobes sweep back and then sharply up, defining thereby the edges of a "throne" or chair-shaped structure. The vertical element forming the back of the "throne" presents its concave surface forwards at right angles to the muzzle. This structure is known as the sella, and it varies in shape from species to species. It is supported behind by a thin wall in the plane of the mid line. This wall appears to act as a supporting buttress for the sella and is known as the connecting process. Its silhouette is highly distinctive from species to species and is a most convenient aid to identification. This wall is mounted on the base of the posterior leaf or lancet, a pointed leaf with complex folds or pockets. The upper part of the lancet is approximately in the same plane as the sella and also perpendicular to the muzzle. The sella and the lancet are thought to focus and concentrate the sound waves which are emitted from the nostrils within the "horseshoe", which may act as a megaphone (Mohres, 1953b).

Both structures probably also act as baffles—particularly the lancet—reducing or controlling the volume of sound reaching the ears directly from the nostrils. However, all species seem to have a narrow channel or unobstructed pathway leading from the nostrils, back past the eyes, to the base of the broad ear conch. Possibly this arrangement allows some form of physiological coordination or comparison to be made between emitted signals and their returning echoes. Frequencies between 60,000—120,000 kilocycles have been recorded for several species, but there is very little modulation and the specific frequency is stable. The emission is relatively long, lasting 10—100 millisecs.

Directional channel, *R. hildebrandti.*

Other anatomical peculiarities are a reduction of the finger phalanges and a fusion of the last cervical first dorsal vertebrae with the first rib and the presternum to form a rigid ring.

These bats appear to be a highly evolved and specialized group but they are nonetheless recognizable as a family in the Eocene.

All species hang by their hindlegs when at rest, their heads hanging down. The wings wrap around the front of the body, while the tail folds back over the rump—this provides a convenient gutter if rain is falling and keeps the fur dry.

They feed on a variety of insects and there are some differences between species in the prey they prefer. This is probably correlated to size and tooth structure. All species are strictly nocturnal.

Most species live in small groups, but solitary bats or pairs are not uncommon and some species congregate in quite large numbers.

A small European species, *R. hipposideros*, is known to have a gestation period of about 50 days. In temperate climates fertilization is delayed during the period of winter and the sperm is stored in a pocket of the vagina until the spring.

Also their sexual associations vary. In some species females form separate "nursery" groups, while in others the females remain with the males. Very little is known about the habits of African species.

A virus has been isolated from the salivary glands of an East African species, *R. hildebrandti*, which has been found to multiply in mosquitoes.

The record for longevity in a wild bat is held by *R. ferrumequinum* in Germany; a bat of this species ringed in 1936 was found again in 1960, at an age of 24 years at least (Eisentraut, 1960).

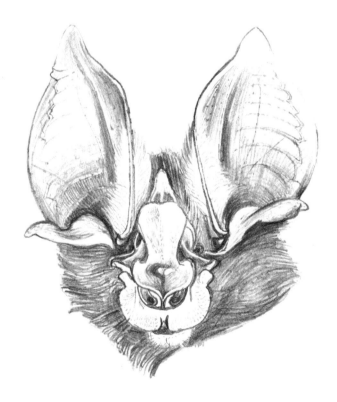

R. ruwenzori.

Rhinolophus ruwenzori

forearm
56—68 mm

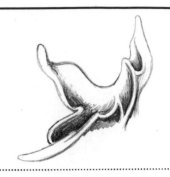

Rhinolophus hildebrandti

forearm
62—67 mm

(metric species)

R. hildebrandti

Rhinolophus fumigatus

forearm
49—60 mm

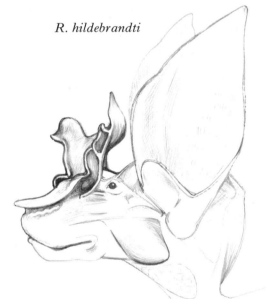

Rhinolophus clivosus

forearm
50—57 mm

(metric species)

R. darlingi

Rhinolophus darlingi

forearm
45—50 mm

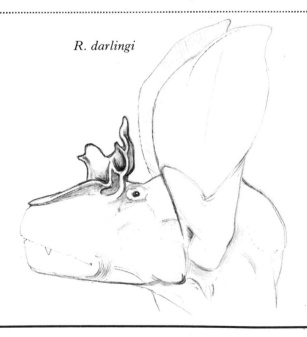

RHINOLOPHUS SPECIES

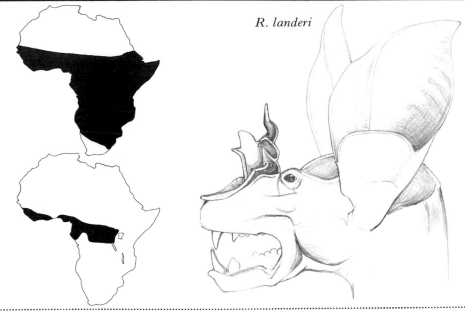

R. landeri

Rhinolophus landeri

forearm
40–48 mm

(metric species)

Rhinolophus alcyone

forearm
49–54 mm

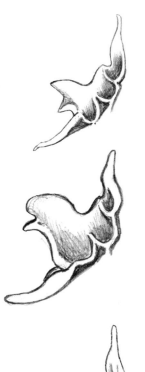

Rhinolophus blasii

forearm
44–48 mm

Rhinolophus simulator

forearm
40–46 mm

Rhinolophus swinnyi

forearm
40–46 mm

Rhinolophus Bats, Horseshoe Bats (Rhinolophus)

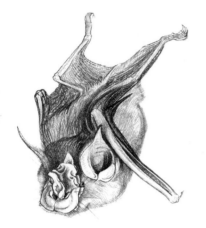

R. hildebrandti.

The species' pictorial key (previous page) lists the measurements, maps their African distribution and depicts silhouettes of the connecting process.

Within *Rhinolophus* there are three pairs of species which are very closely related and are recognizable principally on the basis of their size.

Rhinolophus ruwenzorii is a large species known from the Ruwenzori Mountains. On the Congo (Zaire) side it has been recorded from about 1,000 to 2,500 metres. It appears to be a relic species with a closely related form in Guinea. This distribution pattern for archaic forms found in the two Forest Refuges repeats itself in *Micropotamogale*—even the same disregard for altitude is shared. This species has no close affinity with any other African rhinolophids but seems to belong to an Asiatic group, *R. philippinensis* (Hill, 1942).

Uganda specimens have been taken from caves at medium and high altitudes. They do not appear to be gregarious.

R. hildebrandti is a large, highly evolved species apparently restricted to the eastern half of Africa. It has a large rostrum and massive teeth. This species is often solitary but is also found in small groups, sometimes associated with other species. Kulzer (1959) successfully kept individuals of this species for some months.

R. fumigatus is an open country species of very wide distribution in sub-Saharan Africa. It is a smaller relative of *R. hildebrandti* and, although body sizes overlap, the forearm is consistently smaller in *R. fumigatus*. The cranial swelling or rostrum for the noseleaf is most highly developed in this species, which may be one of the more highly evolved forms of *Rhinolophus*. They roost in the darkest recesses of caves and sometimes roost close together. Aellen (1952) found a colony of about 12, all of which were females. This species may be found in roofs but only when conditions are reasonably cool, dark and moist. The dark roof of a large building in Kampala which sheltered several bat species housed a group of about 5 *R. fumigatus* in a cool corner near a water tank.

R. clivosus is regarded by some authors (Harrison, 1959) as the African form of *R. ferrumequinum*, a Eurasian species ranging from Japan to western

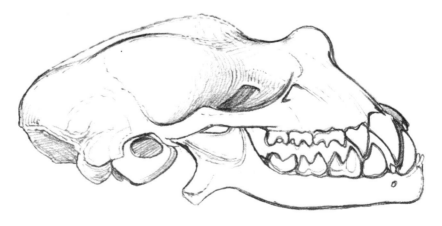

R. fumigatus.

240

R. fumigatus.

Rhinolophus Bats, Horseshoe Bats

(Rhinolophus)

Family Rhinolophidae
Order Chiroptera
Local names
Tai (Kinyaturu) Lumerwa (Karamojong)

Species

Rhinolophus ruwenzorii
Rhinilophus hildebrandti
Rhinolophus fumigatus
Rhinolophus clivosus
Rhinolophus darlingi
Rhinolophus alcyone
Rhinolophus landeri
Rhinolophus blasii
Rhinolophus simulator
Rhinolophus swinnyi

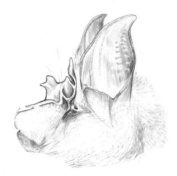

R. darlingi.

R. alcyone.

Europe. It is certainly very closely related.

In Europe *R. ferrumequinum* feeds on spiders, beetles, moths and other insects and it has been seen to alight on the ground. However it usually flutters along slowly with frequent glides at a height of less than 3 metres.

Breeding in Europe involves delayed ovulation and sperm is stored for 6 months in the vagina. There is some segregation of sexes when the females are with young. The young make their first flight at about 22 days. Captives have been kept for nearly two years on a diet of mealworms, moths and water (see Southern, 1964).

R. darlingi is a small species closely resembling *R. clivosus* but with a consistently smaller forearm. In southern Africa it is commonly found in old mining adits and draughty caves. The most northerly record to date is from Banagi, Serengeti, and the species seems to be restricted to Africa south of the Equator.

R. alcyone is a rare forest bat only known from a few localities across the forest belt of Africa. In East Africa the species is only recorded from western Uganda. There are two colour phases, orange and drab brown. Some males have tufts of long stiff hairs in the armpit, presumably used as scent dispersers in sexual behaviour. They have been collected from caves, mines, hollow trees and a hut. Three females from western Uganda were pregnant in mid-June and one was lactating. One birth is recorded in September. I have noticed curious aggregations of fatty tissue over the back of the neck in this species. In late November I captured a pair flying together over water.

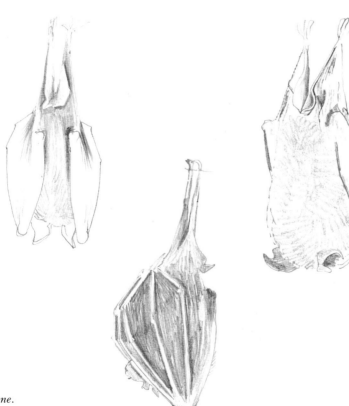

R. alcyone.

R. landeri is a smaller species but closely related to *R. alcyone*. It is very adaptable, being common in a wide range of habitats over a large part of a sub-Saharan Africa.

Kock (1969) found *R. landeri* in the Sudan feeding principally on small Coleoptera, and T. S. Jones (in Rosevear, 1965) observed a number hawking in primary forest, flying very close to the ground; this was an unusual observation in that the bats were active well into daylight. However the forest canopy tends to blur the transition from night to day. These bats are sometimes solitary but more often hang in small groups or open clusters.

R. landeri.

Eisentraut (1940) found some individuals in a Cameroon cave were so lethargic as to allow themselves to be handled. Since *Rhinolophus* are usually most alert and shy and other individuals in the same group had already taken flight, it seems that this tropical species may possess a similar capacity for internal temperature variation to that of the Eurasian *R. ferrumequinum*. Reduction of activity and of body temperature in the European bats has been observed to be the rule, both during seasons when hunting is useless, and during days when weather conditions are unfavourable. Hypothermia is not related to endocrine activity and seems to be an adaptation to allow economy of expanded energy. Brosset and Saint Girons (1969) consider that the essential ecological characteristic of the greater horseshoe bat in Europe is its systematic search for the most favourable micro-climate, using warm roosts when active and cool ones when torpid. These findings will probably turn out to be relevant for African rhinolophid bats.

There appear to be well-defined breeding seasons in this species. Kock (1969) has found that 13 out of 14 Sudanese females carried an embryo at the end of March. He also found males in January that had their shoulder tufts gummy with a yellow secretion. It seems likely that these tufts are of seasonal appearance and may be connected with sexual behaviour. Further study of this fairly common species would be of great interest.

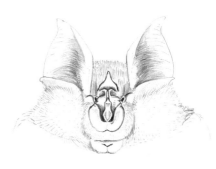

R. landeri.

R. blasii has not been recorded from East Africa but, as specimens have come from the Congolese shore of Lake Tanganyika, it is reasonable to suppose that it will be collected in due course. This species has a peculiar distribution, occurring widely in South-west Europe, South-west Asia, Morocco and Eritrea and then again in southern and central Africa. Is the principal factor excluding it from the Tropics a climatic limitation or is it competition from tropical rhinolophids?

R. simulator is a very small species subject to two colour phases like some other *Rhinolophus* species. Young ones of the orange phase are first white then yellow before acquiring orange fur. There are very few records of this species but it is apparently restricted to the eastern *Brachystegia* woodlands. They have been collected in caves and other shelters, usually in small groups, sometimes together with other species, e.g. *R. clivosus*. Ansell (1960) found a Zambian bat pregnant in October.

R. swinnyi is another very small African form, so far only recorded in East Africa from Zanzibar Island. The ear is relatively small like that of *R. landeri* and the sella is narrow. The skull is distinguished by the large size and width of the cranium compared to the muzzle, this difference in proportion being a concomitant of its diminutive size.

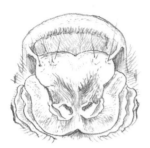

Hipposideros (abae).

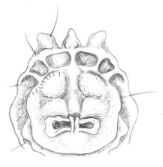

Asellia (tridens).

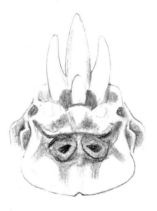

Cloeotis (percivali).

Triaenops (persicus).

Hipposiderid Bats

HIPPOSIDERIDAE Hipposideros
Asellia
Cloeotis
Triaenops

This family is related to the Rhinolophidae and there are numerous similarities between the two.

The structure of the noseleaf is the main distinguishing character. Mohres and Kulzer (1955) testing the ultrasound of the hipposiderid *Asellia* found a different pattern to that of *Rhinilophus*, with a higher frequency at the start and a low frequency burst at the end.

Other differences between the Rhinolophids and the Hipposiderids are the latter's tropical range and greater diversity of forms; there are six genera, four of which are represented in East Africa.

The Hipposideridae are recognized in the Middle Eocene of Europe, and a Miocene humerus, thought to belong to the genus *Hipposideros*, has been found at Songhor in Kenya.

The great age of this family is also shown by a radiation of distinct types or super-species with widely separated representatives living in South-east Asia and in Africa. Hill (1963) has revised the genus *Hipposideros* and defined seven distinct groups, five of which have an African branch.

Illustrations of the form of the noseleaf in the four hipposiderid genera are shown left.

Leaf-nosed Bats

Hipposideros

Within this genus there are seven very distinct super species or species groups, which must have radiated into distinct niches at a very early date because representatives of these groups tend to be found in the tropical Far East and again in Africa. It is interesting that the Miocene fossil *Hipposideros* found at Songhor shows the greatest resemblance with *H. diadema*, the Oriental representative of the *H. commersoni* group.

The five groups represented here are individually treated in the following profiles and reference is made to their Oriental relatives. Hill's revision of the genus contains a detailed analysis and description of the many species and dendrograms of the possible relationship of species within the groups.

Big-eared Leaf-nosed Bat
(Hipposideros megalotis)

**Big-eared Leaf-
nosed Bat
(Hipposideros
megalotis)**

Family Hipposideridae
Order Chiroptera

**Measurements
head and body**
38—44 mm
tail
32—34 mm
forearm
28—32 mm
ear
20—25 mm

This is the most primitive representative of the *Hipposideros* bats. Its large rounded ears are joined at their bases and the noseleaf is less elaborate than in any other species. Known only from the Nakuru area in Kenya and from Ethiopia, this is clearly a relic species restricted to upland areas in north-eastern Africa. There is no related form in Eurasia.

Nothing is known of the biology of this species.

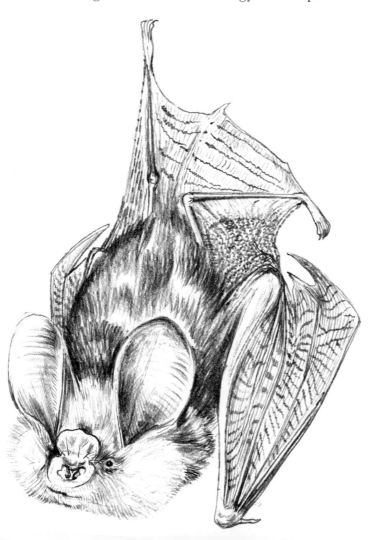

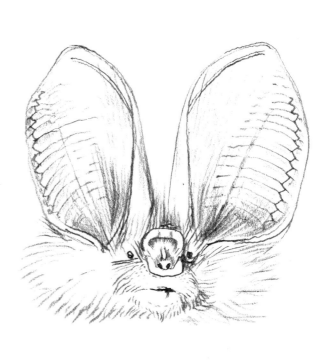

Aba Leaf-nosed Bat
(Hipposideros abae)

This bat, which has red and grey colour phases, is somewhat intermediate in size and form between *H. caffer* and *H. commersoni*, but like the other African hipposiderids it has closer relatives in Asia. The noseleaf, which resembles that of *H. caffer*, is distinguished by three frills surrounding the main leaf. Both sexes have a hairy frontal sac but it is more developed in the males.

H. abae is widely endemic to the northern *Isoberlinia* woodlands and is represented in East Africa only in Uganda. This restriction to the northern woodlands is a rare distribution pattern shared by rather few species of mammals. It may also occur in the pseudo-savanna conditions following the destruction of forest.

This species has conservative habits where roosts are concerned, being rarely found in buildings or roofs; but it turns up in a variety of natural shelters in rocky localities ranging from the darker recesses of large caves (Lang and Chapin, 1917) to shallow holes beneath boulders very close to the ground (Rosevear, 1965). Visitors to their day roosts have found them alert and noisy during daytime. The size of their aggregations seems to depend largely on the limitations of the chosen shelter. Verschuren (1957) in the Congo (Zaire) has found small groups hanging in restricted roosts, or else hundreds that were mixing freely with *H. caffer*. When they hang in fairly dense associations they are, however, sufficiently spaced not to touch one another.

Aba Leaf-nosed Bat (Hipposideros abae)	Family	Hipposideridae	Measurements
	Order	Chiroptera	head and body
			58—69 mm
			tail
			30—40 mm
			forearm
			58—66 mm

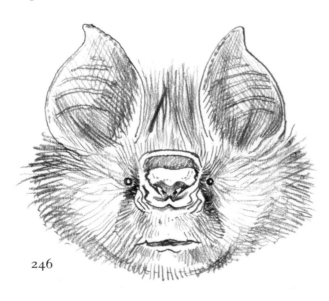

246

Giant Leaf-nosed Bat
(Hipposideros commersoni)

Races

Hipposideros commersoni marungensis	Southern Africa
Hipposideros commersoni gigas	Equatorial Africa

This large bat has various colour phases, from tawny, orange or brown to a greyish tint. The belly is often a pale, neutral colour or white and the shoulder may exhibit a dark contrasting mark. This bat is immediately recognizable by its size. It has characteristic leaf-shaped ears and the broad, relatively simple, noseleaf has lappets round the nostrils. Both sexes possess a frontal gland from which a blunt tuft of hairs protrudes.

This species ranges across most of the warmer areas of Africa and Madagascar. It is not found in the Sahara or in the extreme south. Related species occur in Asia and Australasia. These bats are commonest in woodland and savanna but they are also found in forest and occasionally occur in relatively arid areas.

Populations from the western part of Africa tend to be bigger than those in the south and east and subspecies—formerly species—have been erected on the basis of size. The forearm length of the western types, *H. c. gigas* and *H. c. niangarae* ranges from 96·5 to 116 mm, while the southeastern *H. c. marungensis* ranges from 90 to 104 mm. The populations of intermediate size are mostly from Tanzania, probably constituting a clinal link between the larger bats from the moister tropical belt and the smaller ones from the drier south and east.

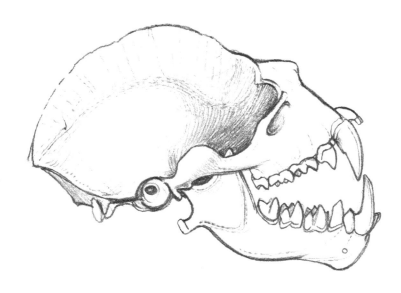

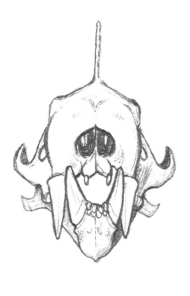

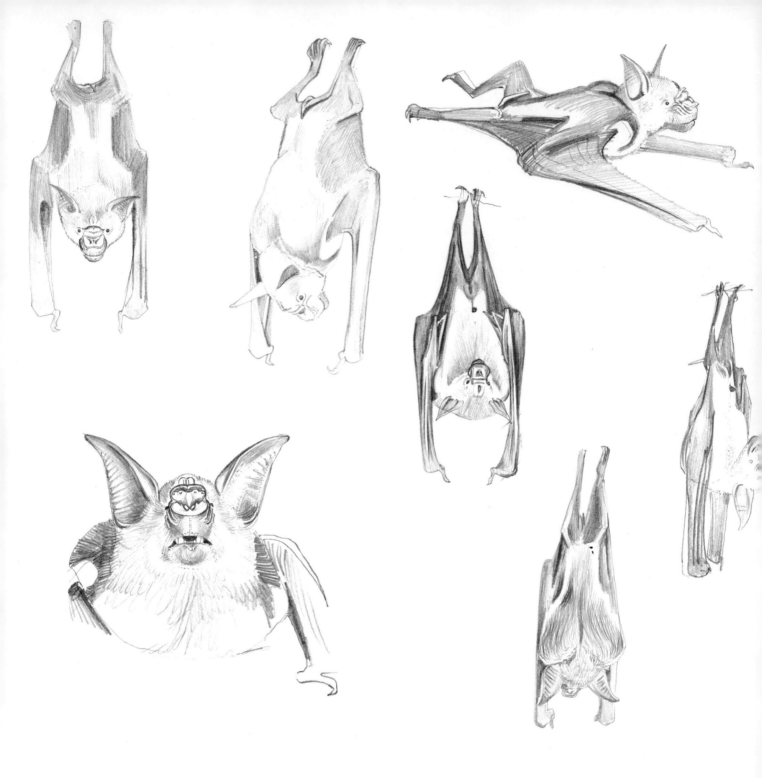

Giant Leaf-nosed Bat
(Hipposideros commersoni)

Family
Order

Hipposideridae
Chiroptera

Measurements
head and body

88—126 mm

tail

30—45 mm

forearm

79—116 mm

weight

74—180 g (ave. 130 g)

They roost in caves, usually in large and sometimes very large groups, possibly numbering hundreds. They have also been collected from hollow trees in forest and even hanging in undergrowth.

They feed on a variety of insects and have even been seen removing beetle larvae from wild figs. Lang and Chapin (1917) report this species gorging itself on flying termites and filling the mouth and cheeks with them. They have been noticed to spit out the hard parts of insects. They hunt at low levels, capturing most of their food in flight, and are sometimes netted over water. They are, however, adept at detecting the finest mist-nets set up in the home cave, flying up vertically along the surface of the net at the last moment.

They crawl about rapidly over the surface of the home roost when disturbed and try to hide, back first, in crevices. Frequent bickering suggests that territorialism (if only over a few square millimetres of space) is a common feature of their behaviour. They make a shrill audible alarm call.

Ansell (1960) reports a full-term foetus from Zambia in October. Juveniles and recently lactating females have been collected in the Southern Highlands in January. Brosset et al. (1969) found a similar season in Gabon, with the young born in October; he states that this species has the long gestation period of five months. Mating would therefore tend to be at the end of the rains and births at the very beginning of the wet season. Lactation lasts for over five months and, although growth is very rapid, the animals do not reach sexual maturity until they are two years old. Fluctuations in fat deposits have been noticed in this species (Monard, 1939).

This species is eaten in some areas of Tanzania where the decline of wild life has made alternative sources of meat scarce.

H. c. gigas
showing back of noseleaf rolled foreward. ·

adult.

subadult.

H. c. gigas.

juvenile.

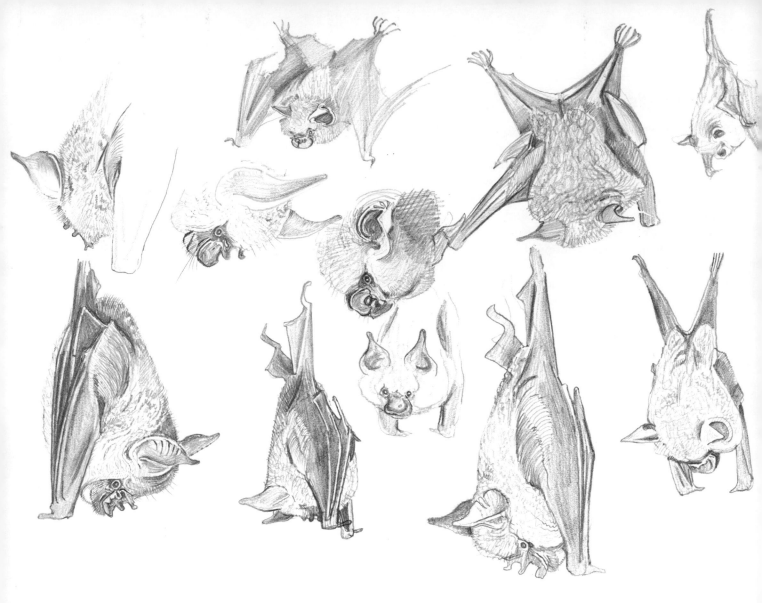

Cyclops Bats
(Hipposideros cyclops) Group

Family	Hipposideridae
Order	Chiroptera

Measurements
head and body
72—87 mm *Hipposideros cyclops*
tail
23—33 mm
forearm
59—71 mm
weight
30—39 g

head and body
89—97 mm *Hipposideros camerunensis*
tail
32 mm
forearm
72—77 mm
weight
32—42 g

Cyclops Bats
(Hipposideros cyclops) Group

These very distinctive bats are limited to the forest belt of Africa and are not closely related to any other African hipposiderids. They are however related to some isolated species in Papua and northern Australia. Hill (1963) considers that these two widely separated hipposiderid groups, displaying considerable similarity and convergence, must share a common origin, however remote. They seem to have highly specialized hearing, the ears and auditory bullae being greatly enlarged, but are otherwise rather primitive. The two species listed here may eventually prove to be a single species when more is known about them. They are indistinguishable morphologically and they share the same habitat, furthermore a large gap between size ranges that has been noted in West Africa (Eisentraut, 1956—1963) is narrowed down to 1 or 2 mm in specimens from western Uganda. However, fully adult *H. cyclops* (*sensu stricto*) of both sexes tend to fall throughout their range within the extremes of measurements given. *H. camerunensis* from the same localities as *H. cyclops* in western Uganda and from very similar roosts are mostly, but not invariably, very much larger; smaller individuals might perhaps be subadult.

No differentiation in habits, habitat or morphology has been noticed and the real relationship between these bats remains to be elucidated.

They can be recognized immediatcly by their long fluffy hair, which is dark brownish grey with a lighter grizzle on the surface. In texture and pattern this hair is like that of *Anomalurus* and has a very bark-like appearance, so much so that the bat becomes very difficult to see when it alights on the trunk of some rough-barked tree—a habit sometimes resorted to when the bat is flushed from its hollow in daylight.

These bats are relatively numerous in the lowland forests of Uganda and western Kenya, particularly in *Cynometra*, but they do not congregate in numbers exceeding about a dozen. These groups are generally predominantly female and Verschuren found an overall sex ratio of about 2 females to one male. When single bats are found they are always males. Roosts tend to be occupied throughout the year. However, they are seldom seen in the open and generally are netted only in light undergrowth beneath a well-formed forest canopy. According to Rosevear (1965) their favourite food is a species of cockroach. These insects are certainly common in its habitat (often coming in to camp lights) and stomach contents of western Uganda specimens reveal a well-crushed mass of brown insect matter with chitinous remains probably of cockroaches or crickets. Pieces of a tettigoniid leaf insect have been found in the mouth of one. It is probable that most of their hunting is at a low level, well beneath the canopy in closed forest. Verschuren (1957) has shown that sphingid moths and cicadas are the principal prey in the Garamba Park. The former are probably taken in flight, the latter are perhaps more likely to be snatched off vegetation. Other foods listed by Verschuren are ants, beetles and woodlice. Large mounds of rejected remains at the base of their hollow trees suggest that most hunting is done within a narrow radius of the shelter and that after capture prey is taken home to be eaten.

They roost in large hollow trees often in association with *Nycteris arge* and *Idiurus zenkeri*. Most museum specimens have been taken from hollow trees, from which they can be smoked out or shot.

There appears to be a well-defined breeding season in this species. In the eastern Congo, Lang and Chapin (1917) found pregnant bats in late January and very young bats in late April. Nearby in Uganda, females have been collected with their false nipples still greatly enlarged but not in breeding condition in June. Males with enlarged testes have been collected in early December, together with females which were still not pregnant. In the Garamba Park Verschuren collected ten pregnant females in February and March and lactating females and young in May. There seems therefore to be a single well-defined breeding season, with the birth of the young coinciding with the arrival of the rains. Lactation lasts two months and Verschuren collected a free-flying juvenile bat with a forearm of 63 mm which had both milk and insects in the stomach.

Male bats have a large pouch beside the anus which is glandular and is capable of emitting a very strong odour. It is lined with rufous hairs and can be turned inside out when the animal is excited, when the bat fans out these hairs it presumably disperses the smell of its gland; as a secondary sexual characteristic it is probably related to courtship and may perform a similar advertising function to the pocket shoulder tufts of male *Epomophorus*. If a male bat is handled after capture, it rapidly opens and closes its pouch. Both sexes have a frontal sac containing fine hairs rather like a minute, pointed sable brush immediately behind the noseleaf.

African Leaf-nosed Bats (Hipposideros caffer) Group

H. ruber.

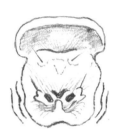

H. caffer.

H. caffer.

African Leaf-nosed Bats (Hipposideros caffer) Group		
Family	Hipposideridae	**Measurements**
Order	Chiroptera	**head and body**
Local names		42—48 mm
Nundu (Kisambaa)		**tail**

Family Hipposideridae
Order Chiroptera
Local names
Nundu (Kisambaa)

Measurements
head and body
42—48 mm
tail
26—39 mm
weight
75—180 g
forearm
43—48 mm   *H. caffer*
skull length
16·7—17·7 mm   *H. caffer*
forearm
48—58 mm   *H. ruber*
skull length
18·9—19·8 mm   *H. ruber*

Types

Hipposideros caffer   Most of Africa outside main forest area.

Hipposideros ruber   From Gambia to Sudan, Ethiopia, Kenya and Angola to Zambia and Tanzania.

These bats belong to a closely related complex of types, which have been treated both as races and species by different authors. Brosset (1966b) thinks that the variety exhibited by populations of this bat suggests that it is in the course of evolution. He also considers that this might be the most numerous mammal type in Africa.

The situation is similar to that of *H. cyclops* in that there are noticeable differences in the measurements of adult bats, particularly in the size of the head, coming from the same localities and even from the same roosts. Very small differences in nasal structure (illustrated by Kock, 1969) have also been observed. The larger form is commonly called *H. ruber* (Noack, 1893; Hollister, 1918; Lawrence, 1964; Kock, 1969) or *H. caffer ruber* (Koopman, 1966; Hayman, 1967). *H. caffer centralis* (Andersen, 1906b) is an intermediate form.

H. ruber.

253

(Another closely related bat, *H. beatus*, is found in the Congo and West Africa.)

The situation is not unlike that of *H. commersoni*, in that the larger type has an equatorial distribution, while smaller forms are found in drier habitats. *H. caffer* (*sensu stricto*) occur in the drier areas to the north and south of the Equator. Bats of intermediate measurements (*centralis* type) range through much of East Africa and the Congo. Both *caffer* and *ruber* are also widespread in this area.

Sometimes two varieties are found together in house rooftops or in caves situated in habitats that are transitional between forested and more open country. Lawrence suggests that *caffer* may be a typically savanna form spreading with the opening up of suitable country, while *ruber*, a forest form (possibly through its adoption of roof roosts), has moved out into dry country. The peculiarities and complexities of East Africa as a faunal overlap zone are demonstrated in this distribution pattern.

H. caffer has broad rounded ears with a triangular point and a relatively simple noseleaf. Red, dark grey and intermediate colour phases are known. In the British Museum collection, about 50% are grey, 20% are red and 30% are intermediate. Albinos have been collected at Dar-es-Salaam (Howell, personal communication).

The *Hipposideros caffer* group are distributed over the entire continent except for the Sahara and Egypt. They also occur in South-west Arabia. The closest relatives of this *Hipposideros* type are the *bicolor* group, ranging across much of tropical Asia, Oceania and Australia. *H. caffer* range over a wide variety of vegetation zones and will shelter readily in houses as well as caves and hollow trees.

They eat many types of small insects. Termites and small Coleoptera have been noted in particular. They leave the day roost immediately after sunset and hunt for up to two hours before resting in some favourite night rest (which is not used during the day). They can nearly fill their stomachs after an hour's hunting. They feed on the wing, flying close to the ground and often picking their prey off the soil or vegetation. As they explore likely surfaces they may briefly hover over crevices. Differences between the hunting habits of male and female bats may appear; both Brosset (1966a) and Kock (1969) found hunting corridors or aerial lanes being used by males only. Their low-flying habits are displayed when confined in a room; they often enter houses at night and when chased they will frequently hug the ground skimming along at ankle level. They may still be flying just before sunrise, but nocturnal activity cycles have not been studied. Brosset found 6 g to be the average weight of insects taken in a night by this species. He estimated that 500,000 individuals were roosting in one Gabon cave; this represents a biomass of bats weighing over 5000 kg and a consumption of 3000 kg of insects, probably harvested within a radius of some 12 km.

This roost was exceptionally large, but many hundreds and thousands have also been noted in large caves in East Africa. I estimated some 1,200 to be present in an immense hollow tree in western Uganda. Small groups will soon colonize new buildings suited to roosting and rapidly increase in numbers. The roosts need to be fairly dark and very hot roofs are generally not favoured by this species. They hang free in close associations but without touching one another. In Uganda, I have found small groups in an open thatched hut, sheltering behind a screen of hornet nests. The selection of roosting sites behind hornets was note-worthy as suitable dark corners without hornets were not tenanted. In southwestern Tanzania I have been put to flight by attacking hornets which nested in the entrance to a small cave containing *H. caffer*.

Day roosts generally have a balance of sexes but associations of one sex have been noticed at the night roosts, which may be related to seasonal or permanent differences in hunting behaviour. Likewise, sexual differences in the laying down of fat have been noted by Brosset. He found female rhythms followed the seasons and he collected fat bats in the wet months between November and May and thinner ones in the dry season, June to August. However, male rhythms were individual and did not follow this clear-cut pattern.

The breeding of *H. caffer* in Gabon has been the subject of an interesting study by Brosset (1968). This species has a single very well defined breeding season with all the births in a colony occurring at once. In Gabon, however, some colonies give birth in March while others in October. A review of evidence from other parts of Africa indicates that in southern latitudes the birth season is about October, while to the north of the Equator it is in March. Brosset suggests that this situation may be due to the different ancestral origins of the populations. As the female is strictly mono-oestrus and the males' short period of sexual activity is synchronized with the females' oestrus, it is perhaps not surprising that once a cycle is established in a population an effective barrier can be erected against inter-breeding with other populations of the same species. Furthermore, some such reproductive barrier might explain the maintenance of distinct "sympatric" populations within the species. Herein lies an interesting field of investigation in relation to the *caffer-ruber* problem.

In the populations investigated so far in equatorial Uganda a "boreal" cycle is apparent, with births in February—March. This occurs during the shorter dry season between the local biannual rains. As gestation lasts about three months, mating is probably late in the November rains. Females are reported to segregate at birth. The young suckle for three months, but are not carried by the female for all this time. They are not sexually mature until their second year. Moulting in these bats is synchronized with the breeding cycle and observers have noted that the females have brighter fur whilst they are pregnant and nursing.

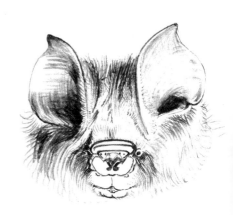

H. caffer.

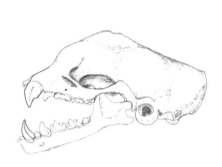

Trident Leaf-nosed Bat
(Asellia tridens)

Family Hipposideridae
Order Chiroptera

Measurements
head and body
50—62 mm
tail
18—25 mm
forearm
45—58 mm

Trident Leaf-nosed Bat
(Asellia tridens)

This small pale bat superficially resembles *H. caffer*, but the noseleaf is distinguished by three blunt knobs on its rear. This structure has some features reminiscent of *H. cyclops* and others of *Cloeotis* (see p. 244).

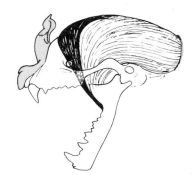 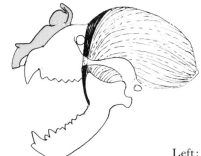

Left: *Asellia tridens.*
Right: *Hipposideros commersoni.*

Asellia has only been recorded in East Africa from Zanzibar, its main habitat being the deserts of northern Africa and the coastal regions of the Red Sea and the Persian Gulf. The predominance of this species in Egypt suggests that it replaces *H. caffer* in these areas. Harrison (1964) found seasonal changes in the populations of these bats in Arabia and concluded that there are distinct possibilities that this bat has migratory tendencies. The occurrence of this species in Zanzibar so very far south of its main range may represent an extension of its coastal distribution or records may happen to be of wandering migrants.

The skull of this species is remarkably different in build from any other hipposiderid, the anterior part of temporal muscles having apparently increased in size so that the bulk of the muscle exerts a vertical rather than a diagonal pull on the coronoid process of the lower jaw. The process is very large and sharply angled. Furthermore, the sagittal crest is most developed in the inter-orbital region. Another peculiarity is that all but two of the lumbar vertebrae are fused into a solid rod. The meaning of these structural specializations has not been investigated but the movement of the sagittal crest looks as though it is a compensatory device for the mechanical weakness that may develop in the narrow inter-orbital area of the skull as the rostrum grows more bulky.

This species is often associated with *Coleura* and *Triaenops* in dark ruins or caves, often beside the sea.

The flight is swift and low and bats in Arabia may fly a kilometre or more from the roost before reaching their feeding area. Roosts in the Middle East may number many hundreds.

There is no information on this species' status or habits in East Africa.

257

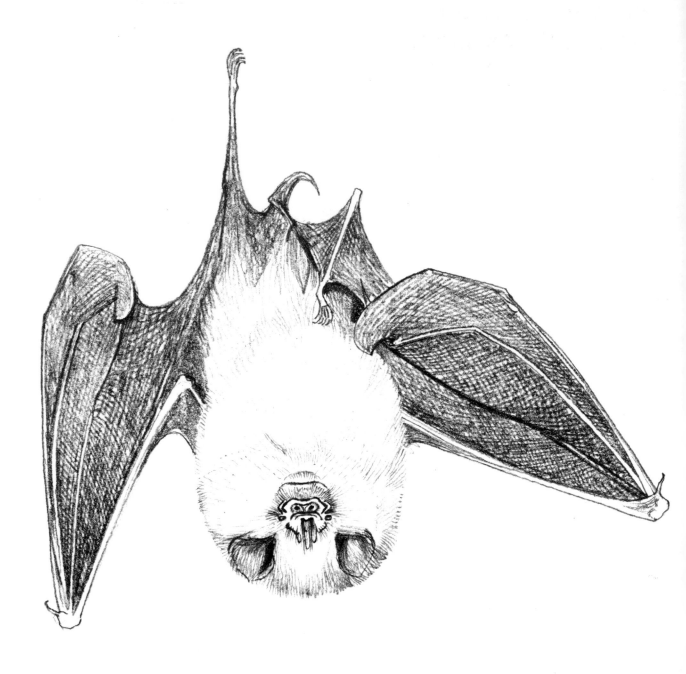

Percival's Trident Bat
(Cloeotis percivali)

Family Hipposideridae
Order Chiroptera

Measurements
head and body
33—50 mm
tail
22—33 mm
forearm
30—36 mm

Percival's Trident Bat
(Cloeotis percivali)

This very small pale bat has some slight resemblance to *H. caffer*, but is recognizable by its diminutive ears, dark wings and the presence of three pointed processes at the back of the noseleaf. There is an orange phase as well as a pale buff type.

Cloeotis is known from the Kenya coast and also from Katanga, Zambia, Rhodesia, Botswana, Transvaal and Swaziland. Very probably it occurs in the intervening areas of Tanzania and Mozambique.

This species is little known, although where it is found it appears to be present in large numbers. The bat was first discovered in 1900 in a coral cave at Takaunga on the Kenya coast. A single individual straying into a beach house in recent years is the only East African record since that time.

In the Transvaal it has been taken from porcupine holes and disused mining adits. It is generally the only bat species in the shelter. Roberts remarked on the narrow entrances to caves in which *Cloeotis* roosts and suggested that this might discourage larger bat species.

The discontinuous distribution is probably a reflection of inadequate collecting. A Zambian specimen identical to the Kenya type supports this supposition.

Nothing has been published on the biology of this species.

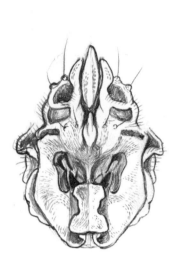

Persian Leaf-nosed Bat
(Triaenops persicus afer)

Family	Hipposideridae	
Order	Chiroptera	

Measurements
head and body
50—57 mm
tail
25—32 mm
forearm
50—55 mm
weight
8—15 g

Persian Leaf-nosed Bat
(Triaenops persicus afer)

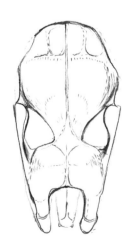

This bat has the most elaborate noseleaf of the East African hipposiderids. Three large spear-shaped leaves are mounted at the back of the noseleaf, which is subdivided into a honeycomb of convoluted cells. A small fourth spear projects from the centre of the noseleaf. The horseshoe has accessory leaflets or scrolls surrounding large nostrils. An elaboration of the ridge between the nostrils forms a figure-of-eight sella-like structure at the front of the muzzle. Presumably correlated with this apparatus, the nasal area of the skull is rather inflated and enlarged so that the rostrum is more developed that in any other hipposiderid. The zygomatic arch appears to have been modified to provide more effective support for the rostrum. It becomes a true strut running directly from the cranium to the mandibular area, loosing the outward flare that is found in other leaf-nosed bats. It has gained strength by becoming tall and narrow—the most economic and efficient form to absorb the postulated mechanical stress. These modifications give *Triaenops* a characteristically long narrow face and skull, somewhat disguised by widely flared ears and a ruff of fur. There are two colour phases, orange and greyish buff. The wing is the narrowest of all hipposiderids. They usually hang like other hipposiderids and the posture depicted here would seldom be adopted.

They generally congregate in great numbers in the rather few caves in which they have been discovered. Until recent years this species was thought to be a strictly coastal species, but its occurrence on the Nile in Uganda, at two localities in the interior of Mozambique and at Kilaguni Lodge in Kenya* suggests that at low altitudes it may extend its range inland. Nonetheless, the coastline of eastern Africa probably provides an optimum habitat. Its northern range in the Persian Gulf is a peripheral extension, as the bat is very rare there and occurs in small numbers.

Kulzer (1959) has described the evening flight of these bats from the Mkulumuzi caves near Tanga. A noisy swarm of bats collected, milling around within the cave's entrance before flying out in a body as if at a given signal. Several swarms followed one another in this way. Thousands of these bats live here and at Ngombeni—near Mombasa—associating principally with *Coleura afra*. In Arabia they are sometimes found mixing with *Asellia*. They fly out in the evening, sometimes before dark and Harrison (1964) has described their flight as delicate and butterfly-like, as they fly low over the ground and bushes.

Very little is known of their biology. Females from Mkulumuzi have been recorded pregnant in December.

* Also recorded recently in Zaire, and from Lake Baringo.

Vesper Bats

Vespertilionidae

This is a very large and successful family of bats. Walker (1964) lists 38 genera and over 270 species. Simpson (1945) lists 25 genera. Vesper bats are subdivided into 6 subfamilies three of which occur in East Africa: the Vespertilioninae, the Miniopterinae and the Kerivoulinae. Ten genera and over thirty species occur in East Africa.

At a casual glance most vesper bats look bewilderingly alike and for certain identification of the more difficult genera and species dental patterns must be examined.

The visual key overleaf has profile drawings of typical species and silhouettes of their skulls, the shapes of which are usually highly characteristic of each genus. The form of the head can often be felt quite clearly even on living bats and it is immediately obvious in the cleaned skull. The pattern of the upper toothrow provides a certain generic diagnosis.

Characteristics common to all vesper bats are: small eyes, separate ears with noticeable tragi and a long tail enclosed in the membrane. Many species have glands on their muzzles.

Fossils of a modern genus, *Myotis*, are known as early as the European Oligocene and the family is clearly very old.

Tate (1942) has discussed the radiation of the Vespertilioninae and defines the primitive vesper bats' skulls as having a moderately full braincase with strong zygomatic arches and a slender snout. The teeth are unreduced with a dental formula $\frac{2.1.3.3}{3.1.3.3} = 38$. Among living forms this formula is found in *Myotis* and Tate regards this genus as being nearest to the stem from which both the Pipistrellini and the Nycticeini developed.

Kerivoulinae and Miniopterinae are less progressive subfamilies that became differentiated at a very early date; both have greatly inflated braincases and *Miniopterus* has lost a premolar.

The more advanced genera have shortened the face and reduced the number of teeth, a tendency which shortens the toothrow and gives more adequate buttressing to the teeth and is presumably a structural improvement allowing of more rapid and effective seizing and chewing of insects. Some genera, particularly the most advanced, have radiated into a large number of species, among which may be species adapted to certain limited ecological conditions or geographic areas as in the very numerous *Eptesicus* and *Pipistrellus*. *Scotophilus* instead has three species, which are structurally almost identical and which range very widely on the continent but which occupy three distinct size classes, one of which is the largest vespertilionid known. On the whole, however, vesper bats are small or medium-sized.

The two mono-specific genera, *Mimetillus* and *Laephotis* appear to have adopted peculiar niches and the former genus has paralleled some of the developments found in the molossids, notably *Platymops*. It roosts in narrow cracks.

The family has a world-wide distribution and occupies all but the very coldest regions and habitats. A very wide variety of situations serve as roosts

262

or shelters for vesper bats and communities may vary between many thousands in caves, to pairs or solitary bats hidden in vegetation. Among some species sexes may segregate after mating and for nursing. In Europe, seasonal migration is well known in some vesper bats; this has yet to be observed in African species.

Most species bear a single young, although *Scotophilus* and *Nycticeius* have twins regularly. Gestation periods range between 40 and 100 days. In the northern hemisphere, both captive and wild vesper bats have been found to live up to 20 years.

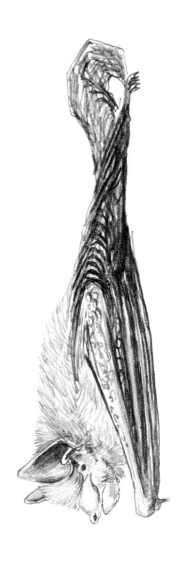

Pipistrellus nanus.

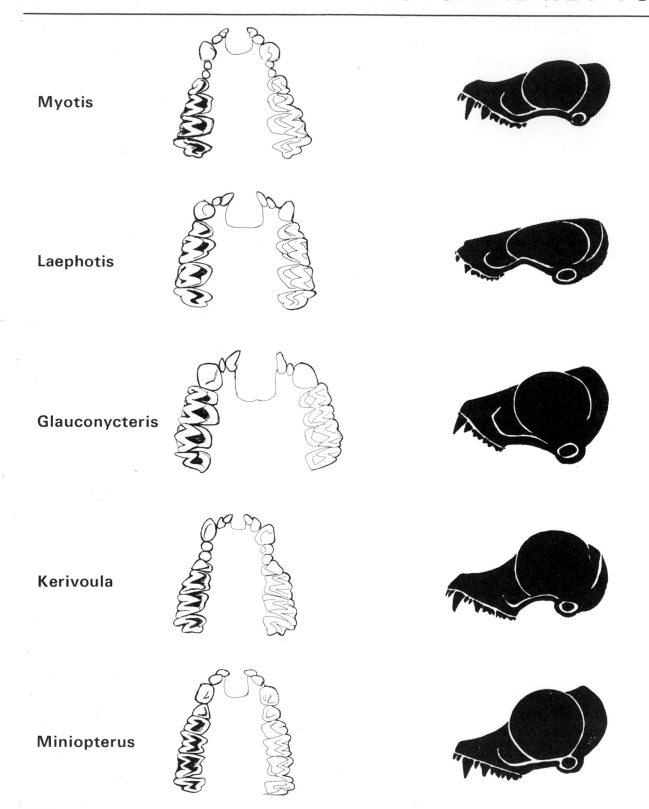

Myotis

Laephotis

Glauconycteris

Kerivoula

Miniopterus

VESPERTILIONIDS

Tooth Formula

$$\frac{2.1.3.3}{3.1.3.3} = 38$$

3 species

Profile on p. 269

$$\frac{2.1.1.3}{3.1.2.3} = 32$$

1 species

Profile on p. 284

$$\frac{2.1.1.3}{3.1.2.3} = 32$$

6 species

Profile on p. 297

$$\frac{2.1.3.3}{3.1.3.3} = 38$$

5 species

Profile on p. 303

$$\frac{2.1.2.3}{3.1.3.3} = 36$$

3 species

Profile on p. 307

Pipistrellus

Eptesicus

Mimetillus

Nycticeius

Scotophilus

Tooth Formula

$$\frac{2.1.2.3}{3.1.2.3} = 34$$

6 species
Profile on p. 273

$$\frac{2.1.1.3}{3.1.2.3} = 32$$

6 species
Profile on p. 277

$$\frac{2.1.1.3}{3.1.2.3} = 32$$

note very short wing
1 species
Profile on p. 279

$$\frac{1.1.1.3}{3.1.2.3} = 30 \text{ or } \frac{1.1.2.3}{3.1.2.3} = 32$$

3 species
Profile on p. 285

$$\frac{1.1.1.3}{3.1.2.3} = 30$$

3 species
Profile on p. 291

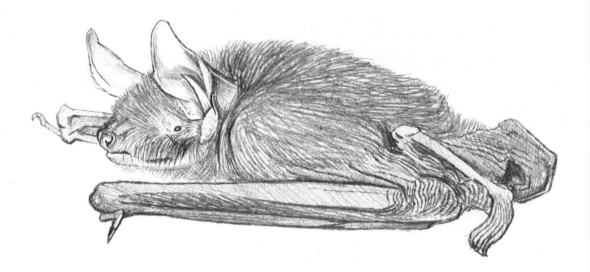

Mouse-eared Bats, Hairy Bats
(Myotis)

Family	Vespertilionidae
Order	Chiroptera

Measurements   *Myotis welwitschii*
head and body
60 mm
tail 55 mm
forearm 52—56 mm
weight 10 g

head and body   *Myotis tricolor*
50—61 mm
tail 45—56 mm
forearm 47—52 mm
weight 8—11 g

head and body   *Myotis bocagei*
50—60 mm
tail 36—42 mm
forearm 36—40 mm
weight 4—10 g

M. bocagei.

M. welwitschii.

M. tricolor.

Mouse-eared Bats,
Hairy Bats (Myotis)

Species

Myotis welwitschii
Myotis tricolor
Myotis bocagei

These small bats are recognizable by their pointed muzzle and numerous teeth, which have sharp pointed cusps. The relatively long toothrow occupies nearly half the length of the skull and the cranium is slightly inflated and without crests.

The genus has been recognized in Oligocene deposits in Europe and, on the basis of its teeth, it is the most primitive genus within the Vespertilionidae. Notwithstanding this, the genus has a world-wide distribution and sixty species are recognized; it is particularly successful in Eurasia. Of the six species found in the Ethiopian region, only the three species listed here are at all widespread and, since these are not particularly common, it is possible that the radiation of other tropical African vespertilionids may have diminished their status and reduced the number of available niches.

Myotis bocagei is the most successful and adaptable species, ranging over many vegetation types and occurring through most of sub-Saharan Africa. The other two species are typical of the southern woodlands and savannas. They seldom approach houses and are usually found roosting in vegetation or in deep caves.

Most *Myotis* species are not particularly fast in flight and do not appear to make the spectacular and rapid changes in direction some pipistrelles are capable of. Some hunt over water, where they can be netted as they fly in low wide circles, and most fly at a height of 1 to 4 or 5 metres from the ground. Insects are caught in flight but, non-flying insects resting on leaves or bark may be taken by European species (Kolb, 1958). Novick (1969) reports that 500 small insects or 150 larger ones may be taken in the course of an hour.

These bats are able to modulate the frequency of their ultrasound over a very variable range, from 100,000 cycles to 20,000 cycles (Peterson, 1964).

They usually hang by the feet, sometimes in dense clusters, but occasionally they cling to the surface of their roost and the European *M. mystacinus* has been found sleeping on its side with folded wings. Experimental release of transported bats has shown that some European *Myotis* species are able to home over several hundreds of kilometres. Limited migrations are also thought to occur in the Middle East.

Although they are often found singly or in pairs, some species are gregarious and may form associations numbering up to 70 individuals. All-male and all-female colonies have been found in various parts of the world. The females often appear to bear their young in segregated groups. Furthermore, *M. mystacinus* male groups are known to leave their year-long roosts for a month or more, and to return with numbers of subadult males, apparently "collected" after a spell spent in the female colonies. Pairs not in breeding

M. bocagei.

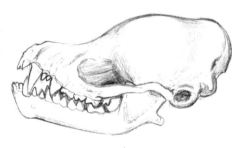

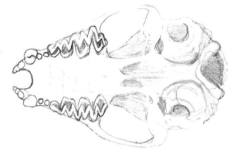

Skulls: *M. bocagei.*

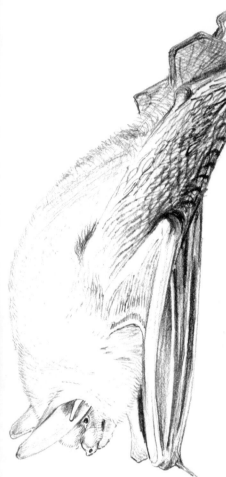

M. bocagei.

condition have also been noticed. In temperate regions *Myotis* is known to store semen in the uterus over the hibernation period. Gestation varies with the species, and may range between 40 and 100 days. A single young is usual, but two have also been recorded. The young mature in one year.

M. welwitschii is easily recognized by the bold black and red pattern of its wings and membranes. It has a scattered distribution through the woodland and savanna areas of southern and eastern Africa. It is rare and nothing is known of its biology.

M. tricolor is a hairy bat, generally dark in colour but often with rusty-coloured hair tips. It cannot be confused with the paler *M. bocagei* as it has a very long forearm in relation to the body.

This species has a wide distribution in southern Africa, extending up the eastern side of the continent as far as Ethiopia. Roberts (1951) describes it congregating in "fair numbers" in wet caves or mining adits, particularly where there is little disturbance. Nothing specific has been recorded on its biology.

Myotis bocagei is the most widely distributed and most frequently collected member of the genus in Africa. It is rather conspicuous because of its bright orange back and pale wings. The Congolese Wabudu are familiar with *M. bocagei*, which roosts in their banana gardens together with the little black *Pipistrellus nanus*; they call it "the big red brother" and liken this bat to the paler individuals in their society (Lang and Chapin, 1917).

They are found throughout a wide range of habitats from forest to semi-desert. One subspecies even occurs in southern Arabia. Darker and lighter forms appear to be related to moister and drier habitats. *M. bocagei* roost in a variety of situations, caves, birds' nests, hollow trees and vegetation; they favour banana trees in some areas, but otherwise avoid villages and houses. They can sometimes be netted over water as they hunt small flying insects. Most of their hunting is done in the earlier part of the evening and then again in the early morning. They eat a wide variety of insect species.

They are frequently found in pairs. A large embryo has been recorded in early January and a small hairless male at the end of June. I have caught a very fat male with large testes in May.

270

M. bocagei.

Pipistrelles and Serotines

Pipistrellus and Eptesicus

The retention of two generic names to describe these bats is only a convenience, for the distinction separating the two groups—namely, the presence of a vestigial premolar in *Pipistrellus* and its absence in *Eptesicus*—has been shown to be inconsistent in an African species of *Eptesicus*, *E. tenuipinnis* (Hayman, 1954), and in a Russian pipistrelle (Kuzyakin, 1950).

Although they represent in all probability only a single group, the lumping together of the very numerous species of pipistrelles and serotines would raise still further problems for their already difficult taxonomy.

Three important authorities, Ellerman and Morrison-Scott (1951), Rosevear (1965) and Hayman (1967) have retained these genera, a course which is followed here.

271

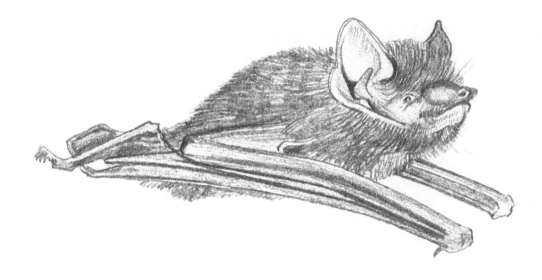

Pipistrelles
(Pipistrellus)

Family Vespertilionidae
Order Chiroptera
Local names

(for *Pipistrellus nanus*) Kahundu
(Lutoro), Belibu (Kuamba),
Nundu (Kisambaa), Lilema
(Kikinga), Kasusu (Kirungu)

Measurements   *Pipistrellus nanus*
head and body   Includes *aero*
36—40 mm
tail 24—41 mm
forearm 25—32.5 mm
weight 2—4 g

head and body   *Pipistrellus nanulus*
41—42 mm.
tail 21—25 mm
forearm 21—25 mm
weight 5—5.5 g

head and body   *Pipistrellus kuhli*
43—55 mm   White border to wing membrane
tail 20—35 mm
forearm 27—35 mm
weight 5—7 g

head and body   *Pipistrellus rueppelli*
43—56 mm   White belly
tail 34—45 mm
forearm 33.5—36.5 mm

head and body   *Pipistrellus rusticus*
37.5—46 mm   Red colour
tail 27—31 mm
forearm 27—30 mm

Pipistrelles (Pipistrellus)

Species

Pipistrellus nanus
Pipistrellus nanulus
Pipistrellus kuhli
Pipistrellus rueppelli
Pipistrellus rusticus
Pipistrellus permixtus

Pipistrelles are difficult to describe as they lack any very striking specializations or characteristics. The ears are not joined at the base, skulls are very small and lack obvious distinguishing features, except when the teeth are examined under a microscope. Forearms are short, less than 40 mm. Dental formula: $\frac{2.1.2.3}{3.1.2.3} = 34$. The species within the genus are probably best distinguished by the shape of the tragus.

The genus is found throughout the world and is particularly successful in Eurasia. Pipistrelles range over a very wide variety of habitats; they often roost in very small spaces where only single or small numbers of animals can find room, although a European species, *P. pipistrellus*, may live in colonies numbering several hundreds. All species feed on very small flying insects, flies, mosquitoes and gnats, which are mostly caught and eaten on the wing. When larger insects are taken, pipistrelles may alight to dismember and eat their prey. The flight is very erratic and fluttering.

The time of emergence has been found to vary with the season in Europe, where the annual activity pattern includes a long winter hibernation and the occurrence of delayed implantation in the females. In most pipistrelles the gestation period is probably about six weeks. The birth of two young is not unusual in some forms.

Pipistrellus nanus, the banana bat, is the commonest and best known African pipistrelle. It is immediately recognizable by little callosities on the feet and thumb. These are undoubtedly adaptive to assist in a firm grip, for its favourite shelter is the slippery furled leaf at the centre of the growing banana tree or the space round the stem of a bunch of bananas. Other roosts are also used, but as this habit is known to small boys in all parts of Africa where bananas grow, naturalists and collectors have been furnished with numerous specimens often prepacked in their natural shelter, for it is a simple matter to close the banana leaf tube with a fold at each end and carry it away.

This species is commonly found solitary or in pairs. It flies at dusk and tends to keep near the ground hunting in the vicinity of its roosts. In common with other pipistrelles, this species has a very wide gape of about 120° (see drawing). Lang and Chapin (1917) noted two features that were associated with this capacity, a process on the chin on which muscles can pull the lower jaw wide open and the great reduction of the coronoid process. These features are shared, but to a lesser degree, by some other bats (i.e.: some *Rhinolophus* and *Nycteris* species). However, I have not found a muscular condition on the chin process.

P. nanus.

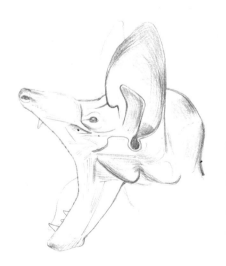

P. nanulus.

P. nanus.

P. nanulus.

P. kuhli.

Breeding is apparently not seasonal, as females with young have been collected at all times of the year.

The form *aero* may be a high altitude isolate of this species, found on the summit of Mt Gargues (Warges), in Kenya.

P. nanulus is readily recognized by its short forearm and also by its very short swollen muzzle. This species has a reddish phase. It ranges from Nigeria to southern Uganda in the forest zone. It hunts over water, flying very close to the surface. I have caught a female with two small embryos in mid-January and a lactating mother in mid-March.

P. kuhli is about the same size but more heavily built than the banana bat, with a stouter skull and teeth. Externally it can be recognized by a white border to the wing.

It is widely distributed in Eurasia and extends across North Africa with scattered records from the eastern side of the continent to the Cape. This species has been recorded from the Hoggar Mountains in the Sahara Desert and also from forest in South Africa. It is clearly very adaptable but in Tropical Africa the presence of *P. nanus* may diminish its opportunities and this may account for its relative rarity in East Africa.

In southwestern Tanzania I once netted a small party of this species, whilst they fluttered around catching small flies off a heavily fruiting peach tree. Two adult females, a male and a subadult animal were caught. Two others were attracted by their cries but avoided getting entangled.

P. rusticus is a rusty-coloured species that appears to be a rare species in tropical Africa. It is recorded from southern Africa and also the Sudan. It resembles *P. kuhli* in having a white margin to the wings but this is very narrow. Start (personal communication) has caught this bat in West Pokot, Kenya.

274

P. rueppelli is easy to recognize by its grey fur and white underside. The face appears to be rather round because the long fur almost hides the muzzle. Collectors have remarked on the conspicuous external penis which is about a quarter of the length of the body.

This is a savanna and dry country species. It is often seen on rivers at dusk. Lang and Chapin (1917) found it fluttering around oil palms growing beside the Bomokandi River and caught a male entering a lighted room.

P. permixtus is a species only known from Dar-es-Salaam and was described by Aellen in 1957. He suggests that this form has Palaearctic or Oriental affinities. This is an interesting record considering the presence of the Oriental *Pteropus* across the Zanzibar channel. There is a possibility that this is a wanderer from Zanzibar or, since Dar-es-Salaam is a great sea port, perhaps a stowaway from further east.

P. rueppelli.

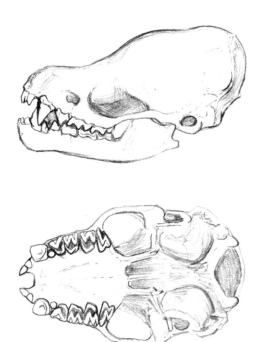

Skulls: *P. nanus.*

Serotines (Eptesicus)

Family Vespertilionidae
Order Chiroptera

Measurements
head and body
37·5—53 mm
tail 26—34 mm
forearm 28—33 mm
weight 5—5·5 g

Eptesicus tenuipinnis
Black body with transparent white wings

head and body
45—59 mm
tail 33—40 mm
forearm 31—38 mm

Eptesicus rendalli
Buff-brown back, off-white below

forearm 46 mm

Eptesicus loveni
Pale brown back,
greyish brown below

Eptesicus capensis group
Small brown bats

forearm 26—29 mm

Eptesicus pusillus

30—32 mm

Eptesicus somalicus

29—35 mm

Eptesicus capensis

Serotines (Eptesicus)

Species

Eptesicus tenuipinnis
Eptesicus rendalli
Eptesicus loveni
Eptesicus pusillus
Eptesicus somalicus
Eptesicus capensis

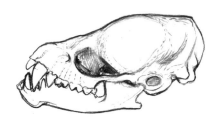

Serotine bats are distinguished by their relatively short ears and by the straight profile of their skull; their crania do not bulge. The possession of one upper incisor on each side separates these bats from the pipistrelles.

Serotines are a very successful group of almost world-wide distribution, occurring in all vegetation zones. Favourite roosts are the roofs of buildings and the interior of hollow trees. They can either hang free from the feet or crawl and cling with their belly pressed to the substrate. Some species can tolerate very hot roofs.

Considering how variable in size serotines are, it is not surprising that their prey also ranges from mosquitoes to beetles and moths. They vary also in their habits and it is difficult to make general remarks on the genus that describe all species.

Their flight is relatively slow and flapping, frequently interrupted by swoops or dives after prey. Small groups may be found hunting near one another and they exchange audible clicks as their paths converge. Insects are caught in flight and, sometimes while these bats are flying, it is possible to hear the rapid crunching of chitin. Clumsy efforts on the part of these bats at catching resting insects have been noted in Europe; serotines landing with outspread wings on the tops of trees and shrubs.

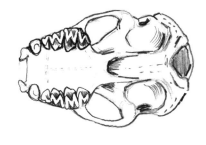

A serotine bat has been known to live for 12½ years in captivity and wild bats are known to have carried rings for well over 10 years.

Although colonies of fifty or more are known, most species are found as solitary animals or in small groups. Most forms that have been studied have been found to be conservative about their roosts and deep deposits of dung are not unusual.

Skulls: *E. tenuipinnis.*

Known predators are owls and hawks. In species from the temperate regions female serotines store quiescent sperm throughout their winter hibernation. When the young are born, nursery colonies are formed from which males are absent. One or, occasionally, two young are born, naked and blind. The infant is dropped from an inverted hanging position into the interfemoral membrane, where it is licked clean and nudged into the wing fold and thence onto the body. At one week the infant's eyes open and its size is already more than half the adult's. At this time the mother begins to "park" her young and, although she will still answer the young's piping call, she is less attentive towards it. At two weeks the young are fully haired and start combing their fur with their claws, they also start to practice flapping their wings. They fly for the first time at 3 weeks of age, at which time the perma-

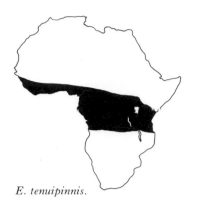

E. tenuipinnis.

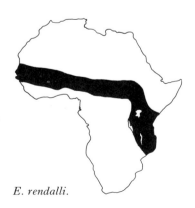

E. rendalli.

nent dentition has replaced the milk teeth. The young animal is able to fend for itself by the time it is one month old.

E. tenuipinnis is very easily recognized by its white, transparent wings and tiny black-brown body.

This species ranges across tropical Africa and is principally a forest species. Bats of this species have been recorded in groups of half a dozen or more, roosting in hollow trees, but they are often to be seen hawking over water or around lights to which they are attracted by the dazzled insects; occasionally they have been known to fly into lit rooms. So transparent are their wings that, when they are watched against a clear sky, only their little dark bodies appear to be hurtling through the air.

I have caught 3 lactating females in Uganda during January, and males with very large testes during June, October and November. At this time the males' facial glands are very large and exude a greasy orange secretion, which tints the entire animal, including its wings, with a pinkish colour. The hair on the face seems to be less dense at this time, suggesting that it may either fall out or be rubbed off (see drawing, left).

E. tenuipinnis have been recovered from the crops of bat-hawk.

E. rendalli can be distinguished by the presence of a basal cusp to the upper incisor and by its buff-brown back and dirty-white belly. Like *E. tenuipinnis* it has pale wings but they are less transparent.

This species ranges through woodland and savanna from Malawi and Mozambique across Tanzania and Kenya to West Africa (Gambia).

E. rendalli frequently enters lit rooms at night in pursuit of moths. They have been found roosting in thick vegetation and in the roofs of thatched houses, even clinging to a single strand of projecting grass (Lang and Chapin, 1917). These authors also remarked on their seasonal invasion of houses after bush-fires had deprived them of natural cover at the beginning of the dry season.

They hunt early in the evening, skimming low with a whirring flight as they catch small insects. They are alert bats and will fly to another roost when disturbed. They are generally found in pairs or as single individuals.

E. loveni is only known from the type locality of Mt Elgon. It is distinguished by bicuspid inner, upper incisors and a pale colouring. Hayman (1967) thinks it possible that this may be an African relative of the European *E. serotinus*.

E. capensis is a term of convenience for a complex of bats which may turn out to include several species. These are all small bats, with a forearm of less than 35 mm and which do not have the swollen cranium that is so characteristic of many other small vespertilionids.

These bats range throughout Africa south of the Sahara and Madagascar. They can be found in all habitats; they shelter in houses, in rocks and tree cavities. Kock (1969) notes that *E. somalicus* bats use a regular flight-path and alight frequently to rest.

There is a very widespread and tiny form, which may be a subspecies or a distinct species (forearm 26—29 mm) and that has been described under two names, *E. minutus* and *E. pusillus*. Another form is known as *E. somalicus*; bats belonging to this type are distinguished by longer hair and a less developed nuchal crest and they appear to be sympatric with *E. capensis* in

South-west Africa.

E. somalicus is common in Uganda, where I have caught a pregnant female with two large foetuses in January.

The taxonomic problems concerning this group have been discussed by Rosevear (1962—1965), Koopman (1965) and by Hayman (1967) and the need for a complete revision of the group has been felt by all workers on African bats.

E. somalicus, resting posture.

Moloney's Flat-headed Bat
(Mimetillus moloneyi)

Family	Vespertilionidae
Order	Chiroptera

Measurements
head and body
50—60 mm
tail
26—33 mm
forearm
26·5—30 mm
weight
6—11·5 g

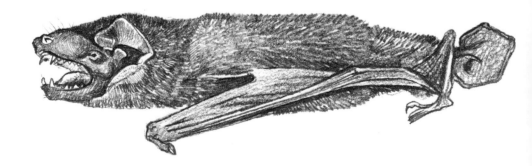

Moloney's Flat-headed Bat (Mimetillus moloneyi)

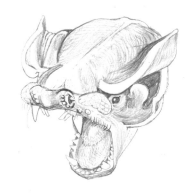

This bat is very easily recognized, both in flight and in the hand; its most obvious features are the extreme shortness of the wing and the flattening of the head and body. The fur is dark and velvety and often appears to be greasy in texture like that of some molossids. The legs are short and stoutly built. Like other Vespertilionidae, *Mimetillus* has very large glands on the muzzle (see dissection); these are largest in sexually active animals.

The flat skull and body appear to be adaptive specializations similar to those found in the molossid, *Platymops*, which lives inside fine cracks in shale and rock; *Mimetillus* instead roosts in cracks beneath the bark of dead trees.

These bats range across the tropical belt of Africa, reaching as far south as Angola, Zambia and southern Tanzania. They were formerly thought to be an exclusively forest species, but recent collecting has shown that they also occur in wooded country and at high altitudes—2,300 m.

Mimetillus prey on small flying insects, notably termites, pursuing them at dusk, often while it is still quite light. The flight is very fast and a whirring of wings can be heard when a bat passes overhead. The rapid flapping must be maintained to keep the relatively heavy little body in the air and it would be interesting to have a measure of the wing-beat frequency. *Mimetillus* flies either in a direct and straight line or in wide shallow arcs, but appears to be incapable of sudden turns. It is usually seen in relatively open spaces, above the forest canopy, along exposed mountain ridges or along river courses but has also been netted in forest over stagnant pools. I have watched captives and have noticed that when they are prepared for flight they spread their wings before dropping. If the space proves inadequate, they drop to the ground and then scramble up again. Trying to scramble upwards before take-off appears to be a definite behaviour pattern and may be related to the need for an elevated launching point in such a short-winged bat. Their roosts prob-

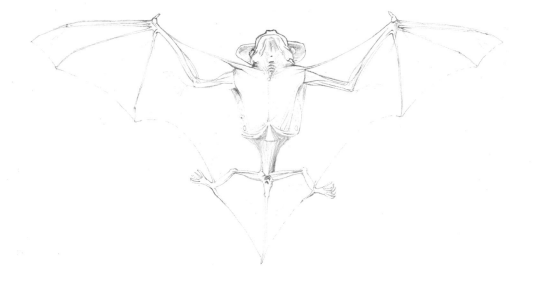

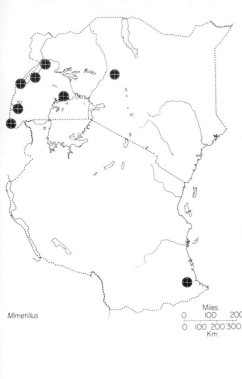

Mimetillus

ably allow a vertical drop and clear flyway; Brosset (1966b) records them as being at a height of about 8—12 metres. Rosevear (1965) notes that *Mimetillus* have been collected from roofs in Sierra Leone villages, where it seems, they have adapted to the recent built-up conditions within the forest zone.

Mimetillus live in small colonies of 9—12 individuals in well established roosts under dead bark. They return to their roosts every 10 or 15 minutes for a rest, a habit that is probably related to their tiring rapidly because of the frequency of their wing-beats.

I have netted several *Mimetillus* and my first impression was that they groomed themselves feverishly and continuously. However, this may have been due to the acquisition of parasites from the molossid bats they were confined with. They used their hindfeet to scratch all over the body except

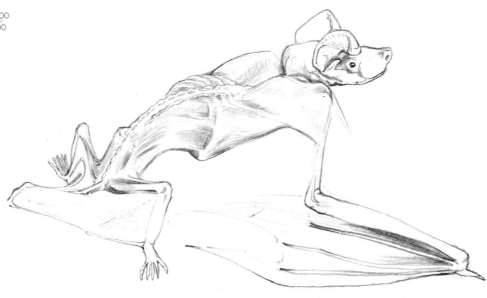

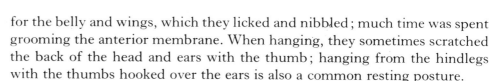

for the belly and wings, which they licked and nibbled; much time was spent grooming the anterior membrane. When hanging, they sometimes scratched the back of the head and ears with the thumb; hanging from the hindlegs with the thumbs hooked over the ears is also a common resting posture.

When in the open, my captives showed every sign of being uncomfortable; they moved sideways and backwards as though seeking a crevice in which to crawl. The down-pointing tip of the tail seems to serve as a sensitive probe. Whenever the animals walk on a level surface they carry the body well clear of the substratum and sometimes walk or posture with the head raised and the body at a steep angle; the forearm may be almost perpendicular at such times. When walking or running, the hind claws splay out sideways and all their movements are very fast and jerky. When resting, or sometimes in response to a sudden noise, *Mimetillus* can flatten themselves; the ears fold down and the animals become compact little pancakes. As they raised themselves from this flattened position they reminded me of deck-chairs being unfolded!

These little bats have very wide gapes and my captives were extremely fierce and made rapid snaps at an approaching pencil or finger. They also struck out with quick little jabs of the forearm, while making an audible squeak. They feed on flying termites and ants. One animal I netted had the head and jaw of a flying ant firmly embedded in the corner of its mouth, and the insect had severely damaged the bat's cheek.

It is possible that the larger hornbills, Bucerotidae, may be important predators of *Mimetillus*, for hornbills spend much time removing and probing loose bark and I have actually seen one of these birds killing and eating a bat. Chapin (1932) retrieved *Mimetillus* from the stomach of a bat-hawk, *Machaerhamphus*. It had been swallowed without the least mutilation. Other bats from the crops and stomachs of these birds have been found with deep wounds and Lang and Chapin (1917) collected a *Mimetillus* just recovering from two severe wounds in the shoulder and neck, which they suggested were probably bat-hawk injuries.

The calendar shows ten records from Uganda, indicative of a biannual breeding season with births in February or March and again in August. The young, therefore, are likely to be born at the end of dry spells and would start fending for themselves during the rains, just when insects are most abundant.

Active male *Mimitellus*.

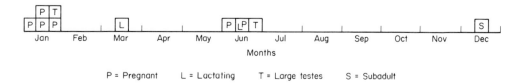

P = Pregnant   L = Lactating   T = Large testes   S = Subadult

De Winton's Long-eared Bat
(Laephotis wintoni)

| **Family** | Vespertilionidae |
| **Order** | Chiroptera |

Measurements
head and body
45—51 mm
tail
41—46 mm
forearm
35—37·5 mm

De Winton's Long-eared Bat (Laephotis wintoni)

This large-eared, tawny brown bat has a short face and very small canines. Its dentition is characterized by an extreme reduction and upward migration of the canines and incisors. The cheek teeth appear to be the only functional teeth and it is possible that this bat has a specialized diet or manner of feeding. It is also possible that the peculiarly raised and reduced incisors and canines create a channel which is linked in some way to the beaming of ultrasound.

The twenty or so specimens known to science have all been taken from under the bark of dead trees, a micro-habitat which, in the case of *Mimetillus* and some molossids, has led to the flattening of the skull; the skull of this species, however is long and narrow. Only three specimens have been collected to date in East Africa, all from Kenya, but the species is also known from Angola, Katanga, and Zambia and may eventually prove to be more widely distributed in the savannas and woodlands of central Africa. I have been told of small, long-eared bats that roost under bark, from Uganda (north of Mt Elgon), and also from north of Lake Rukwa in Tanzania; these might be *Laephotis*.

284

Laephotis seem to roost singly or in pairs but, although rare are probably locally abundant in suitable localities; Hayman (1957) reports that Mr A. Lips collected 13 individuals from 2 villages in Katanga.

Further information on this species is needed; data on its echo-location system, roosting postures and its feeding and hunting behaviour would be particularly interesting.

Twilight Bats

Nycticeius

The genus *Nycticeius* includes two subgenera, *Scotoecus* and *Scoteinus*. It is a group with a very wide range being found in tropical Asia, Australasia and also America. The genus is distinguished by a blunt muzzle, a relatively short tragus in a round ear, four or five upper cheek teeth and one upper incisor.

N. (S.) hirundo.

Evening or Twilight Bats
(Nycticeius (Scotoecus))

Family	Vespertilionidae
Order	Chiroptera

Measurements

$N. (S)$ *hirundo*
Dark-winged forms

head and body
54—62 mm
tail 28—38 mm
forearm 33—36 mm

$N. (S)$ *h. hindei*

head and body
61—68 mm
tail 32—40 mm
forearm 37—38 mm

$N. (S)$ *h. albigula*

head and body
46—63 mm
tail 30—35 mm
forearm 30—33 mm

$N. (S)$ *h. artinii*

head and body
50 mm
tail 30—32 mm
forearm 30—31 mm

$N. (S)$ *albofuscus*
Light-winged form

Evening or Twilight Bats (Nycticeius (Scotoecus))

Species

Nycticeius (Scotoecus) hirundo
Nycticeius (Scotoecus) albofuscus

The subgenus *Scotoecus* is distinguished from *Scoteinus* by blunt, short and rounded tragi and by the greater breadth of the skull's rostrum. The upper canines are peculiarly flattened on their front surface. The significance of these dental features is not known, as there is little information on the biology of *Scotoecus*. The subgenus is represented by two species, both restricted to Africa; *N. (S.) hirundo* is a slightly larger bat with a pale belly and dark wings. It is known to have dark grey and reddish colour phases. *N. (S.) albofuscus* has a dark belly and pale wings.

Three forms of dark-winged *Scotoecus* have been recorded in East Africa: *N. (S.) h. hindei* from widely scattered localities in Uganda, Kenya and Tanzania, *N. (S.) h. artinii* from dry localities in northern and western Kenya and *N. (S.) h. albigula* from western Kenya and Uganda. It is possible that more than one species is involved but the group awaits a thorough revision. I am indebted to Anthony Start for the measurements of the various forms.

N. (S.) hirundo hindei appears to have a well-defined breeding season in western Uganda where I have collected two pregnant females in early March; both were carrying two foetuses, one in each horn of the uterus. A lactating female and a juvenile were caught together in May; two males were also caught in May. In this species the penis is very long.

N. (S.) albofuscus.

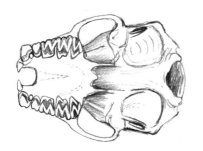

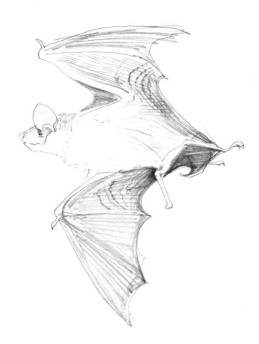

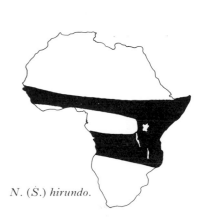

N. (S.) hirundo.

287

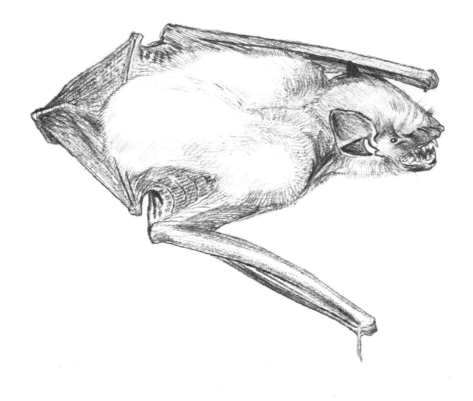

**Schliefen's
Twilight Bat**

**(Nycticeius (Scoteinus)
schlieffeni)**

Family Vespertilionidae
Order Chiroptera

**Measurements
head and body**
40—51 mm (males)
46—56 mm (females)
tail
26—37 mm
forearm
29—35 mm
weight
6—9 g

Schliefen's Twilight Bat (Nycticeius (Scoteinus) schlieffeni)

N. Scoteinus is distinguished from *N. Scotoecus* by its more pointed tragus with a straight anterior margin and by the rostrum being narrower. The upper canine has a more rounded cross-section. The colour is variable and white-bellied specimens are known.

The single species ranges through all the drier open parts of Africa and parts of Arabia. This bat roosts in houses and huts and has been taken from crevices in branches (Verschuren, 1957), where it was found together with two species of *Tadarida*. The flight of this bat is very erratic and some numbers have been seen together, emerging before dusk and flying about between the trees. (Roberts, 1951.)

Kock (1969) watched this species in the Sudan, while it was drinking from small expanses of water shortly after sunset. It feeds mainly on moths and beetles, and individuals were well fed within half an hour of emerging from their roosts. Some beetles, *Bolboceras* and *Geotrupes* species, were not eaten by a captive, which was unable to cope with their hard chitin. Soft insects were bitten while still being held in the interfemoral membrane.

Kock found that, on cool, wet nights, these bats slept or only woke for very brief periods. When well fed, a captive could go without food for 5 days.

S. gigas skull.
S. nigrita (below).

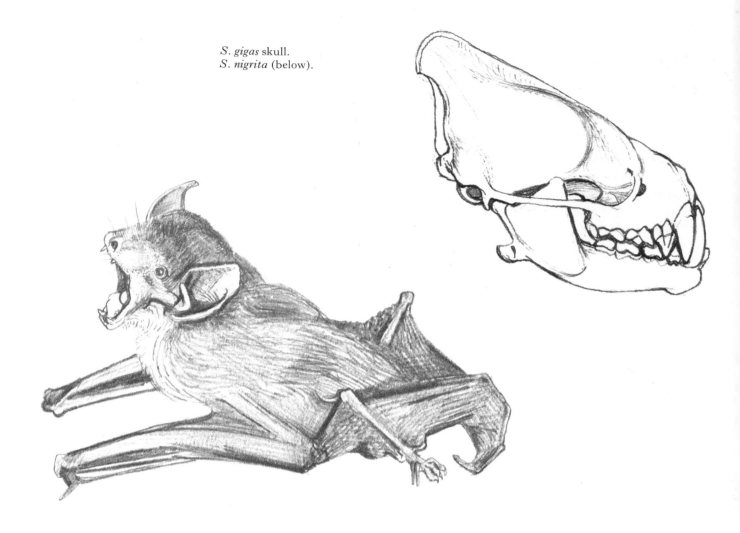

Scotophilus Bats, House Bats
(Scotophilus)

Family   Vespertilionidae
Order   Chiroptera

Measurements
head and body
112—117 mm
tail 68—78 mm
forearm 70—80 mm

Scotophilus gigas

head and body
72—80 mm (males)
80—89 mm (females)
tail 50—55 mm
forearm 50—65 mm
weight 20—40 g

Scotophilus nigrita

head and body
56—70 mm
tail 37—55 mm
forearm 43—50 mm

Scotophilus leucogaster

Scotophilus Bats, House Bats (Scotophilus)

Species

Scotophilus gigas
Scotophilus nigrita
Scotophilus leucogaster

Scotophilus are robust bats with blunt heads and long tapering tragi. There is a conspicuous swollen gland inside the corners of the mouth and the posterior molar teeth are greatly reduced and simplified in pattern.

The two smaller species are very alike and the situation is similar to that in other groups of bats where there appears to be a larger and smaller form without any clear structural difference being apparent.

Scotophilus gigas has not yet been recorded from East Africa but is known from various vegetation zones in the Sudan, the Congo, Malawi and Mozambique and will probably turn up somewhere in the intervening country. Only a very few specimens have been collected and these vary widely in fur colour between dark brown and white with brown wings. Rosevear (1965) records *Scotophilus gigas* as having been collected in a house, flying over a clearing, over a river bank and from the hollow trunk of a doum palm, *Hyphaene*. He points out that practically nothing is known of its habits. Kock (1969) believes that this species may be carnivorous and kill other bats, citing as evidence the absence of any other species in the same roosts.

S. gigas.

Scotophilus nigrita, the yellow house bat, is by contrast a common and widely distributed species familiar to most collectors of African bats. It ranges through the forest, woodland savanna and semi-arid zones of sub-Saharan Africa and has been recorded in the southeastern tip of Arabia. It roosts in roofs, hollow trees and caves. Roberts (1951) found it in woodpecker and barbet holes. This species is not easily disturbed and even after handling it often returns to the same roost. It can tolerate very hot temperatures but the type of roost occupied in any single locality is probably influenced to some considerable degree by competition with other species of bats.

All observers have agreed that this bat is almost exclusively an eater of small beetles although it is quite a catholic feeder in captivity. Kock (1969) found that captives would eat most types of insects but emphatically avoided the cantharid beetles, *Mylabris* and *Cylindrothorax*. He successfully fed them geckos. In common with other observers, he found they attacked another species of bat, *Nycteris*, put in the same cage and the viscera of one victim were eaten.

They emerge at dusk and appear to be very effective hunters. I have caught *S. nigrita* with 2 or 3 grammes of beetles already in their stomachs before the last traces of daylight had disappeared from the sky. Kock records them with full stomachs after an hour's hunting and described their nocturnal activity as being a series of intermittent hunting bouts interspersed with pauses for digestion and rest. They will hunt undeterred by rain. They drink flying at speed along chosen flightways, sweeping close to the water's

S. nigrita.

S. nigrita.

surface. They are more easily caught in nets near the roost as their routes are well established, but probably avoid nets more easily while hunting. They may return to the day roost within an hour of leaving it at dusk or they may use temporary resting spots. They are able to hang free but often prefer to huddle in groups clinging with their feet and thumbs. Captives scuttle about their cage until they can back into some dark corner with their tail and hind legs protected and the head hanging down. They shriek when caught, and threaten with a wide gape, exposing large pink glands in the corners of their mouths. Kock found the temperature of freshly caught bats struggling in the net to be 39°—40°C. By contrast, sick, cold bats measure 22°—24°C, at which temperature they are torpid. At 29°C they are lethargic but can

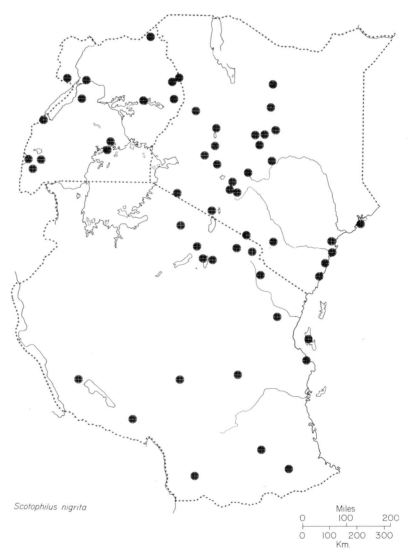

Scotophilus nigrita

Miles
0   100   200
0 100 200 300
Km.

shiver themselves up; at 31°—33°C they can just manage to take off and glide clumsily to the ground, but they are only fully active at temperatures about 34°C. Although sometimes found singly or in small groups, *S. nigrita* is more usually colonial, 20—30 and even up to 80 bats sleeping in one roost,

292

often densely clumped in favoured places with single outliers scattered in less favourite spots. They do not associate closely with other species and, in Kampala, two neighbouring roofs are respectively occupied by *Tadarida* species and *Scotophilus*. The former are under very exposed corrugated iron, the latter under somewhat cooler terracotta tiles. Temperature is unlikely to be the only decisive factor, as *Scotophilus* may be found elsewhere under equally hot tin roofs. Kock noticed that in the Sudanese Nuba Mountains—in savanna country—*Scotophilus* were the dominant house bats, but that molossids occupied similar situations in the roofs of houses along the Blue Nile in a drier area. He found these bats present throughout the year and very attached to their roosts.

There is none the less a possibility of local migrations by this species. This suggestion was first made by the collector Willoughby Lowe, who found *S. nigrita* absent for some months of the year in Darfur (Thomas and Hinton, 1923). Along the edge of the Sahara the range of this species may well show a seasonal fluctuation and Kock has pointed out that the species' limits in this zone may be determined to some extent by a compromise between the increasing energy demands of higher temperatures and the evaporation rates on the one hand and the decreasing availability of food on the other. In East Africa, there is no firm evidence of migrations or seasonal movements, although there are some months in which I have been unable to net them and others when they have been numerous. Kock has noted a seasonal fluctuation in the deposition of fat during the latter part of the rains, at which time Sudanese bats have a definite birth season (July). I have recorded 12 pregnancies between January and March and this is clearly the main birth season. Males with greatly enlarged testes and facial glands have been caught in May and July and again in October and November, suggesting a second birth season between August and September. Kock thinks that delayed implantation may occur in this species. The females tend to be larger than the males

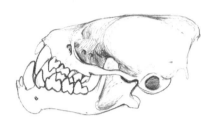

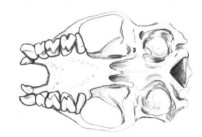

S. nigrita.

and Kock suggests that there might be a greater development of the swollen muzzle glands in male bats. However, I have found females with equally large glands while they were in breeding condition and also in early pregnancy and the glands may in fact be connected with courtship in both sexes as all the males I have examined show a correlation between the size of their testes and the glands: the larger the testes, the larger the glands.

This species twins with unusual frequency and I have always found both foetuses in the right horn of the uterus. Verschuren (1957) noticed a peculiarity in juveniles of both sexes, which have two pairs of rudimentary nipples. These disappear with growth but the occurrence is interesting in a species habitually bearing more than one young. This phenomenon may be compared with that found in the American vespertilionid, *Lasiurus*, which has the rare distinction of four functional mammae and which bears two to four young.

I have had the opportunity to watch the early development of this species. A female caught on February 20 gave birth to 2 naked, but very active, young two days after her capture. When I first saw what had happened, the mother had eaten the cauls and was on the ground, in a corner of her cage licking the young. The infants, however, were still attached by their umbilical cords. These broke or were bitten by the mother as they struggled to cling to her belly. Then the mother lay in a corner, suspended by one hind leg but resting her body on the right forearm, while the left wing sheltered the young under its half furled membrane. When she squeaked the young appeared to respond, but I could hear no sound from them. From the very beginning there was a lot of mutual mouth to mouth licking, which might have been an anticipation of weaning behaviour. After the first day the mother hung herself up on the upper walls of her cage but always maintained the tension in her legs, keeping them bent and so making her wings and body into some sort of tent for the young. Every evening the mother became very active, at which time she would regularly urinate. For the first three days the young clung tightly to

294

her belly, firmly fastened to the nipples. On the third night after their birth one of the young detached itself from the mother at about 8.30 p.m. just when the mother was most active. This was the first occasion either infant released its tenacious hold on the mother. From then on the young would occasionally leave the mother for short periods during the day, and would either hang quietly beside her or climb over her but only if she kept quite still. They were silent for long periods at a stretch but at times there were sudden prolonged bursts of shrill squeaking in which both the mother and the young participated. Unfortunately the young were accidentally damaged a week after their birth and died soon after.

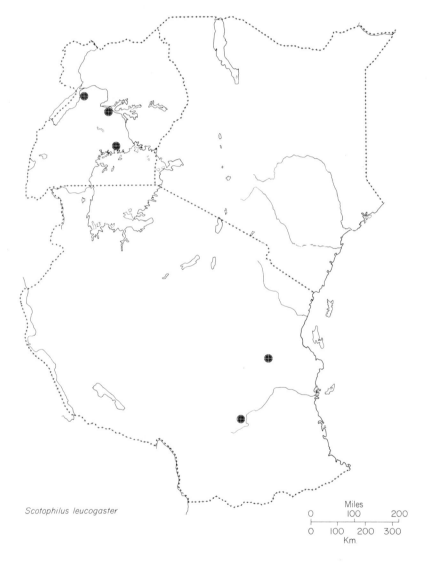

Scotophilus leucogaster

Miles
0   100   200
0   100   200   300
Km

Scotophilus leucogaster is a smaller bat than *S. nigrita* and is usually drab brown on the back. It has a dirty, whitish belly instead of a bright orange or yellow one like the larger species. I have caught one pregnant female in February and three in March and males with greatly enlarged testes and facial glands in May and July. When examined all four pregnant females carried two embryos.

S. leucogaster.

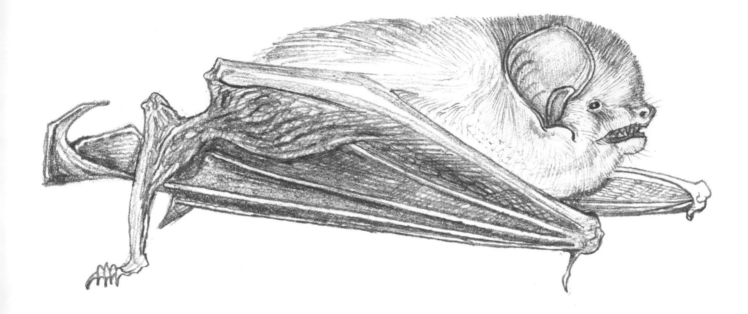

G. humeralis.

Butterfly Bats
(Glauconycteris)

Family   Vespertilionidae
Order   Chiroptera

Measurements
head and body

53—65 mm

tail 41—45 mm
forearm 41—45·5 mm
weight 9—15 g

head and body

53—60 mm

tail 44—50 mm
forearm 39—45 mm
weight 8·5—15 g

Glauconycteris variegata
Yellowish bat with strongly reticulated
pattern over the wings

Glauconycteris gleni
A white bat with a trace of reticulation
on the interfemoral membrane

Butterfly Bats (Glauconycteris)

head and body 63 mm **tail** 49 mm **forearm** 46—47 mm	*Glauconycteris superba* Black bat with white markings on shoulder
head and body 45 mm (*beatrix*) **tail** 43 mm (*beatrix*) **forearm** 35—40 mm	*Glauconycteris humeralis* Brown bat often with pale shoulder spot
head and body 45—55 mm **tail** 45—55 mm **forearm** 40—44 mm	*Glauconycteris argentata* Pale bat with pale brown wings

Glauconycteris are generally recognizable by their prettily marked wings or body or by the domed braincase behind a very short, broad muzzle.

Glauconycteris is an African genus of bats with a closely related genus, *Chalinolobus,* in Australasia. They have blunt, broad faces with widely spaced nostrils and beautiful conch-shaped ears, that flare out widely from the corners of the mouth where they connect with a peculiar lobe on the cheek.

The butterfly bats range over most of Africa south of the Sahara but are typical of the more heavily wooded or forested areas.

They roost in trees, inside buildings, amongst leaves or palm fronds or against bark. They are most commonly collected coming to lights after small insects and not infrequently fly into lighted rooms. They occasionally make audible cries.

The pale colouring on the two commonest species shows up strongly in lamp-light and their marbled pattern has probably earned the genus its popular name. Notwithstanding this, not much is known of their habits. They usually live in small groups and are known to have one or two young.

Glauconycteris variegata, the butterfly bat, is the best known species. The beautiful reticulated pattern of the wings is evidently designed to imitate leaves, for they sometimes exhibit unusual behaviour when bullied, half opening the wings to display the pattern to best advantage and refusing to fly or give any sign of life (see drawings overleaf). This species is characteristic of woodland and is known from Ghana to the Sudan and northern Kenya southwards to Zululand, South-west Africa and Angola. Lang and Chapin (1917)

G. variegata.

297

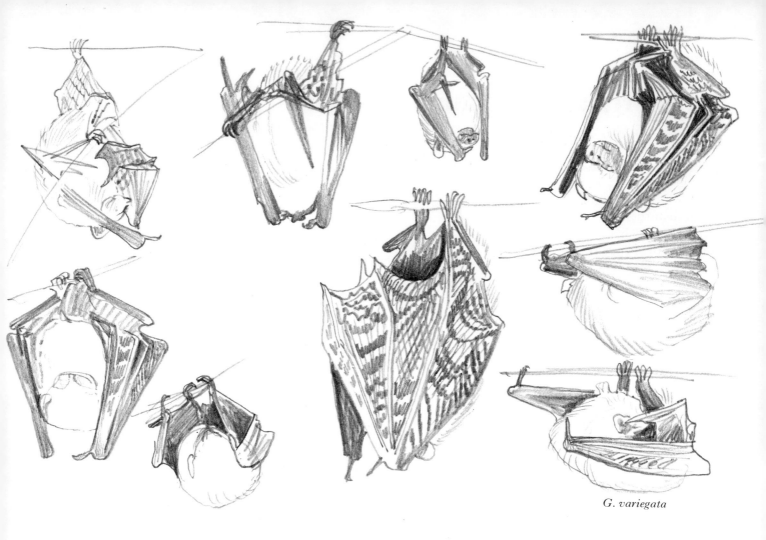

G. variegata

found this species roosting in pairs on trees, amongst bunches of leaves and also in groups numbering more than ten, huddled together in the thatch of huts. They also noticed these bats leaving their roosts two hours before sunset on overcast days. They fly with much changing of speed, direction and altitude and come down to lights chasing various small insects, I have found their stomachs to contain moths. Over the Uganda—Congo border in Upper Uelle, Lang and Chapin found some females with full-term foetuses in March and April. I have caught sexually active males in southern Uganda during July and October.

G. gleni.

Glauconycteris gleni is a white bat with a trace of greyish brown veining on the interfemoral membrane and between the feet and the elbow, so that the white is not so conspicuous when the animal is at rest. The fur next to the skin is dark but it is white-tipped on the face and chest, graduating to brown on the rump. Originally known from Sango Bay Forest and Kampala it first brought itself to the attention of Robert Glen, the collector, by coming after moths attracted to a light. I have seen this bat flying in daylight, after it had been disturbed by birds in dense vegetation some five or six metres from the ground. It seems to represent a unique and very rare species, somewhat intermediate between *G. variegata* and *G. argentata*. Glen collected two lactating females in January. This species has been described by Peterson and Smith (1973).

G. gleni.

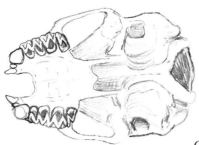

G. variegata

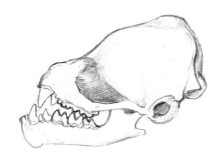

Glauconycteris argentata is a silvery butterfly bat that resembles the former species to which it is probably quite closely related in size and in the pale coloration of its fur. The wings, however, are pale brown and show a trace of reticulation only near the body. The range of this species includes moist forest, at both high and low altitudes, and also relatively open country. It has not yet been found in areas with less than 900 mm annual rainfall. It has a scattered distribution in East Africa and through the woodlands and savannas fringing the Congo basin. The most westerly records are from the

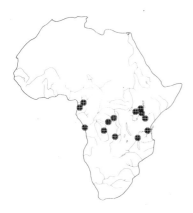

G. argentata.

G. superba.

299

Cameroons. Several naturalists have described this species' habit of colonial roosting on the fronds of various palm and *Dracaena* species. Several bats cling with all four limbs in compact rows to single leaves of the fronds facing out from the midrib. These little colonies may number 12 to 32 individuals, all clinging to one frond about 7 metres above the ground. Loveridge (1937) records four out of eight females collected in the Kenya uplands as carrying young ones in March.

Glauconycteris superba, the pied butterfly bat, is a rare lowland forest species which has been recently found in western Uganda. Hayman (1967) has suggested that this form may link *G. superba* with *G. alboguttatus.* The sketches show the white markings on the back and emargination of the ears. This bat was netted over a stagnant pool in the heart of Budongo Forest. Nothing is known of its biology.

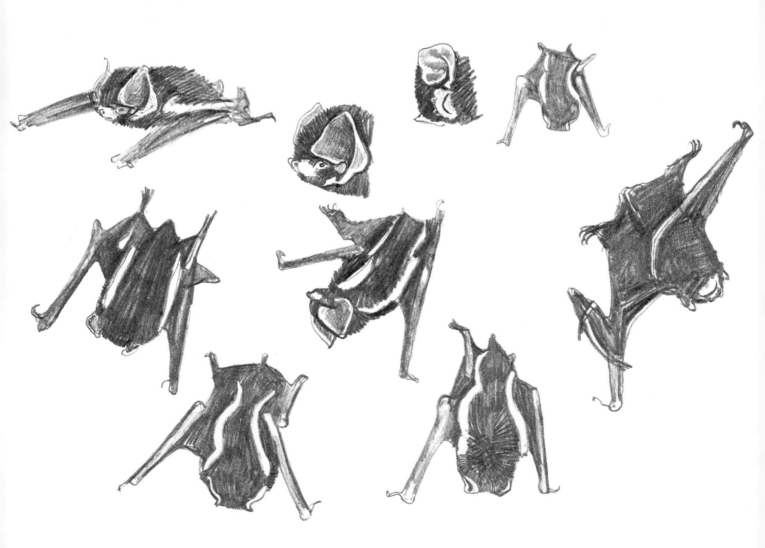

Glauconycteris humeralis, Allen's spotted butterfly bat, is a brown bat sometimes bearing paler spots on the shoulders. An unspotted form, *G. beatrix*, known from Uganda is possibly only a variety of the same bat. These are little known bats restricted to the forest zones of Africa and extending into West Uganda.

 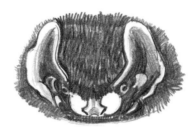 

Above and opposite, *G. superba*.

Woolly Bats, Kerivoula Bats

(Kerivoula)

Family Vespertilionidae
Order Chiroptera

Measurements
head and body
45—51 mm
tail 41—48 mm
forearm 34—39 mm

Kerivoula argentata

head and body
37—39 mm
tail 36—40 mm
forearm 30—32 mm

Kerivoula harrisoni

head and body
39—41 mm
tail 43—45 mm
forearm 32—34 mm

Kerivoula smithi and *Kerivoula cuprosa*

Woolly Bats, Kerivoula Bats (Kerivoula)

Species

Kerivoula argentata
Kerivoula harrisoni
Kerivoula smithi
Kerivoula cuprosa
Kerivoula africana

K. argentata : Frosted reddish back, white underside and hairy fringe to the tail membrane.
K. harrisoni : Similar to above but smaller and more frosted on the back.
K. smithi and *K. cuprosa* : Bats of similar size and grizzled fur. Separated on dental pattern.
K. africana : Dark brown fur, grey tipped.

Woolly bats are immediately recognizable by their curly hair, funnel-shaped ears with a long pointed tragus, a long sharp muzzle and a very highly domed cranium. Dental formula: $\frac{2.1.3.3}{3.1.3.3} = 38$.

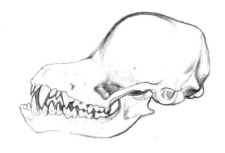

K. argentata.

Woolly bats are very little known in spite of being a most distinctive group (the only genus within the subfamily Kerivoulinae) and in spite of a very wide distribution. They range through most of the Oriental region and much of sub-Saharan Africa. Although some species seem to be tied to forest and none have been collected from really arid habitats, it is difficult to be certain of the ecological limitations on species while records and observations are so scattered.

The grizzled or frosted colouring which has been described as resembling that of the common hare in quality and texture but with shining tips to the hair, seems to have a cryptic function, and roosting sites most favoured by these bats generally seem to match their coat colour and texture very closely.

303

Old birds' nests, thatch, usnea lichen, shrivelled foliage and hollow branches are recorded roosts.

They usually emerge late in the evening to feed on small insects and hunt close to the ground, flying in arcs or circles with a weak fluttering flight.

Kerivoula are generally found singly or in small groups. Two juvenile *K. harrisoni* found sheltering with a single adult imply that two young are possible for this species. Seven Philippine *Kerivoula* are described by Walker (1964) performing daylight aerobatics together, interspersed with pauses at their roosting sites in foliage. A Uganda *Kerivoula argentata* was killed in the radiator of a car as it fluttered about in daylight over a road.

K. argentata.

K. smithi.

K. harrisoni.

K. argentata is a species of the southern savannas, the majority of species having been collected from the woven grass nests of weaver birds, Ploceidae (typically savanna birds). They are generally chestnut red on the back with a white underside but paler colour phases are known. Shortridge (1934) records a clump of four *K. argentata* clinging close together and hanging under the eaves of a thatched hut. The site is a common one for the nests of stinging wasps and the resemblance of the bats to a wasps' nest may afford them some special protection. A female with young attached has been flushed from a weaver's nest in dry acacia country.

K. harrisoni is a smaller darker species related to the former. Members of this species have been found actually sheltering behind a wasps' nest inside an old sunbird's nest (Roberts, 1951). The same author records that one of these extraordinarily well-camouflaged bats remained still in a weaver's nest that had been collected, and was only noticed when it emerged at night into the museum laboratory.

This bat is distributed over most of eastern Africa. Hayman (1967) suggests that further study may show that *K. muscilla*, a West African and Congolese form, may belong to this species.

K. smithi is a rare bat known from a very few widely scattered records from Nigeria, the Cameroons and the eastern Congo. The single East African record is from Garissa, Kenya. It is distinguished from the following species by its long upper incisors and unicuspid outer lower incisors.

K. cuprosa has short upper incisors and trilobed outer lower incisors. (Harrison, 1957a.)

Lang and Chapin (1917) describe this species as resembling *Pipistrellus nanus* but with longer frizzled hair looking as though it was singed. A female with very large young was found inside a hollow branch.

K. africana, described by Dobson in 1878, is only known from the type specimen from the "east coast of Africa (Zanzibar)". It has peculiar inner upper incisors each with distinct outer cusps.

Long-fingered Bats
(Miniopterus)

Family   Vespertilionidae
Order   Chiroptera

Measurements   *Miniopterus minor*
head and body
47 mm
tail 41 mm
forearm 35—42 mm
skull 12—13 mm

head and body   *Miniopterus inflatus*
64 mm
tail 48—55 mm
forearm 45—50 mm
skull 16—17·1 mm
weight 11—14 g

head and body   *Miniopterus schreibersi*
50—63 mm
tail 49—60 mm
forearm 42—47 mm
skull 14·5—15·5 mm
weight 6—10 g

Long-fingered Bats (Miniopterus)

Species

Miniopterus minor
Miniopterus inflatus
Miniopterus schreibersi

The measurements, particularly those of the skull, are the only reliable key to species.

The long-fingered bats belong to the subfamily, Miniopterinae. They have a most distinctive structure and formation to the wing, a high-domed skull somewhat resembling that of *Kerivoula* and a dental formula $\frac{2.1.2.3}{3.1.3.3} = 36$.

Long-fingered bats range through most of the tropical Old World: Africa, the Palaearctic and the northern areas of Australasia. They live in all habitats except desert and do not occur in the Sahara. They have been recorded from sea-level up to 2,300 m altitude (in the Cameroons).

They are primarily cave bats, preferring to roost in very dark holes and crevices well away from any light. They congregate in large groups or clusters, occasionally hanging free but, more usually, clinging to one another or hanging against the walls.

They feed mostly on high-flying insects including small beetles. Although they hunt rather high in the air they are recognizable, as they tend to be early fliers and resemble martins or swallows in their very rapid flight with abrupt swoops and changes of altitude and direction. Within the home cave they are able to swerve round corners and artificial obstacles with unabated speed.

Recoveries of ringed specimens of *M. schreibersi* in Europe have revealed that these bats can live up to at least nine years (Caubère and Caubère, 1948), and Dwyer (1966b) has estimated a maximum life-span of fifteen years.

M. minor appears to be mainly restricted to the tropical coasts of Africa, the islands off Africa in both Indian and Atlantic Oceans and to the lower reaches of the Congo River. Like its close relatives, it is a gregarious cave bat and appears to differ in size only.

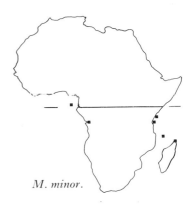

M. minor.

M. inflatus.

M. inflatus.

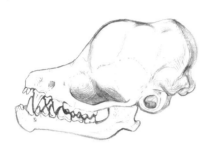

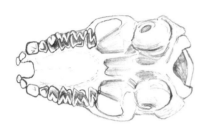

M. schreibersi.

M. inflatus is a tropical African species, that sometimes co-exists with the very widespread *M. schreibersi* but is distinguished by its larger size. Within these two species there is a considerable overlap of all measurements, excepting those of the skull, the skull measurements are, therefore, regarded as the most reliable indicator of the species. Within each size-class the upper and lower ranges tend to be very close or to overlap. These present a difficult problem for the field-worker who has to assign individuals to a species. Hayman (1967) summarizes the taxonomic problems and the associated literature.

M. schreibersi is found from Australasia, South-east Asia and southern Europe down to Africa and to the Cape. It is one of the most successful of all bats. It often roosts in tens of thousands, sharing dark caves with many other species of bats.

The social and reproductive behaviour of this species in tropical Africa deserves attention, for much interesting information has accrued from other areas. In Europe and eastern Australia, colonies hibernate or greatly reduce activity during the winter. European *M. schreibersi* mate before their winter sleep and the ovum is fertilized. However, the development of the embryo is minimal during hibernation, and the period of pregnancy therefore becomes greatly extended; mating taking place in September and October and parturition in June (Courrier, 1927), in Australia mating is in May and parturition is in December (Dwyer, 1963a). In the New Hebrides, 15 degrees south of the Equator, *M. schreibersi* have a well-defined breeding season with mating in September and births in December. The diagram below correlates the birth seasons in these three study localities and illustrates how *Miniopterus* manages to achieve a mating and birth season at the most favourable times by retarding the development of the embryo.

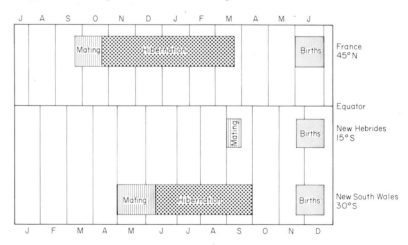

The breeding cycle in 3 populations of *M. schreibersi* (from Courrier, 1927; Dwyer, 1963).

Matthews (1942) found that, in South African populations, the ova were released from the left ovary but that the embryo developed in the right horn of the uterus. Dwyer (1966b) found periodic changes in Australian populations of *M. schreibersi* that might be sequential events in the breeding cycle as much as responses to changes in the environment. Partially discrete breeding populations are made up of many mobile colonies; the focus for one of these populations tends to be a "maternity" colony, which may attract females over a radius of 300 km. Where there are large cave systems with ideal physiological conditions for breeding, the numbers of animals in a colony can approach 50,000 bats. Within such colonies there are very large clusters which have a definite composition, but the presence in one cave of all classes of bats tends to hide the existing social structure.

During the late stages of pregnancy, the females form "maternity" clusters which last as a recognizable entity until the young are able to fly and fend for themselves; altogether a period of about 5 months. The females then mate again, passing through an adult mating colony as part of a general movement away from the "maternity" colony. In Australia, this may involve flying to another cave but it is also possible for the movement out of the "maternity" cluster and into the mating cluster to take place within a single cave system. Dwyer notes that the females are the mobile element, and observed that, whereas males were seldom found to fly for more than 60 km, marked females had been recovered after flights of more than 160 km.

When the females desert their young for the males, the juveniles form distinct clusters. In Australia, these juvenile groups are sometimes found in sites that are otherwise seldom occupied. There must be a learning problem for these young bats, and it is possible that their presence in roosts, that would normally be inferior to those already established, may be significant. The dispersion of inexperienced animals contrasts with the more stable

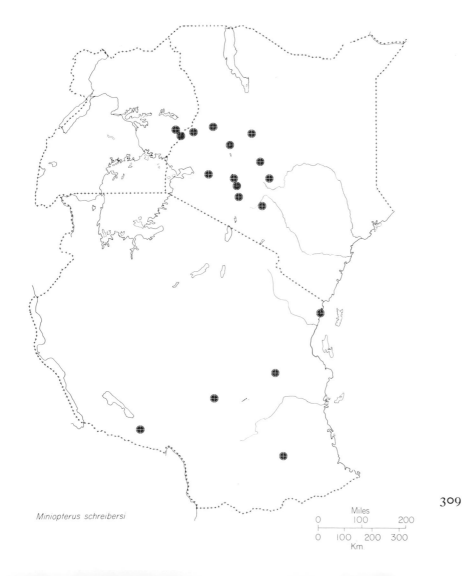

Miniopterus schreibersi

Miles
0   100   200

0   100 200 300
Km.

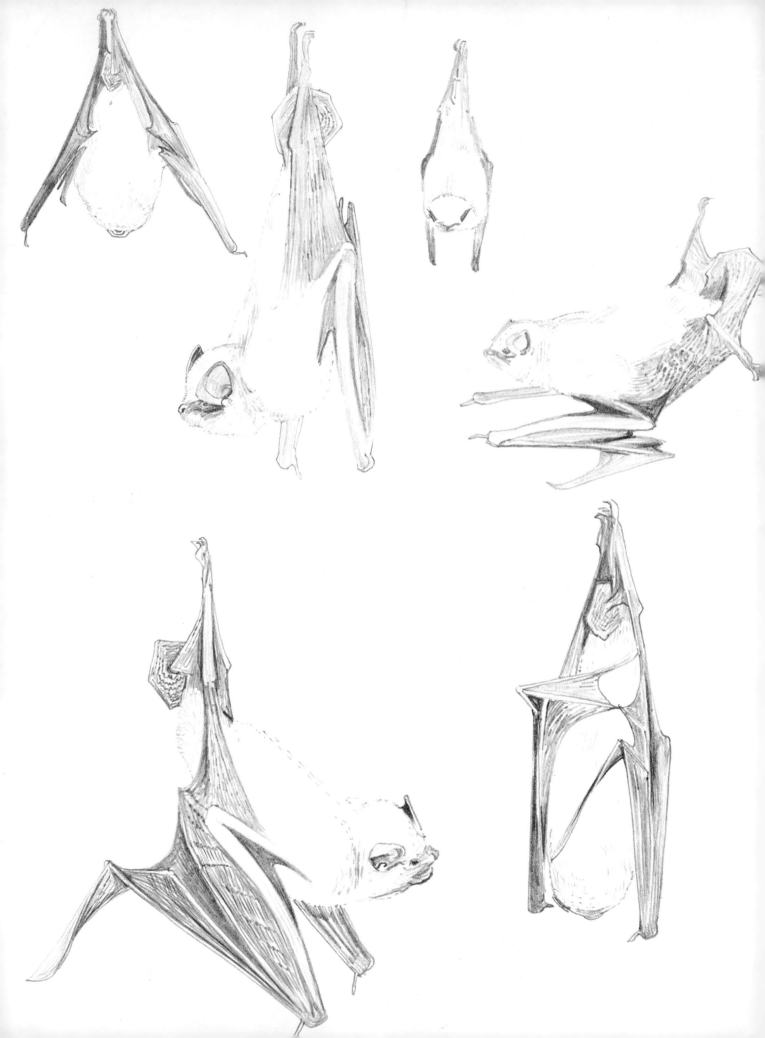

habits of adults that are faithful to particular roosts, the juveniles, therefore, may be a key factor in maintaining gene-flow in a single species that is dispersed over half the globe. Dwyer asks

"Do these individuals learn the location of this new breeding site merely by an initial accident consequent upon exploratory surveys? Or do they perhaps join experienced individuals and learn socially, from them?"

Dwyer found that just over half the young survived the first year, and that most of the casualties had died through falling off the roof. He thought that the actual number falling was probably related to the total number of animals present in the cluster. The age structure of a colony was estimated at 55·5% adults, 28% juveniles and 16·5% yearlings.

That mothers and young recognize each other, and that the former only care for their own young is suggested by the following observations by Dwyer, who watched a few adult females remaining in a juvenile cluster after the evening exodus;

"they were all carrying small young. One female was observed to leave the cluster, the attached young was clinging tenaciously to both cluster and 'parent' but no other juveniles attempted to attach to the adult bat".

Later a female and young were captured

"The juvenile was removed from its 'mother' and placed on the cave wall near the entrance. The adult bat was then released. For ten minutes this female remained in the small entrance chamber and made several flights past the calling juvenile, on two occasions hovering briefly within a few inches of it. No attention was paid to the juvenile by any other female bats that were liberated in this chamber during the same period."

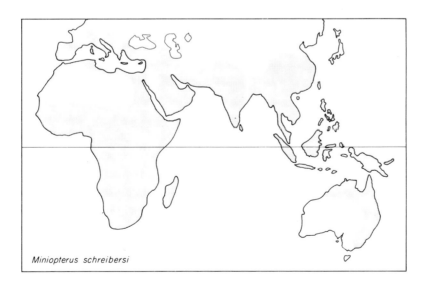

Miniopterus schreibersi

KEY TO MOLOSSIDS

Otomops   Forearm 62—73 mm   Large bats, long-winged. Ears very large joined to snout

Platymops   Forearm 30—36 mm   Skull flattened, gular sac, warty forearm

Myopterus   Forearm 33—37 mm   Ears widely separated. Ears and wings white, underside white. 4 upper cheek teeth, M3 much reduced

Tadarida   Forearm 30—66 mm   Ears separate or conjoined

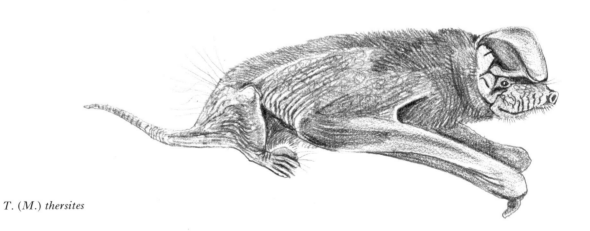

T. (M.) thersites

Molossid Bats, Free-tailed Bats

Molossidae

Molossid bats are a very distinctive group, immediately recognizable by the tail being free beyond the rather narrow interfemoral membrane. The snout is rather dog-like and the ancient Greek for mastiff, "molossus", is the derivation of the family name.

The fur has a velvety and sometimes slightly greasy texture. The ears are thick and generally have a rather angular structure of folds and lappets. A tragus is generally present and in many species the compact structure of the ears is reinforced by a cartilaginous bridge joining them across the forehead and snout. Vaughan (1966) pointed out that the ear form of molossids is related to their characteristic continuous swift flight. Their long, narrow wings are thought to be the most highly evolved among bats; the scapular articulation in particular, represents an advanced condition (see Revilliod, 1916). Control of the wing membrane is improved by well-developed muscles and connective tissue. In contrast to the wings the hind legs are short and stout and these bats never hang free but instead cling firmly with all four limbs to the substrata of their shelters. Some species are capable of considerable agility as they run on their padded wrist-joints and muscular little legs. The unspecialized condition of the hind legs led Anthony and Vallois (1913) to suggest that molossids might resemble the most primitive morphological type among living bats. It seems instead that they have retained legs that allow considerable mobility and versatility while developing a very refined organ of flight in the forelimbs.

In one respect, however, all molossids have highly specialized feet; the outer toes are rather thicker than the others and carry a remarkable fringe of hairs. The outermost hairs are long and fine and probably have a sensory role similar to that of the whiskers of many other mammals. The edges of the toe carry quite different hairs, which are short, stout and with spoon-shaped tips; similar hairs occur on the muzzle of many molossids. Almost all published descriptions of molossids draw attention to this conspicuous peculiarity and many have suggested a function. Lang and Chapin remark:

> "These curious hairs have undoubtedly a tactile function as one might surmise from seeing the bats moving and spreading their toes. Their tiny sharp claws guided by these tactile brushes are able to take advantage more rapidly of even the slightest unevennesses in rock or wood to which they cling."

Later, discussing a possible role as a comb they say

> "bats scratch or comb themselves with perfect ease, all over the belly or about the head and can also reach their backs, should any parasite be caught even momentarily between such spoon hairs, molossida are certainly adroit enough to dispose of it readily."

They have this to say of the rare molossid, *Myopterus*:

> "Minute hairs of this kind may be seen in the same position in all the molossids we collected, and they are of regular occurrence on the first and fifth toe of this

Ears and head of *Otomops* from above to show streamlining.

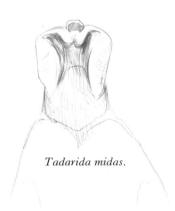

Tadarida midas.

313

family. In *Myopterus albatus* they are especially large on the upper lip and always turned upward in the direction of the nostrils, the lower ones being hooked, whereas most of the upper ones bear a spatulate tip. Such a unique tactile broom right in front of the mouth should be more effective than ordinary vibrissae and this species could well be called the 'brush-lipped bat'."

Rosevear (1965) writes . . .

"it has been suggested as with so many similar unexplained developments in the mammalia, that they play some role in sexual attraction or stimulation. There is a lengthy paper on these spatulate hairs in the Molossidae by Jablonowski (1899). Another unexplained organ is the gular sac which occurs here and there in this family, as in the Emballonuridae."

Like the Emballonuridae some molossid species are very firmly attached to a particular spot in the roost and the presence of stains suggests that these places are marked by glandular secretion. Glands are found on various parts of the body as well as on the throat, in the genital region, on the top of the head and on the face. The facial glands in the muzzle of *Tadarida mexicana* have been investigated by Werner (1966), who has found that two types of glands are present; sweat glands are concentrated in two oval masses each in the lateral margins of the upper lip, while sebaceous, holocrine glands are scattered wherever there are hair follicles and these glands are always in association with hairs. Werner suggests that the odour might serve to assist other bats in the location of day-time retreats. The molossid gular gland has been investigated by Horst and Wimsatt (1964). The gular gland is highly developed in males of the American genus *Molossus*. Females have a vestigial gland and the latter authors found that even after ovarioctomy and treatment with testosterone there was no change in the gland's size or activity. On the other hand, in males the gland atrophied after castration but maintained itself if the animal was treated with testosterone. The gland is a complex of sebaceous units surrounded by an aggregate of sudoriparous glands within a capsule of dense connective tissue. Horst and Wimsatt concluded that in this genus the gland serves some role in reproduction. Glandular secretions in the molossids probably also assist in species recognition, roost and territory recognition and probably as a deterrent to other bat species. Judging from the greasy appearance of the fur in some individuals it may also help to condition or scent the fur, which is frequently and carefully groomed with the hairy hind legs and muzzle. It is most probable that detailed studies of sexual and territorial behaviour in molossids will reveal that glandular secretions and the spoon hairs are linked structures, the latter serving as scent dispensers. This seems to be particularly likely for the muzzle, where the spoon hairs grow from an area that is probably glandular in all molossid species.

Of all bat species there are few that smell as strongly as molossids and, as this is offensive to some other species, it may help them to monopolize roosts that might otherwise attract other animals—including predators.

Molossids are distributed throughout the warmer latitudes of the world and are tolerant of very high temperatures. Their ability to withstand heat may be the principal cause of their widespread tenancy of hot tin roofs.

Many species are highly successful in the tropical world. Molossid pre-eminence seems to have been acquired at an early date and the type is very stable. This is demonstrated by a singularly well-preserved fossil from the Lower Miocene in France; this species, *Tadarida stehlini*, belongs to a con-

temporary genus and shows that molossids of twenty million years ago were scarcely, if at all, inferior to living forms.

The capacity for sustained rapid flight gives these bats a greater range than many other types. It has been shown that American species migrate; some African species may also be shown to do so when more is known about them. They are able to fly at high altitudes and they probably feed mainly on

T. pumila.

those insect types that fly freely in the open air. These insects are usually strong fliers and the molossids may largely fill the nocturnal niche of swallows and swifts. In America, *Tadarida* have been recorded flying 3,000 m above the ground.

The behaviour of African molossids is still a neglected study, although some species are accessible in considerable numbers in the roofs of private and public buildings all over the continent. Other information on breeding, and activity is presented in the profiles that follow.

Bini Free-tailed Bat
(Myopterus whitleyi)

Myopterus whitleyi can be recognized by its widely separated ears with large tragi, white wings and underside, 4 upper cheek teeth and a greatly reduced last upper molar. It differs from the other molossids in having a smooth rounded nose with rough but unwrinkled lips that are bristly along their upper margins.

Bini Free-tailed Bat **(Myopterus whitleyi)**	**Family** Molossidae **Order** Chiroptera	**Measurements** **head and body** 56—66 mm **tail** 25—33 mm **forearm** 33—37 mm **skull** 17—19 mm

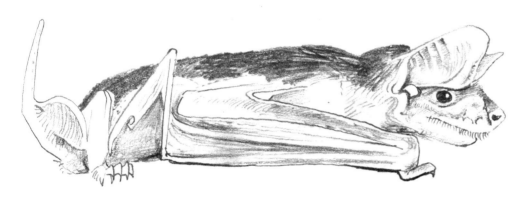

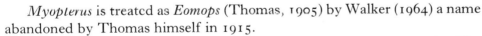

Myopterus is treated as *Eomops* (Thomas, 1905) by Walker (1964) a name abandoned by Thomas himself in 1915.

It ranges within the forest zone from Ghana to southern Uganda. The only East African specimens known came from Entebbe (De Beaux, 1922a) where, in spite of intensive collecting having been carried out since, it has never been seen again.

This bat does not appear to be gregarious and in West Africa individuals have been found roosting in plantain leaves or clinging to the trunk of trees.

KEY TO TADARIDA

Tadarida (Mormopterus)

forearm
38 mm

Small size,
ears separate.
4 upper cheek teeth.
Upper incisors
widely separated.
M3 not reduced.
Palate margins enlarged

M3

Tadarida (Tadarida)

forearm
45—66 mm

Ears separate.
5 upper cheek teeth.
M3 not reduced.
Large palate margins

Tadarida (Chaerophon)

forearm
35—52 mm

Ears joined.
5 upper cheek teeth.
M3 slightly reduced.
No palate margin

Tadarida (Mops)

forearm
27—66 mm

Ears joined.
5 upper cheek teeth.
M3 reduced.
Little palate margin

Wrinkle-lipped Bats (Tadarida)

Less specialized than the other molossid genera, *Tadarida* contains 21 species already recorded in East Africa and probably other species will be discovered in future. The subgenera *Mormopterus*, *Mops*, *Chaerephon* and *Tadarida* are primarily based on the relative development of the last molar. Although a few species can be easily identified on external features, an examination of the teeth is essential to correct identification.

The wrinkle-lipped bats are a very homogeneous group, but the degree of sociability, the choice of roosts and their breeding habits seem to vary from species to species. Furthermore, although all species have a strong smell, some are distinguished by peculiar scent glands in the males: at the base of the penis in *T. (Mops) demonstrator*, below the root of the tail in *T. (Chaerephon) cistura*, on the chin in *T. (T.) ansorgei* and on the forehead in *T. (Chaerephon) major*, implying that sexual and social life in these species has developed peculiar patterns. Most species have rather long narrow wings and seem to feed at relatively high levels.

Aspect ratio 11. Wing loading ratio $1\cdot40 \dfrac{N}{M_2}$.

T. pumila.

There are, however, some very interesting differences in proportion among these bats; for instance, *Tadarida pumila* and *Tadarida nanula* although weighing almost the same have quite different proportions (see drawing, opposite), the latter species has a very short forearm and the former a very long one. The roosting habits of the two species go some of the way to explain the difference, *nanula* squeezes into cracks in bark in dense bunches, while *pumila* does not make a habit of living in cracks (although it may occasionally use them to hide from disturbance). In any case, the altered wing proportions must affect flight very considerably and it would be interesting to know what other ecological and behavioural differences are correlated with shorter or longer wings.

Aspect ratio $7\cdot83$. Wing loading ratio $1\cdot43 \dfrac{N}{M_2}$.

T. nanula.

Like all molossids, Tadarida are quadrupedal bats, they never hang free but scuttle into crannies for shelter after landing on the nearest exposed surface to the roost. Several species have a food preference for moths and beetles.

The gregarious species embark together for the evening flight, launching themselves into the air simultaneously, as if by a signal. In America, they have been shown to migrate and the periodic appearance and disappearance of African *Tadarida* species in some localities suggests that the same may occur here.

Some species of *Tadarida* colonize roofs very readily and their tolerance for very high temperatures under corrugated iron is most remarkable. They are frequently considered a nuisance because of their strong smell and it is reported that bats in the roof can cause some people to suffer from headaches, probably an allergic reaction.

Gestation periods of *Tadarida* may resemble those of American species, about 80 to 90 days. The young are born naked but very active and they grow fast. They are never carried in flight, but are left in the roost until they are able to fly. The females of an American species can reproduce at the age of 7 months and this may apply to most African species as well. Some *Tadarida* species that have been adequately studied are known to breed twice a year in

Tadarida pumila and *T. condylura.*

the tropical zone, often with well synchronized timing. Optimum feeding conditions for the young when they are about to fly appear to be one of the factors related to their reproductive timing.

Captive *Tadarida* are very hardy and are not difficult to keep once the problem of feeding them is solved.

Tadarida Free-tailed Bats (Tadarida (Tadarida))

Species

Tadarida (Tadarida) lobata
Tadarida (Tadarida) africana
Tadarida (Tadarida) fulminans
Tadarida (Tadarida) aegyptiaca
Tadarida (Tadarida) ansorgei

Tadarida Free-tailed Bats

Tadarida (Tadarida)

Family Molossidae
Order Chiroptera

Measurements
head and body oo mm
tail oo mm
forearm 57—66 mm

Tadarida (Tadarida) lobata
Large ears joined to nose, narrow skull, white underside

head and body 86 mm
tail 61 mm
forearm 63—66 mm

Tadarida (Tadarida) africana
Large bat, broad skull 23·5—25·7 mm

head and body 85 mm
tail 57 mm
forearm 57—62 mm

Tadarida (Tadarida) fulminans
Smaller, more lightly built bat, skull 22—23 mm

head and body 65—70 mm
tail 30—45 mm
forearm 45—53 mm

Tadarida (Tadarida) aegyptiaca
Flat brain case. Lower canines separate at base, skull 19—21 mm

head and body 65 mm
tail 36 mm
forearm 46·5—47·5 mm

Tadarida (Tadarida) ansorgei
Elevated brain case. Skull 18—20 mm. Chin and throat black around small gular gland

The members of this subgenus have forearms measuring between 45—66 mm, separate ears and well-developed palate margins. The cusps of the last upper molar have an N pattern and, except for *T. (T.) aegyptiaca*, all forms have the bases of the lower canines nearly touching.

T. (T.) lobata is striking for its very large ears, which are rather translucent as are the wings. This species has been the subject of a paper by Peterson and Harrison (1970). Until recently, there was only one known specimen, collected by Sir Frederick Jackson at Turkwel, Suk in 1890, but since that date two more specimens have been caught; one from the Cherengani Hills and one from Hatfield, Rhodesia. These localities are about 2,400 km apart, and

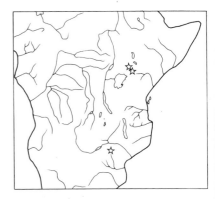

T. (T.) lobata.

T. (T.) africana.

it is likely that the species occurs in various places in the intervening country.

T. (T.) africana is a very rare bat that has only been recorded from 5 localities, which are very widely scattered over the eastern half of the continent; southern Ethiopia, Kenya (Mt Menengai), the eastern Congo, Malawi, Tanzania and the Transvaal.

T. (T.) fulminans ranges from Madagascar across East Africa to the eastern Congo, Ruanda, Zambia and Malawi. It has been found in caves, but appears to be uncommon and little known.

T. (T.) aegyptiaca is a very widespread species, common throughout the Middle East, across North Africa to Algeria and down to the drier parts of East Africa to the Cape.

This species lives in colonies, in caves or buildings, where it lodges in crevices. Where there is adequate space, hundreds of bats may be found together. They do not usually associate with other species, although Shortridge (1934) found small numbers sheltering in tree cracks together with *Eptesicus capensis*. Roberts (1951) suggests that these bats may be responsible for the spreading of bedbugs, *Cimex*.

T. (T.) ansorgei is known from scattered eastern and central Africa localities, from the southeastern Sudan and western Uganda to Angola and Rhodesia. It has been collected in the same caves as *T. fulminans* in Rhodesia. Lang and Chapin described a very large colony in a rock cleft. The bats in this colony refused to leave their roost during daylight in spite of smoking and shooting.

T. (T.) fulminans.

T. (T.) aegyptiaca.

T. (T.) ansorgei.

Chaerephon Free-tailed Bats (Tadarida (Chaerephon))

The members of this subgenus are of medium size with a forearm range of 35—52 mm. Behind the joined ears there may be a tuft in the males. They have 5 upper cheek teeth and no margin to the palate, the cusps of the last molar form an abbreviated N shape.

Tadarida (C.) major usually has a white belly-stripe. The peculiar pocket on the forehead may have a glandular function and the long hairs on the male act as scent dispensers. The skull is somewhat flattened.

This is primarily a northern savanna species, known to range from Mali to the southern Sudan, northeastern Zaire, Uganda as far south as Dar-es-Salaam in Tanzania. It roosts in rock clefts, trees and buildings. Lang and Chapin (1917) found only two males in a colony of thirty and caught no males at all in a large series collected in the Congo. They noticed that each individual bat was very attached to its nesting crevice and returned to it as soon as disturbance ceased; furthermore they never flew into the open, as other species of molossids do when disturbed. They also remarked on the strong smell of well-established roosts. All the East African localities recorded for this species are in rocky areas of savanna.

Peterson (1971) has shown that *Tadarida (C.) bemmeleni cistura*, formerly considered a species, is the East African representative of *bemmeleni* a West

Species

Tadarida (C.) major
Tadarida (C.) bemmeleni
Tadarida (C.) nigeriae
Tadarida (C.) hivittata
Tadarida (C.) aloysiisabaudiae
Tadarida (C.) pumila
Tadarida (C.) chapini

Above: *T. c. pumila.*

323

Chaerephon Free-tailed Bats

Tadarida (Chaerephon)

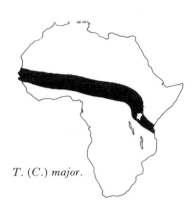

T. (C.) major.

Family Molossidae
Order Chirptera

Measurements
head and body
65—70 mm
tail 34—36 mm
forearm 42—44 mm

Tadarida (Chaerephon) major
Lappet-lidded pocket on forehead

head and body
65—70 mm
tail 37—40 mm
forearm 46—48 mm

Tadarida (Chaerephon) bemmeleni cistura
Glands beside root of the tail

head and body
61—72 mm
tail 34—45 mm
forearm 44—50 mm

Tadarida (Chaerephon) nigeriae
Long white hairs on flanks and under wings. Conspicuous crest in males

head and body
65—81 mm
tail 35—45 mm
forearm 46—52 mm
weight 14—26 g

Tadarida (Chaerephon) bivittata
White spots or stripes on head

head and body
70—80 mm
tail 37—42 mm
forearm 48—52·7 mm

Tadarida (Chaerephon) aloysiisabaudiae
Elongated skull

head and body
54—74 mm
tail 25—43 mm
forearm 35—43·3 mm
weight 10—13·5 g

Tadarida (Chaerephon) pumila
Ribbon of white hairs on underside of wings. Short crest. Colour of wings and underside variable

head and body
oo—oo mm
tail oo—oo mm
forearm 34—39 mm

Tadarida (Chaerephon) chapini
Males have long bicoloured crest between the ears. Underside light coloured

T. bemmeleni.
⌀ *bemmeleni.*
☆ *cistura.*

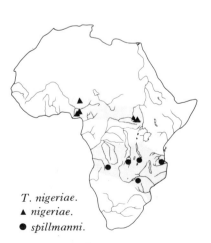

T. nigeriae.
▲ *nigeriae.*
● *spillmanni.*

African forest bat. This species has caudal glands forming distinct swellings on each side of the root of the tail in the male. The tips of the wings are nearly transparent and the ribbons of white fur below the armpits are conspicuous. Start (1966) caught two females in a state of exhaustion swimming in a pool in the Rift Valley. Others have been caught over dammed valleys in forest and one was extracted from a hole among rocks near Kampala.

The scattered records of *Tadarida (C.) nigeriae* present a rather curious distribution pattern, but this may be due to a scarcity of collectors. Rosevear points out that all the northern records of this bat are from Guinea woodlands and the distinct southern race, *T. (C.) nigeriae spillmanni*—distinguished by its white wings—is recorded from the *Brachystegia* belt. It seems therefore, that this is a woodland-adapted bat. It has been found sheltering under bark and its rather flat skull might be related to this habit.

324

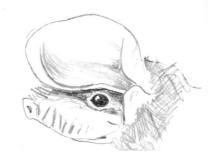

Tadarida (C.) bivittata has been shown by Hayman and Harrison (1966) to be a distinct species. It ranges down the eastern half of the continent from Eritrea to Zambia and might be found in any part of East Africa. I have netted this species on the edge of forest in western Uganda and Start (personal communication) has caught it in dry country near the Lakes Hannington and Baringo. One Uganda female was found to be pregnant in March.

Tadarida (C.) aloysiisabaudiae is a uniformly brown bat with dark wings. Its principal peculiarity is an elongation of the muzzle but there are no data available that might throw light on the adaptive significance of this feature. It is a forest species and has recently been shown to range from Ghana to western Uganda (Peterson, 1972). It has been netted over forest rivers, but no details of its biology have been published. Peterson (1967) discussed sexual dimorphism in its dentition and in the basisphenoid pits.

Tadarida (C.) pumila contains forms that were formerly treated as two distinct species, *pumila* and *limbata*. These are now recognized as the manifestation of a high degree of dichromatism (Hayman, 1967). Some have white patches on the underside and light wings (*limbata* type), others are wholly black (*pumila* type) and intermediate forms are not uncommon. This species has a fringe of pale bristles on the outermost toes. It is a very common and widely distributed species, made more familiar through having colonized roofs in almost every African town. It is found throughout sub-Saharan Africa as far south as South-west Africa and Natal. It also occurs on Madagascar and on some islands in the Indian Ocean.

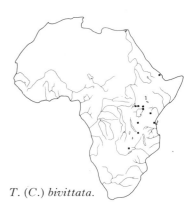

T. (C.) bivittata.

The natural roosts of this species are hollows and crevices in trees and it is commonly found in the crowns of various species of palms. In the roofs of houses they tolerate high temperatures. Mutere recorded a range of 17—39°C and a humidity range of 38%—94% in the roof of the Entebbe Virus Research building, where a well-established colony lives. In the Sudan, Kock (1969) found that these bats exhibited a slight preference for the north side of a building at a time when this side was shady and protected from wind.

Like other molossids, *T. pumila* is a fast flyer and most of its hunting is done in open sky above the tops of trees and buildings, up to 70 or more metres high. It may, however, descend to lower levels and Lang and Chapin report it hawking along narrow channels in coastal mangrove swamps. It eats a wide variety of small, soft-bodied insects. Kock reports these bats flying off 15 minutes after sunset in the Sudan, but in Uganda, small batches start leaving the roost immediately after sunset; they swoop out together, having assembled in a line at the entrance of their shelter. Kock noticed that the forehead tufts of the males formed a recognizable silhouette as they flew out into the evening sky. This tuft is only displayed when the ears are spread, erecting the fold that bridges the inner margins of the ear.

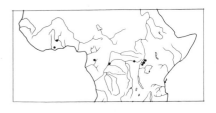

T. (C.) aloysiisabaudiae.

As well as ultrasound *T. pumila* has an audible repertoire of excited squeaks and screams, the shrillest of which are uttered when the animal is hurt or handled. Short periods of audible "excitement" can be heard in the late evening and are reminiscent of the vocal choruses of swifts. Sometimes they can actually be seen flying together with swifts and, in some localities, they live in close association with the palm swift, *Cypsiurus parvus*; this bird feeds on much the same sort of insects, uses the same roosting trees and also falls prey to the same enemy as these bats, *Machaerhamphus* the bat-hawk.

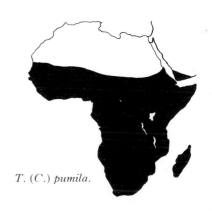

T. (C.) pumila.

I have watched the bat-hawk catch *T. pumila* at dusk on two occasions and exactly in the same spot and in the same way but a year apart. On both occasions the bird was flying low and fast between buildings and trees at about 10 metres height. Its passage coincided with the flight of a batch of *pumila* from the eaves of a three-storey house. The bird flung itself upwards, falling momentarily upside down as its talons came to the fore, seized a shrieking bat and then continued to fly with unabated speed. On one occasion it disappeared with the bat in its claws, on the other it seemed to swallow it in flight, at least this was the impression it gave in the diminishing light. Those who have attempted to catch these bats as they emerge from their shelter, will know how crucial the timing of the strike is. The hawk's low altitude and the timing of its evening flight are probably essential to the rapid slaughter of this type of bat. Chapin (1932) describes shooting a bat-hawk at dusk which had already swallowed 4 bats of 4 different species. The hawk was killed after what must have been a very short spell of hunting. Frequent sightings of bat-hawks at the same time and place suggest that these birds may take a regular harvest from bat colonies.

Development may have favoured this species, for it takes regularly to roofs where it may assemble in large numbers. Tree roosts, on the other hand, naturally tend to limit numbers. Verschuren (1957) noted six bats living in a narrow crevice in the swamp tree, *Mitragyna stipulosa*. In palm groves, the total of bats in the area may be considerable, although individual trees only harbour a few. Where they are found in large numbers, bats may crawl on top of one another and pack into small spaces. When disturbed they fly out into the open.

There have been several studies on the breeding biology of *pumila*. Marshall and Corbet (1959) have shown that on the Equator in south Uganda there is continuous breeding and this was confirmed by Mutere who pointed out, nevertheless, that there are peaks that coincide with rainfall peaks. Kock (1969) found that in the Sudan 72% of the females were pregnant in October and had no evidence of a second season. Harrison (1958) found evidence of two brief breeding seasons following one another in Kenya. There may therefore be gradients in the seasonality of breeding and further information from different localities would be interesting. It is odd that this species should be able to breed continuously, while *T. condylura, T. thersites,* and *T. nanula,* which share the same habitat, should be strictly seasonal breeders.

Tadarida (C.) chapini has been collected from west Uganda and Kenya in recent years and its distribution appears to be less restricted than early records suggested.

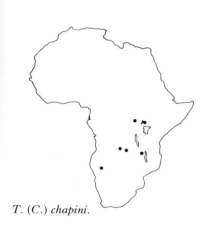

T. (C.) chapini.

326

Mops Free-tailed Bats (Tadarida (Mops))

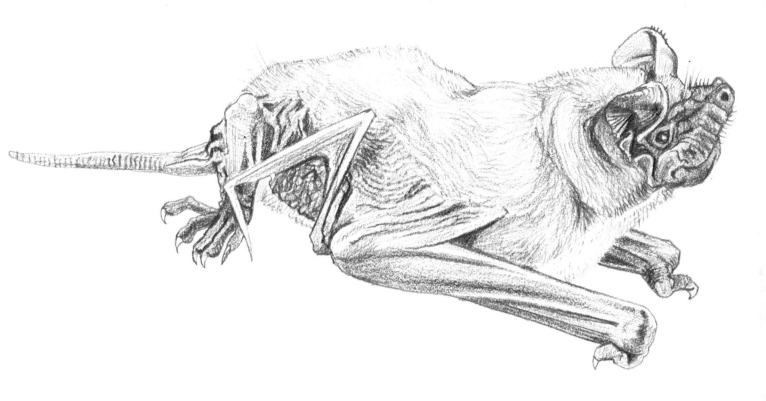

Species

Tadarida (Mops) nanula
Tadarida (Mops) brachyptera
Tadarida (Mops) midas
Tadarida (Mops) congica
Tadarida (Mops) trevori
Tadarica (Mops) demonstrator
Tadarida (Mops) condylura
Tadarida (Mops) niveiventer
Tadarida (Mops) thersites
Tadarida (Mops) leonis

This subgenus is probably the most advanced branch of *Tadarida* containing many of the most successful species in the genus. It is characterized by the reduction of the last molar to a V pattern, the ears are joined over the top of the head and there is little palate margin.

T. (M.) nanula is the smallest bat in the genus. However, its body size and weight are almost identical to *T. (C.) pumila*, it is only the short wing and forearm that make it appear smaller. When settled, this is a very compact little bat, as it folds its wings away tightly, a habit well suited to its roosting sites, which are generally hollow cracks in tree trunks. There is a wide range of colouring in this species, from pale russet to almost black on the back, and from orange to a white-tipped dark grey underneath.

This species is common in Uganda and extends as far as western Kenya; it occurs over most of the Congo basin. Although primarily a forest species it is also found in dry, open and lightly wooded country. These bats live in colonies and have been found in thatch and in house interiors as well as in trees. Rosevear reports an all-female colony, in which every individual was either about to give birth or had newly-born young.

Above: *T. c. condylura*.

327

Mops Free-tailed Bats

(Tadarida (Mops))

Family	Molossidae
Order	Chiroptera

Measurements
head and body
52—60 mm
forearm 29—31·5 mm
weight 7—15 g

Tadarida (Mops) nanula and
Tadarida (Mops) brachyptera
Smallest species

head and body
98—121 mm
tail 40—56 mm
forearm 58—66 mm
weight 40—61·5 g

Tadarida (Mops) midas
Large species

head and body
82—92 mm
tail 39—50 mm
forearm 57 (56—58) mm
weight 42—64 g

Tadarida (Mops) congica
Large skull, 25 mm
Forest species

head and body
83—90 mm
tail 34—42 mm
forearm 53 (51—54) mm
weight 46 g

Tadarida (Mops) trevori
Savanna species

head and body
71—73 mm
tail 37 mm
forearm 42—44 mm

Tadarida (Mops) demonstrator
Scent gland on penis

head and body
66—78 mm
tail 45—51 mm
forearm 45—50 mm
weight 30—40 g

Tadarida (Mops) condylura
High sagittal crest

head and body
72—84 mm
forearm 44—47 mm

Tadarida (Mops) niveiventer
Dark crown and white underside

head and body
68—76 mm
forearm 38—41 mm
weight 22—29 g

Tadarida (Mops) thersites
Blackish with sensory hairs on naked
rump
Tadarida (Mops) leonis
Red with creamy underside

T. (M.) nanula and *T. (M.) brachyptera.*

In Uganda, this species appears to have 2 birth seasons a year. Fertilization occurs in January and the births late in March or in April. The timing of the births is well synchronized, as most females are in the same condition at any one time. The second season occurs between July and September.

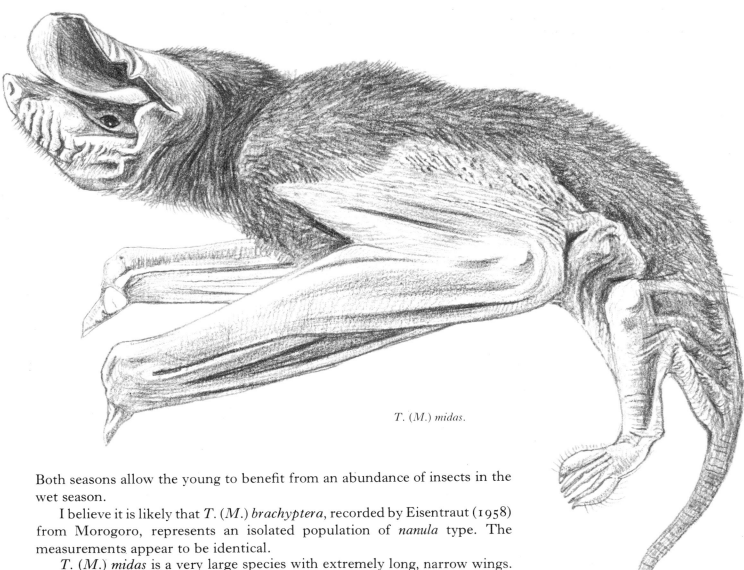

T. (M.) midas.

Both seasons allow the young to benefit from an abundance of insects in the wet season.

I believe it is likely that *T. (M.) brachyptera*, recorded by Eisentraut (1958) from Morogoro, represents an isolated population of *nanula* type. The measurements appear to be identical.

T. (M.) midas is a very large species with extremely long, narrow wings. It has very broad ears that are bridged across the forehead. The fur has a silky bloom and both dark and red phases are known. This is a woodland and savanna species, ranging right across Africa from Senegal to Eritrea and from the Congo through Zambia, Malawi and Botswana to Madagascar. Although, at present, only recorded from Uganda, it will probably turn out to be widespread in East Africa. Lang and Chapin (1917) remarked that this species might make local migrations, since they saw these bats in the northeastern Congo only at the end of the dry season, at which time small, hard-shelled Coleoptera were common. These proved to be their favourite food and the bats were seen dodging and diving after the beetles. They appeared at about dusk and flew some 20—40 metres above the River Dungu. Verschuren (1957) has described their flight as resembling that of *Taphozous* with infrequent wing beats. They seem to fly at medium heights except when coming down to drink. No drinking timetable has emerged from my netting programme and catches over water have been scattered throughout the night. For example, on a June night three came to drink together at 8.10, one at 8.40, one at 9.02, one at 11.45 p.m., one at 00.05, three between 00.45 and 4.45 a.m. and one at 5.45 a.m. Verschuren never saw them fly or land near the home

T. (M.) midas.

329

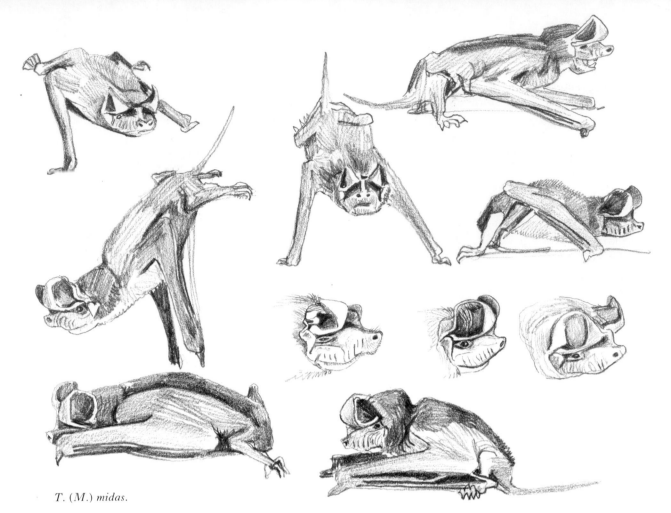

T. (M.) midas.

T. (M.) congica.

shelter and he believes they fly all night without interruption. He likens their departure and arrival flight at the home roost to that of swifts. The air whistles over their wings as they sweep up and dive into their holes without attaching themselves to the tree trunk. Like other Molossids they seem to fly together in small groups and I have twice caught pairs coming in to drink together.

In southern Uganda I have found this species common between January and May and, in spite of keeping up a systematic netting programme throughout the year, I have only two other records, one for September and one for October. This species makes a very rapid, sharp series of clicks when flying. These sounds are quite unlike those made by other molossids I have heard and identified.

Verschuren (1957) has found well-established roosts of this species in long, narrow cracks in trees, notably tall emergents such as *Vitex* and *Parinari* species. The sheltering crack is generally no wider than will allow a bat to squeeze through, and is situated on an exposed trunk or limb, so that flight in and away from the tree is unhampered. Verschuren noticed that colonies of 10—20 bats left their roost at dusk and returned as a group at dawn after having foraged well away from the roost. He also found that females outnumbered males by about 4 to 1.

In southern Uganda, these bats seem to have well-defined breeding seasons; five females caught in January were all in the same condition of early pregnancy. A female caught in March was lactating, and one caught in June still showed traces of milk glands and enlarged teats. A male caught in June

330

had large testes. The capture in October of a lactating female that had recently given birth shows that there probably is a second birth season.

T. (M.) congica resembles the much commoner *T. condylura* and also *T. (M.) trevori*. This is a large forest species living in hollow trees. It is known from Cameroon, eastern Zaire and western Uganda (see Peterson, 1972).

In western Uganda, I shot and netted a large series of these bats. Almost all the adult females in this series were very recently pregnant in early March. Three subsequent efforts to collect, at different times of the year and in the same locality, never revealed a single other animal of this species. About 200 km to the north-west and sixty years earlier, Lang and Chapin recorded six females of this species with large embryos in September. These records suggest that this species has a biannual breeding season and that it may be subject to regional movements or migration. On the night I collected them, these bats were hunting just before dark and were coming down to drink at 7.30 p.m.

T. (M.) trevori is a very similar bat to *congica* but is smaller and more of a savanna species. Although formerly treated as a subspecies of *congica*, Peterson (1972) has convincingly demonstrated its distinctness. It is restricted to Uganda and northeastern Zaire (Congo).

T. (M.) demonstrator is distinguished by a scent gland at the base of the penis. It has dark wings and frosting on the brown back. The underside is generally light. A narrow white band of fur runs along under the wing membrane following the line where it joins the body. Also this species seems to have a rather restricted distribution in the Sudan, northeastern Congo and Uganda. I have netted this species in western Uganda, early in the evening.

T. (M.) condylura has a high sagittal crest on the skull of the male. The fur is usually dull brown, but may range between grey and near black—a red phase is also known. The furry area on the back is rectangular.

This is a common and widespread species, ranging from Gambia to Somalia and to the Cape. It lives in all habitats from forest to dry bushland but is not found in desert or semi-desert. It colonizes roofs very readily and this habit undoubtedly accounts for the numerous distribution records and perhaps implies an increasing abundance at the expense of other species with more conservative roosting habits. Cracked and hollow trees are the natural roosts used by this species and in these situations colonies may number only 10 to 20 individuals, but much larger colonies, numbering several hundreds, are common in roofs. The animals pack close together in crevices and corners and may occasionally share a roof with other *Tadarida* species, notably *T. pumila* in Uganda. There is a slight preponderance of female animals in the colonies and Mutere (1969) measured a temperature range of 18°—50°C and humidity fluctuation of 35%—90% under one roof.

These bats are readily netted over water, generally together with *T. pumila* and *T. thersites*. In the course of a systematic netting programme over a pool in a wooded valley, I never caught a single specimen of this species; yet I managed to catch some specimens over fishponds a few miles away and in more open country.

In southern Uganda this species has two breeding seasons, with mating occurring in April—May and then again in November—December. The young are born in February—March and July—August. Growth is very rapid

T. (M.) demonstrator.

T. (M.) demonstrator.

T. (M.) condylura.

T. (M.) niveiventer.

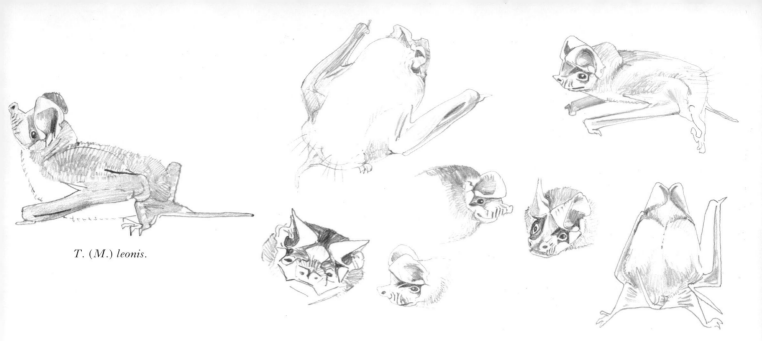

T. (M.) leonis.

T. (M.) thersites.

and Mutere believes that a bat born in the previous season might be able to breed in the following one. He also noted that implantation in this species always takes place in the right horn of the uterus. Only one young is born at a time. The weight at birth is about 10 g and the animals are virtually mature by the time they weigh 20 g.

T. (M.) niveiventer is a close relative of T. condylura. The only differences are a dark head, pale underwings and the possession of deep basisphenoid pits. This species appears to be restricted to the "miombo" woodland zone. Hayman (1967) suggests that the southern Tanzanian, T. (M.) angolensis orientis (Allen and Loveridge, 1942), may represent this species.

T. (M.) thersites is a dark-winged bat, usually almost black on the back, although it may occasionally be found to be reddish brown. There are two clumps of long, sensory vibrissae on the rump (see drawing, facing page). The sides of the body are black and the belly is a paler brown. When the sleek fur of the back is wet it resembles gun metal.

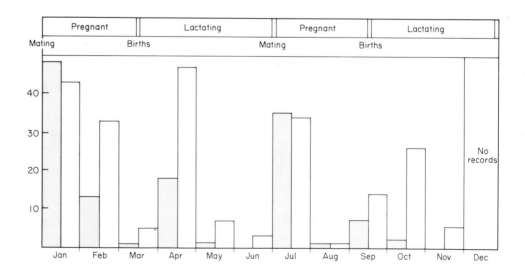

T. (M.) thersites : a record of catches near Kampala. Males, shaded columns; females, white columns. The condition of adult females is shown above.

This is a forest species, found from Sierra Leone to Uganda. It lives in colonies in hollow trees and occasionally in roofs.

In southern Uganda, I have netted *thersites* over a pool in a forested valley. Very large numbers were caught on some nights and few on others. An analysis of the results suggests that the catches were influenced by the condition and sex of the animal, by the state of the weather and by the presence or absence of flying termite swarms. It is possible that there were large-scale movements of these bats—perhaps the males—which influenced the catches when they were out of the area. However, some *thersites* were always about. Some variation in the amount of time spent netting must upset the statistical picture: nonetheless, the catches reveal something of the biology of the species and the data are presented below, left.

Altogether, nearly twice as many females as males were caught and the former mostly during the earlier part of their pregnancy or while they were lactating. During the birth period, or at other times catches were low. Both sexes were caught together in equal numbers only at the time of fertilization or at the very beginning of pregnancy. At other times males were caught in much smaller numbers. As there are two seasons a year, this meant that there were two peak catches of males per year.

The only females that were not pregnant with the others were immature, clearly the offspring of the preceeding breeding season; their measurements, however, were often identical to those of the pregnant bats. It is possible that these bats may be able to breed at the age of $3\frac{1}{2}$ months but I think it unlikely.

T. (M.) leonis is a red bat with a pale peach-coloured underside. It is a true forest species that is very closely related to *T. thersites*. I have netted this species deep in forest while it was flying over a small pool in heavy rain.

T. nanula.

T. (M.) thersites.

T. (M.) thersites with ears closed down

333

Natal Wrinkle-lipped Bat Tadarida (Mormopterus) acetabulosus

This is a small species (forearm 38 mm) with the ears separate and distinguishable from all other *Tadarida* by possessing only 4 upper cheek teeth and having widely separated incisors. Other diagnostic features are large palate margins and a well-developed back molar. It is dark brown above and below.

This species has not yet been recorded in East Africa and is primarily a Malagasy animal but, as it has been found as far north as Ethiopia and as far south as Natal, it may be expected to turn up also in East Africa. The subgenus is also found in tropical America.

Peterson (1965) has pointed out the anatomical resemblance this species has to *Platymops*.

Nothing is known of the species' biology.

Natal Wrinkle-lipped Bat Tadarida (Mormopterus) acetabulosus	**Family**	Molossidae
	Order	Chiroptera

Flat-headed Bat
(Platymops setiger)

Races

Platymops setiger setiger   Near Lake Rudolph
Platymops setiger macmillani   Southern Kenya

Flat-headed Bat
(Platymops setiger)

| **Family** | Molossidae |
| **Order** | Chiroptera |

Measurements
head and body
50—60 mm
tail
22—26 mm
forearm
29—36 mm

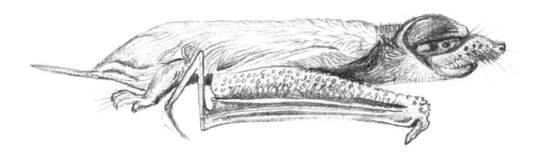

This is a highly specialized bat with a very flattened head and body. It is an extreme case of adaptation to a very peculiar niche. The use of very narrow crevices as shelter enhances safety from predators and allows the colonization of localities in which few other bats can compete. The body form of *Platymops* is an example of anatomical modification due to the physical properties of the environment. Some degree of flattening can be observed in several other bats living in rock crevices, notably the vespertilionid *Mimetillus*, but the molossidae have the largest number of types exhibiting this tendency and *Platymops* presents.a culmination of the trend.

Peterson (1965) describes the skulls of the flat-headed bats as being essentially of the *Tadarida* type and he has shown that these bats are fundamentally closer to *Tadarida* and particularly to the subgenus *Mormopterus* than to any other known genera. In this last subgenus, the structure of the ears is very close to that of *Platymops*.

Apart from head and body there are other extraordinary features, some of which appear to be directly related to living in narrow cracks. The feet, the thumb and the muzzle carry long, very fine sensory hairs, presumably allowing for an easy assessment of space in every direction. The tail, as well as supporting a short membrane and fat deposits at its base, also performs a sensory role: the tip of the tail is held aloft to gauge the height of the crevice roof and of the space behind. The tips of the ears and wings also appear to assist the bat feel its ceiling. Live bats will scuttle about with the tip of the wing protruding behind the elbow, and with the third phalanges of the third and fourth fingers raised to probe the space above. A slight change in the typical molossid arrangement of the finger grouping occurs in *Platymops*; the fifth finger has a longer distal phalanx, which folds in at the elbow towards the body. When the entire wing is folded into its resting position, the tip of this finger joins the others in slotting into the armpit but when the long fingers assume the "ready" position outside the membrane behind the elbow, it remains separate. The forearms are covered in peculiar little warts. Both sexes have a gular sac, which may be responsible for the strong smell of their roosting cracks.

The upper lip of *Platymops* is very odd: it is swollen laterally and thickly grown with very stiff, short bristles. Shorter bristles also occur on the lower lip. This growth of bristles is embedded in a mass of muzzle tissue that renders the lip largely immobile—in contrast to most other molossids, which have relatively mobile lips. These bristles are quite different to the long fine vibrissae that also occur on the upper lip, and if the bristles have a sensory function it must be of a different nature to that of the long delicate "feelers".

This genus is restricted to the drier stony areas of the northeastern Sudan,

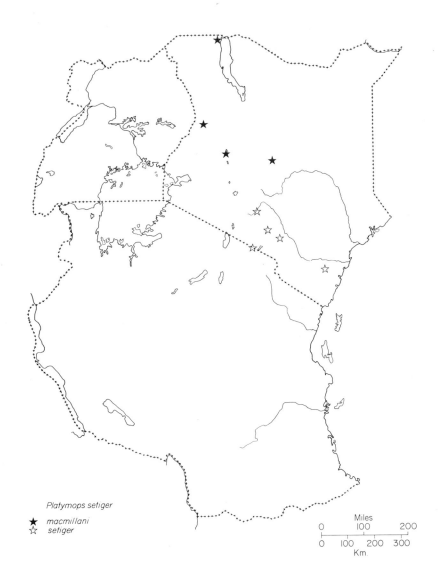

Platymops setiger

★ macmillani
☆ setiger

Miles
0   100   200

0   100   200   300
Km.

southern Ethiopia and Kenya. *Platymops* have been seen flying over water holes and marsh "with rapid erratic flight at heights of 9 metres or less" (Walker, 1964) and Williams (1967) records them roosting, both singly and also in small scattered colonies, in the rock fissures of granite hills. He found them emerging at dusk with a direct and rapid flight. Small beetles have been recorded as items of their diet. The thick fat deposits around the rump are probably symptomatic of a seasonal fluctuation in their food supply. Harrison (personal communication) found them co-habiting with scorpions.

337

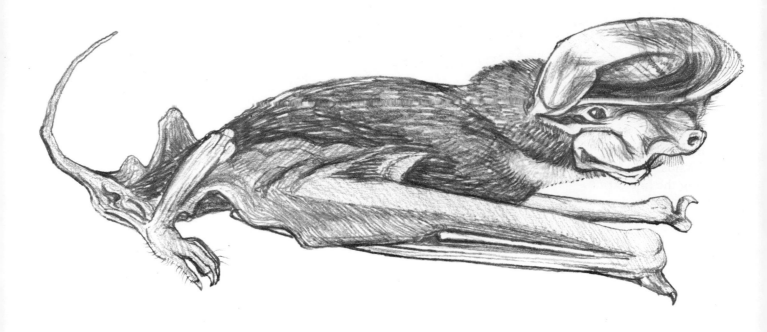

Otomops Bat, Giant Mastiff Bat

(Otomops martiensseni)

Family	Molossidae
Order	Chiroptera

Measurements
head and body
88—110 mm
tail
44—50 mm
forearm
62—72 mm
weight
31—38·5 g

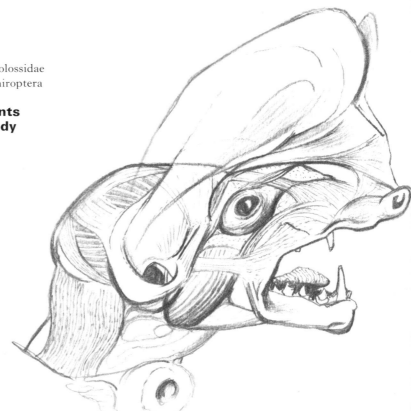

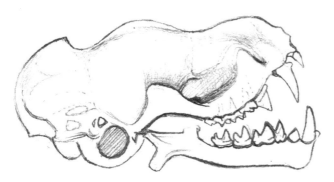

Otomops Bat, Giant Mastiff Bat (Otomops martiensseni)

These are large bats with very long ears which are attached along the whole length of the head. They have a narrow white margin around the fur of the back. Their proportions are very slender compared to those of other molossids. The body is long and streamlined with the extraordinary ears following the line of the slipstream perfectly. The slightly horny consistency of the pinnae gives them a certain rigidity and thus avoid folds or vibrations that might interfere with sonar reception. The tragus and antitragus have virtually disappeared. Instead, the flange behind the antitragus, running forward from the back corner of the ear, has developed in such a way that it can either seal or open the very wide entrance to the ear canal. This arrangement follows the streamlining of the ear and seems to serve two purposes: to deflect the air stream which would otherwise eddy about in the enormous meatus and to reinforce the ear's rigidity (see drawing, in margin).

Otomops showing ear flap closed (upper), and open (lower).

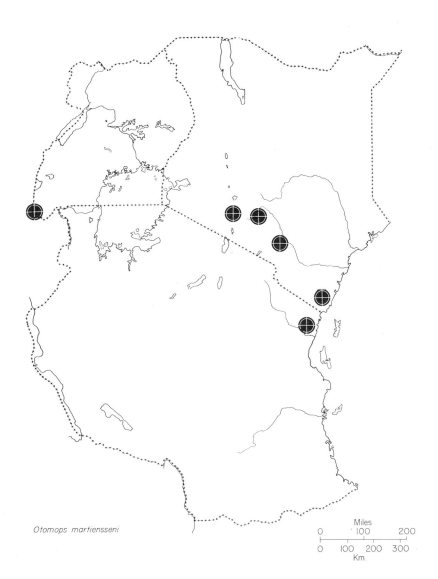

Otomops martiensseni

Miles
0   100   200

0   100   200   300
Km.

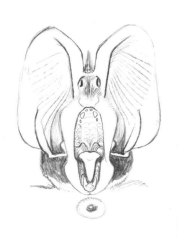

The upper lips have flanges on either side of the mouth, which are the exaggeration of a feature found in species of *Tadarida*. They probably function somewhat like a megaphone and perhaps help to direct and control ultrasonic impulses emitted through the mouth.

The pattern on the back of *Otomops martiensseni* is curious, it has every appearance of being designed for display; both fur and skin are black or dark brown but there is a white margin to the fur on the back. In most mammals exhibiting conspicuous contrasts these usually accentuate some feature or else serve to break up the body outline; in this case, a very symmetrical geometric figure is formed and a cryptic function can be dismissed. The eyes of *Otomops* are of medium size for the family and are certainly fully functional, but intra-specific displays involving visual responses would be very strange in a nocturnal, cave-dwelling animal.

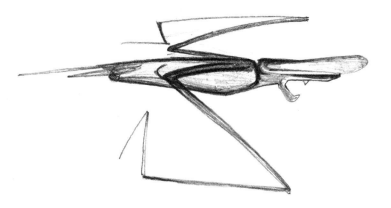

A large gland, marked by a conspicuous circular skin pocket on the throat, is present in both sexes. The gland, the form of the ears and lip flanges and the great length and narrowness of the wings suggest that *Otomops* is a very highly evolved and specialized type of molossid, as all these features are extreme exaggerations of the trends that are perceptible in the genus *Tadarida*.

Several species of *Otomops* are distributed throughout the Oriental Tropics to New Guinea. *Otomops martiensseni* is found over the eastern half of Africa, from the eastern Congo to Kenya and south to Angola, Natal and also Madagascar. It has been found in western Uganda and in the highlands of Kenya and it lives in a variety of ecological zones ranging from semi-arid habitats to montane forest, from sea level to 2,000 m. The few colonies that have been found contain many hundreds of bats packed close together. The lava tunnels on Mt Suswa, in Kenya, shelter particularly large numbers.

Otomops showing streamlining and wing proportions.

These bats can be seen flying in the evening very fast and in a straight line; their wings have a high aspect ratio, built for speed and they should be capable of feeding at considerable distance from their roosts. Alpine swifts, with similar shaped wings, *Micropus melba* and *Micropus aequatorialis*, are known to feed at distances many hundreds of kilometres from their nesting and roosting haunts. Large colonies of *Otomops*, like the one found on Mt

340

Suswa, are unlikely to find a year-long, adequate source of food in the vicinity of Suswa, a very dry and barren area in the dry season, so that very long foraging flights are a distinct possibility. However, visitors to the caves thought that the numbers of bats in the cave appeared to fluctuate a great deal and seasonal migrations during periods of food shortage are also possible.

This bat is currently being studied by Mutere and it will be interesting to learn more about its habits and behaviour.

Otomops martiensseni in flight (from film by U. Norberg).

Bibliography

Insectivores

Allanson, M. (1934). Seasonal variations in the reproductive organs of the male hedgehog. *Phil. Trans. Roy. Soc.* B**223**.

Allen, G. M. (1939). A checklist of African mammals. *Bull. Mus. Comp. Zool. Harvard* **83**.

Allen, J. A. (1917). The skeletal characters of *Scutisorex*. *Bull. Amer. Mus. Nat. Hist.* **37**.

Allen, J. A. (1922). The American Museum Congo Expedition Collection of Insectivora. *Bull. Amer. Mus. Nat. Hist.* **47**.

Ansell, A. D. H. (1969). *Petrodromus tetradactylus* at Ngoma. *Puku Occ. Papers (Dept. Game Fish., Zambia)* **5**.

Ansell, W. F. H. (1960). "Mammals of Northern Rhodesia," 31. Government Printer, Lusaka.

Ansell, W. F. H. (1964). Captivity behaviour and post-natal development of *Crocidura bicolor*. *Proc. Zool. Soc. London* **142**.

Bateman, J. A. (1959). Laboratory studies of the golden mole and the mole rat. *Afr. Wildlife* **13**.

Benson, S. B. (1939). Concealing coloration among some desert rodents of the southwestern United States. *Univ. Calif. Publ. Zool.* **40**.

Bourlière, F. (1955). "The Natural History of Mammals." G. Harrap, London.

Boys Smith, J. S. (1967). Notes on the behaviour of a hedgehog. *J. Zool.* **153**, pt. 4.

Broom, R. (1902). On the organ of Jacobson in the elephant-shrew (*Macroscelides proboscideus*). *Proc. Zool. Soc. London*.

Broom, R. (1916). On the structure of the skull in *Chrysochloris*. *Proc. Zool. Soc. London*.

Broom, R. (1938). Note on the premolars of the elephant-shrew. *Ann. Transval Mus.* **19**.

Brown, J. C. (1964). Observations on the elephant-shrews (Macroscelididae) of Equatorial Africa. *Proc. Zool. Soc. London* **143**.

Brown, J. C. (1971). The description of mammals. I: The external character of the head. *Mammal Review* **1** (6).

Brown, L. (1968). A note on the eagle-owl, *Bubo lacteus*. *J. E. Afr. Nat. Hist. Soc.* **26**.

Burton, M. (1955). Elephant-shrews. *Illus. Lond. News* **226**.

Butler, P. M. (1939). Studies on the mammalian dentition. Differentiation of the post-canine dentition. *Proc. Zool. Soc. London* B**109**.

Butler, P. M. (1956). The skull of *Ictops* and the classification of the Insectivora. *Proc. Zool. Soc. London* **126**.

Butler, P. M. (1969). *In* "Fossil Vertebrates of Africa." Vol. I. (Leakey, L. S. B. ed.), Academic Press, London, New York.

Butler, P. M. and Hopwood, A. T. (1957). Insectivora and Chiroptera from the Miocene rocks of Kenya Colony. *Fossil Mammals Afr.* **13**.

Cabrera, A. (1925). "Genera Mammalium (Insectivora)," Madrid.

Campbell, B. (1938). A reconsideration of the shoulder musculature of the

Cape golden mole. *J. Mammal.* **19**.

Clark, W. E. Le Gros (1928). On the brain of the Macroscelididae. *J. Anat.* **62**.

Clark, W. E. Le Gros (1932). The brain of Insectivora. *Proc. Zool. Soc. London.*

Copley, H. (1950). "Small Mammals of Kenya." Highway Press, Nairobi.

Corbet, G. B. (1966). Macroscelididae. *In* "Preliminary Identification Manual for African Mammals" (Meester, J., ed.). Smithsonian Institution, Washington.

Corbet, G. B. (1970). Patterns of subspecific variation. *In* "Variation in Mammalian Populations," Symp. Zool Soc. London, No. 26. (Berry, R. J. and Southern, H. N. eds), Academic Press, London.

Corbet, G. B. and Hanks, J. (1968). A revision of the elephant-shrew family (Macroscelididae). *Bull. Brit. Mus. Nat. Hist.* **16** (2).

Corbet, G. B. and Neal, B. R. (1965). The taxonomy of the elephant-shrews of the genus *Petrodromus*, with particular reference to the East African coast. *Rev. Zool. Bot. Afr.* **71**.

Crowcroft, P. (1957). "The Life of Shrews." Reinhardt, London.

Davis, D. H. S. (1958). Notes on some small mammals in the Kalahari Gemsbok Park with special reference to those preyed upon by barn owls. *Koedoe* **1**.

Deanesly, R. (1934). The reproductive cycle of the female hedgehog. *Phil. Trans. Roy. Soc.* B**223**.

Delany, M. J. (1971). The biology of small rodents in Mayanja Forest, Uganda. *J. Zool. London.*

De Beer, R. G. (1929). The development of the skull of the shrew. *Phil. Trans. Roy. Soc.* B**217**.

Dimelow, E. J. (1963). The behaviour of the hedgehog *Erinaceus europaeus* in the routine of life in captivity. *Proc. Zool. Soc. London* **141** (2).

Dobson, G. E. (1882). A Monograph of the Insectivora, Systematic and Anatomical. London 1882—1890 (1892).

Dollman, G. (1915—16). On the African shrews belonging to the genus *Crocidura. Ann. Mag. Nat. Hist.* **8**. 15—17.

Durrell, G. (1953), "The Overloaded Ark." Faber and Faber, London.

Edwards, J. T. G. (1957). *In* "Handbook on the Care and Management of Laboratory Animals (*Erinaceus*)." London University Federation Animal Welfare.

Ellerman, J. R. and Morrison-Scott, T. C. S. (1951). Checklist of Palaearctic and Indian mammals 1758 to 1946. British Museum (Nat. Hist.) London.

Ellerman, J. R., Morrison-Scott, T. C. S. and Hayman, R. W. (1953). Southern African Mammals 1758—1951: A Reclassification. British Museum (Nat. Hist.) London.

Eloff, G. (1951). Adaptation in rodent moles and insectivorous moles and the theory of convergence. *Nature* **168** 4281.

Evans, F. G. (1942). Osteology and relationships of the elephant-shrews (Macroscelididae). *Bull. Amer. Mus. Nat. Hist.* **80**.

Fiedler, W. (1953). Die Kaumuskulatur der Insectivora. *Acta Anat.* **18**.

Filhol, H. (1892). Note sur un insectivore nouveau. *Bull. Soc. Phil.* **4** (8).

Fitzsimons, F. W. (1920). "The Natural History of South Africa". Vol. IV. "Mammals." Longmans, Green and Co., London.

Forcart, L. (1942). Beiträge zur Kenntnis der Insectivorenfamilie Chrysochloridae. *Rev. Suisse Zool.* **49**.

Frechkop, S. (1931). Note préliminaire sur la dentition et la position systématique des Macroscelidae. *Bull. Mus. Roy. Hist. Nat. Belg.* **7**.

Friant, M. (1935), La morphologie des dents jugales chez les Macroscelididae.
Proc. Zool. Soc. London.

Gasc, J. P. (1963). La musculature cephalique chez quelques insectivores.
Mammalia **27** (4).

Gerard, P. (1923). Etude sur les modifications de l'uterus pendant la gestation
chez. *Nasilio brachyrhynchus. Archs Biol., Paris* **33**.

Godet, R. (1951). Contribution à l'éthologie de la taupe. *Bull. Soc. Zool.
France,* **76**.

Goodman, M. (1963). Serum protein comparisons by two-dimensional
starch-gel electrophoresis and immunodiffusion plate reactions applied
to the systematics of primates and other mammals. *Proc. XVI Int. Congr.
Zool.* **4**.

Gould, E., Negu, S. N. C. and Novick, A. (1964). Evidence for echolocation
in shrews. *J. Exp. Zool.* **156**.

Grassé, P. P. (1955) (ed.). Ordre des Insectivores. Anatomie et reproduction.
Affinités zoologiques des diverses familles entre elles et avec les autres
ordres de mammifères. *In* "Traité de Zoologie." Vol. 17, Mammifères,
Systématique. Masson et Cie, Paris.

Gregory, W. K. (1910). The orders of mammals. *Bull. Amer. Mus. Nat.
Hist.* **27**.

Gregory, W. K. (1920). Studies in comparative myology and osteology. IV:
A review of the lacrymal bone of vertebrates with special reference to that
of mammals. *Bull. Amer. Mus. Nat. Hist.* **42**.

Guth, C., Heim de Balsac, H. and Lamotte, M. (1959, 1960). Recherches sur
la morphologie de *Micropotamogale lamottei* et l'évolution des Potamogalinae. *Mammalia* **23**, **24**, 2.

Haeckel, E. (1866). "Generelle Morphologie der Organismen," Vol. 2. Georg
Reimer, Berlin.

Harris, R. A. and Duff, K. R. (1970). The hedgehog. *Animals* **3** (13).

Heim de Balsac, H. (1936). Biogéographie des mammifères et des oiseaux de
l'Afrique du Nord. *Bull. Biol. Fr. Belg.* Suppl. 21.

Heim de Balsac, H. (1954). Un genre inédit et inattendu de mammifère
(Insectivore Tenrecidae) d'Afrique Occidentale. *Compt. Rend. Acad. Sci.
Paris* **239**.

Heim de Balsac, H. (1956a). Morphologie divergente des Potamogalinae
(Mammifères Insectivores) en milieu aquatique. *Compt. Rend. Acad. Sci.
Paris* **242**.

Heim de Balsac, H. (1956b). Un soricide inédit et aberrant du Kasai exige la
création d'un genre nouveau. *Rev. Zool. Bot. Afr.* **54** 1—2 (Oct.).

Heim de Balsac, H. (1957a). Insectivores de la famille des Soricidae de
l'Afrique Orientale. *Zool. Anzeiger* **158**, 7/8.

Heim de Balsac, H. (1957b). Insectivores Soricidae du Mt Cameroun. *Zool.
Jahrbuch.* **85** (6) (Dec.).

Heim de Balsac, H. (1958a). Mammifères insectivores *In* "La Reserve Inte-
grale du Mt Nimba." Mem. Inst. Fr. Afr. Noire.

Heim de Balsac, H. (1958b). Persistance inattendue de certains caractères ancestraux dans la denture des Soricidae. *Compt. Rend. Acad. Sci. Paris,* (3 November).

Heim de Balsac, H. (1959). Premières donnée sur la repartition générale, l'écologie et la variation morphologique du genre *Paracrocidura* H. de B. en Afrique Centrale. *Rev. Zool. Bot. Afr.* **57**.

Heim de Balsac, H. (1966a). Faits nouveaux concernants l'évolution craniodentaire des Soricines (Mammifères Insectivores). *Compt. Rend. Acad. Sci. Paris* **263**.

Heim de Balsac, H. (1966b). Evolution progressive et évolution regressive dans la denture des Soricinae (Mammifères Insectivores). *Compt. Rend. Acad. Sci. Paris,* **263**.

Heim de Balsac, H. (1966c). Contribution à l'étude des Soricidae de Somalie. *Mon. Zool. Ital.* **74** (Suppl. Dec.).

Heim de Balsac, H. (1967). Faits nouveaux concernant les *Myosorex* de l'Afrique Orientale. *Mammalia* **31** (4).

Heim de Balsac, H. (1968a). Contribution à l'étude des Soricides de Fernando Po et du Cameroun. *Bonn. Zool. Beitr.* **19**, 1—2.

Heim de Balsac, H. (1968b). Recherches sur les *Soricidae* de l'ouest de l'Afrique. *Mammalia* **3**.

Heim de Balsac, H. (1970). Précisions sur la morphologie et les biotopes de deux *Soricidae* d'altitude endemique du Kilimanjaro. *Mammalia* **34** (3).

Heim de Balsac, H. and Barloy, J. (1966). Révision des Crocidures du groupe *Flavescens—Occidentalis Manni*. *Mammalia* **30** (4).

Heim de Balsac, H. and Bourlière, F. (1955). Ordre des Insectivores, Systématique. *In* "Traité de Zoologie" (Grassé, P. P., ed.). Vol. 17, Mammifères, Systématique. Masson et Cie, Paris.

Heim de Balsac, H. and Lamotte, M. (1957). Evolution et phylogenie des Soricides africains. *Mammalia* **20** (2) June 1956, and **21** March 1957.

Heim de Balsac, H. and Mein, P. (1971). Les musaraignes mammifères des Hypogees de Thebes, *Mammalia* **35**, 2.

Heim de Balsac, H. and Verschuren, J. (1968). Insectivores. *Explorat. Parc Natn. Garamba* Inst. Parcs Natr. Congo Kinshasa.

Heim de Balsac, H. and Vuattoux, R. (1969). *Crocidura douceti* H. de B. et le comportement arboricole des Soricidae. *Mammalia* **33** (1).

Henckel, K. O. (1928), Das Primordialcranium von *Tupaiia* und der Ursprung der Primaten. *Z. Anat. Entwickl.* **86**.

Herter, K. (1957). Das Verhalten der Insectivoren. *In* "Handbuch der Zoologie" (Helmche, J. G., von Lengerken, H. and Starck, D., eds.), Vol. 8. Gruyter, Berlin.

Herter, K. (1963). Igel Ziemsen: Wittenberg Lutherstadt. (English ed. 1965 "Hedgehogs," London Phoenix House.)

Hill, J. E. (1938). Notes on the dentition of a jumping shrew, *Nasilio brachyrhynchus*. *J. Mammal.* **19**.

Hoesch, W. and von Lehmann, E. (1956). Zur Säugetier-Fauna Südwest-Afrikas. *Bonn. Zool. Beitr.* **7**.

Hollister, U. (1916). Shrews collected in the Congo Expedition of the American Museum. *Bull. Amer. Mus. Nat. Hist.* **35**.

Hollister, N. (1918). East African Mammals in the United States National

Museum. *Bull. U.S. Natn. Mus.* **99** (3).

Hoogstraal, H., Huff, C. G. and Lawless, D. K. (1950). A malarial parasite of the African elephant shrew. *J. Natn. Malaria Soc.* **9**.

Lang, H. (1923). Notes on the glands of *Elephantulus*. *J. Mammal.* **4**.

Laurie, W. A. (1971). The food of the barn owl in the Serengeti National Park, Tanzania. *J. E. Afr. Nat. Hist. Soc.* **28** (125).

Loveridge, A. (1922). Notes on East African Mammalia—other than horned Ungulates—collected and kept in captivity 1915—19. Part II. *J. E. Afr. Uganda Nat. Hist. Soc.* **17**.

Loveridge, A. (1933). Field notes in Allen and Loveridge: Reports on the scientific results of an expedition to the southwestern highlands of Tanganyika Territory, II: Mammals. *Bull. Mus. Comp. Zool. Harvard*, **75**.

Loveridge, A. (1937). Field notes in Allen and Lawrence: Scientific results of an expedition to rain forest regions in East Africa, III: Mammalia. *Bull. Mus. Comp. Zool. Harvard*, **79**.

Loveridge, A. (1942). Field notes in Allen and Loveridge: Scientific results of a 4th expedition to forested areas in East and Central Africa, I: Mammals. *Bull. Mus. Comp. Zool. Harvard*, **89**.

Marlow, B. J. G. (1955). Observations on the herero musk shrew *Crocidura flavescens herero* St Leger, in captivity. *Proc. Zool. Soc. London* **124**.

Marshall, F. H. A. (1911). The male generative cycle in the hedgehog: with experiments on the functional correlation between the essential and accessory sexual organs. *J. Physiol. London* **43**.

McDowell, S. B. (1958). The greater Antillean insectivores. *Bull Amer. Mus. Nat. Hist.* **115** (3).

McKenna, M. C. (1963). Primitive Paleocene and Eocene Apatemyidae (Mammalia, Insectivora) and the primate-insectivore boundary. *Amer. Mus. Novit.* **2160**.

McLaughlin, J. D. and Henderson, W. M. (1947). The occurrence of foot and mouth disease in the hedgehog under natural conditions. *J. Hyg. Camb.* **45**.

Mayr, E. (1954). Notes on nomenclature and classification. *Syst. Zool.* **3**.

Meester, J. (1958a). Variation in the shrew genus *Myosorex* in southern Africa. *J. Mammal.* **39** (3).

Meester, J. (1958b). The fur and moults of the shrew *Myosorex caffer*. *J. Mammal.* **39** (4).

Meester, J. (1959). Adaptation of *Crocidura hirta* Peters to variation in moisture conditions in southern Africa. *Koedoe* **2**.

Meester, J. (1961). A taxonomic revision of southern African *Crocidura* (Mammalia Insectivora). *Ann. Mag. Nat. Hist.* **13** (4).

Meester, J. (1963). The genera of African shrews. *Ann. Transv. Mus. Ser.* 13, **4**.

Meester, J. (1965). The origins of the southern African mammal fauna. *Zool. Afr.* **1** (1).

Misonne, X. (1959). Une nouvelle capture de *Potamogale ruwenzorii*. *Mammalia* **23** (4).

Patterson, B. (1965). The fossil elephant shrews, Macroscelididae. *Bull. Mus. Comp. Zool. Harvard* **133**.

Pearson, O. P. (1946). Scent glands of the short-tailed shrew. *Anat. Rec.* **94**.

Pellegrini, M. S. (1969). Aspetti strutturali ed ultrastrutturali del fegato di riccio *Erinaceus europaeus* durante il ciclo annuale. *Arch. Ital. Anat. Embriol.* **74** (4).

Peters, W. C. H. (1852). Naturwissenschaftliche Reise nach Mossambique in 1842—48. *Zool. Säugetierk.* (Vol. I. Macroscelididae, Berlin.)

Pitman, C. R. S. (1942). "A Game Warden Takes Stock." J. Nisbet and Co., London.

Pocock, R. I. (1912). On elephant shrews. *Proc. Zool. Soc. London* **142**.

Poduschka, W. and Firbas, W. (1965). Das Selbstbespeicheln des *Erinaceus europaeus* steht in Beziehung zur Funktion des Jacobsonschen Organes.

Portmann, A. (1952). "Animal Forms and Patterns," Faber and Faber, London.

Ranson, R. M. New laboratory animals from wild species. Breeding laboratory stock of hedgehogs (*Erinaceus europaeus*). *J. Hyg. Camb.* **41**.

Rahm, U. (1960a). Acquisition d'une dépouille de *Potamogale ruwenzorii*. *Folia Scient. Afr. Centr.* **6** (2).

Rahm, U. (1960b). Note sur les spécimens actuellement connus de *Micropotamogale ruqenzorii* et leur répartition. *Mammalia* **24** (4).

Rahm, U. (1961). Beobachtungen an der ersten in Gefangenschaft gehaltenen *Mesopotamogale ruwenzorii*. *Rev. Suisse. Zool.* **68** (1).

Rahm, U. (1966). Les mammifères de la forêt équatoriale de l'est du Congo. *Ann. Mus. Roy. Afr. Centr. Sér.* 8, *Sci. Zool.* **149**.

Rahm, U. and Christiaensen, A. R. (1963). Les mammifères de la région occidentale du Lac Kivu. *Ann. Mus. Roy. Afr. Centr. Sér.* **8**, *Sci. Zool.* **118**.

Rathbun, G. (1973). The golden-rumped elephant shrew. *A.W.L.F. News*, **8**, No. 3.

Romer, A. S. (1945). "Vertebrate Palaeontology", Chicago University Press.

Roux, G. H. (1947). The cranial development of certain Ethiopian insectivores and its bearing on the mutual affinities of the group. *Acta Zool. Stockl.* **28**.

Saban, R. (1954). Phylogenie des insectivores. *Bull. Mus. Natn. Hist. Nat. Paris* **26**.

Schaerffenberg, B. (1942). Zur bidugie des Maulwurfs, *Z. Säugetierk.* **14**.

Schulte, H. von (1917). A note on the lumbar vertebrae of *Scutisorex* Thomas. *Bull. Mus. Nat. Hist.* **37** (29).

Setzer, H. W. (1956). Mammals of the Anglo-Egyptian Sudan. *Proc. U.S. Natn. Mus.* **106**.

Shriner, H. W. (1903). Adaptation to aquatic, arboreal, fossorial and cursorial habits in mammals. *Amer. Nat.* **37**.

Simonetta, A. M. (1967). A new golden mole from Somalia with an appendix on the taxonomy of the family Chrysochloridae (Mammalia Insectivora). *Mon. Zool. Ital.* (*N.S.*) **1**.

Simpson, G. G. (1931). A new classification of mammals. *Bull. Amer. Mus. Nat. Hist.* **59**.

Simpson, G. G. (1940). Studies on the earliest primates. *Bull. Amer. Mus. Nat. Hist.* **77**.

Smithers, R. H. N. (1966). "The Mammals of Rhodesia, Zambia and

Malawi." Collins, London.

Stephan, H. and Banchet, R. (1960). The brain of *Chlorotalpa stuhlmanni* and *Chrysochloris asiatica*. *Mammalia* **24** (4).

Swynnerton, G. H. (1959). A tentative grouping of the species of the genus *Crocidura* Wagler 1832 occurring in Africa. *Durban Mus. Novit.* **5**.

Swynnerton, G. H. and Hayman, R. W. (1951). A checklist of the land mammals of the Tanganyika Territory and Zanzibar Protectorate. *J. E. Afr. Nat. Hist. Soc.* **20** (1).

Thomas, O. (1910). Description of a new species of shrew *Sylvisorex somereni*. *Ann. Mag. Nat. Hist.* **6** (July).

Toschi, A. (1949). Note ecologiche su alcuni mammiferi di Olorgesaili. *Zool. Appl. Caccia* Suppl. 2.

Van der Horst, C. J. (1944). Remarks on the systematics of *Elephantulus*. *J. Mammal.* **25**.

Van der Horst, C. J. (1946). Biology of reproduction in the female of Elephantulus. *Trans. Roy. Soc. S. Afr.* **31**.

Van der Horst, C. J. (1950). The placentation of *Elephantulus*. *Trans. Roy. Soc. S. Afr.* **32**.

Van der Horst, C. J. (1954). *Elephantulus* going into anoestrus: menstruation and abortion. *Phil. Trans. Roy. Soc.* B**238**.

Van der Klaauw, C. J. (1929). On the development of the tympanic region of the skull in the Macroscelididae. *Proc. Zool. Soc. London*.

Verheyen, R. (1951). "Contribution à l'Etude Ethologique des Mammifères du Parc National de l'Ipemba." Inst. Parcs Natn. Congo Belge, Brussels.

Verheyen, W. N. (1961). Recherches anatomiques sur *Micropotamogale ruwenzorii Bull. Soc. Roy. Zool. Anvers* **21** (22).

Verheyen, W. and Verschuren, J. (1966). Rongeurs et Lagomorphes. *Explorat. Parc Natn. Garamba, Miss H. de Saeger,* **50** Inst. Parcs Natn. Congo Belge, Brussels.

Verschuren, J. (1958). Ecologie et biologie des grands mammifères. *Explorat. Parc Natn. Garamba.* Inst. Parcs Natn. Congo Belge, Brussels.

Vesey-Fitzgerald, D. F. (1962). Habitat notes on Central African species of *Crocidura*. *Mammalia* **26** (2).

Watson, J. M. (1951). The mammals of Karamoja (VI). *Uganda J.* **15** (1).

Wendt, H. (1956). "Out of Noah's Ark," Weidenfeld and Nicholson, London.

Witte, G. de and Frechkop, S. (1955). Sur une espèce encore inconnue de mammifère africain, *Potamogale ruwenzorii* sp. n. *Bull. Inst. Roy. Sci. Nat. Belg.* **31** (84).

Wortman, J. L. (1920). On some hitherto unrecognised reptilian characters in the skull of the Insectivora and other mammals. *Proc. U.S. Natn. Mus.* **52**.

Zlabek, K. (1938). Le masseter des insectivores. *Arch. Anat. Hist. Embryol.* **25**.

Bats

Aellen, V. (1952). Contribution à l'étude des Chiroptères du Cameroun. *Mem. Soc. Neuchat. Sci. Nat.* **8**.

Aellen, V. (1954). Description d'un nouvel *Hipposideros* (Chiroptera) de la Côte d'Ivoire. *Rev. Suisse Zool.* **61**.

Aellen, V. (1957). Les Chiroptères africains du Musée Zoologique de Strasbourg. *Rev. Suisse Zool.* **64**.

Aellen, V. (1959). Chiroptères nouveaux d'Afrique. *Archs Sci. Geneve.* **12**.

Aellen, V. and Perret, J. L. (1956). Mammifères du Cameroun de la collection J. L. Perret. *Rev. Suisse Zool.* **63**.

Allen, G. M. (1911). Bats from British East Africa. *Bull. Mus. Comp. Zool. Harvard* **54**.

Allen, G. M. (1914). Mammals from the Blue Nile Valley. *Bull. Mus. Comp. Zool. Harvard* **58**.

Allen, G. M. (1939). A checklist of African mammals. *Bull. Mus. Comp. Zool. Harvard* **83**.

Allen, G. M. (1940). "Bats". Harvard University Press, Cambridge, Mass., U.S.A.

Allen, G. M. and Lawrence, B. (1937). Scientific results of an expedition to rain forest regions in East Africa, III: Mammalia. *Bull. Mus. Comp. Zool. Harvard* **79**.

Allen, G. M. and Loveridge, A. (1927). Mammals from the Uluguru and Usambara Mountains, Tanganyika Territory. *Proc. Boston Soc. Nat. Hist.* **38**.

Allen, G. M. and Loveridge, A. (1933). Reports and scientific results of an expedition to the southwestern highlands of Tanganyika Territory, II: Mammals. *Bull. Mus. Comp. Zool. Harvard* **75**.

Allen, G. M. and Loveridge, A. (1942). Scientific results of a 4th expedition to forested areas in East and Central Africa, 1: Mammals. *Bull. Mus. Comp. Zool. Harvard* **89**.

Allen, H. (1891). On the wings of bats. *Proc. Phil. Acad. Nat. Sci.*

Allen, L. A. Lang, H. and Chapin, J. P. (1917). The American Museum Congo Expedition Collection of Bats. *Bull. Amer. Mus. Nat. Hist.* **37**.

Al Robaae, K. (1968). Notes on the biology of the tomb bat, *Taphozous nudiventris. Säugetierk. Mitt.* **16**.

Anciaux de Faveaux, F. M. (1958). Speologica Africana. Chiroptères des grottes du Haut Katanga (Congo Belge). *Bull. Inst. Fr. Afr. Noire* **A20**.

Andersen, K. (1905). On some bats of the genus *Rhinolophus*, with remarks on their mutual affinities and descriptions of 26 new forms. *Proc. Zool. Soc. London.* **3** and **4**.

Andersen, K. (1906a). On the bats of the *Hipposideros armiger* and *commersoni* types. *Ann. Mag. Nat. Hist.* **17** (7).

Andersen, K. (1906b). On *Hipposideros caffer* Sund and its closest allies, with some notes on *H. fuliginosus. Temm. Ann. Mag. Nat. Hist.* **17** (7).

Andersen, K. (1906c). On some new or little known bats of the genus *Rhinolophus* in the collection of the Museo Civico Genoa. *Ann. Mus. Civ. Stor. Nat. Giacomo Doria* **42** (3), III.

Andersen, K. (1907). Chiropteran notes. *Ann. Mus. Civ. Stor. Nat. Giacomo Doria* **43** (3).

Andersen, K. (1910). On some species of the genus *Epomops. Ann. Mag. Nat. Hist.* **5** (8).

Andersen, K. (1912). Catalogue of the Chiroptera in the collection of the

British Museum, I. Megachiroptera.

Anderson, A. B. (1949). Small mammals of the Southern Sudan. *Sudan Notes and Records* **30** (2).

Anthony, R. (1912). Contribution à l'étude morphologique général des caractères d'adaptation à la vie arboricole chez les vertébres. *Ann. Sci. Nat. Zool.* **9**.

Anthony, R. and Vallois, H. (1913). Considérations anatomiques sur le type adaptif primitif des Microchiroptères. *Int. Monatsch. Anat. Physiol.* **30**.

Ansell, W. F. H. (1960). Some fruit bats from northern Rhodesia, with the description of a new race of *Epomophorus gambianus* Ogilby. *Rev. Zool. Bot. Afr.* **61**.

Ansell, W. F. H., Benson, C. W. and Mitchell, B. L. (1962). Notes on some mammals from Nyasaland and adjacent areas. *Nyas. J.* **15** (1).

Baker, H. G. and Harris, B. J. (1957). The pollination of *Parkia* by bats and its attendant evolutionary problems. *Evolution* **11**.

Baker, H. G. and Harris, B. J. (1958). Pollination in *Kigelia africana* Benth. *J. West Afr. Sci. Ass.* **4** (1).

Baker, H. G. and Harris, B. J. (1959). Bat pollination of the silk-cotton tree *Ceiba petandra* L. Gaertn (sensu latu) in Ghana. *Niger. Field*.

Blackwell, K. (1967). Breeding and handrearing of fruit bats. *Int. Zoo Yearbook* **7**.

Braestrup, F. W. (1933). On the taxonomic value of the subgenus *Lophomops* (Nyctonomine bats), with remarks on the breeding times of African bats. *Ann. Mag. Nat. Hist.* **10—11**.

Brosset, A. (1962). Bats of central and western India, *J. Bombay Nat. Hist. Soc.* **59** (1).

Brosset, A. (1966a). "La Biologie des Chiroptères," Collection 3: "Les Grands Problèmes de la Biologie." Masson et Cie, Paris.

Brosset, A. (1966b). Les Chiroptères du Haut Ivindo, Gabon. *Biologica Gabonica*, **2**.

Brosset, A. (1968). Permutation du cycle chez *Hipposideros caffer* au voisinage de l'equateur. *Biologica Gabonica*, **4**, 4.

Brosset, A. and Dubost, G. (1967). Chiroptères de la Guyane Française. *Mammalia* **31**.

Brosset, A. and St Girons, M. (1969). Les chiroptères du Gabon. *Biologica Gabonica* **5**.

Brosset, A. and Vuattoux, R. (1968). Redécouverte du "rat volant" de Daubenton, *Myopterus senegalensis* Oken en Côte d'Ivoire. *Mammalia* **32**.

Britton, P. (1972). Rough-wing swallow attacking bat. *E. Afr. Nat. Hist. Soc. Bull.* (Feb.).

Caubere, B. and Caubere, R. (1948). L'essaim des Chiroptères des grottes du Queire. *Mammalia* **12**.

Chapin, J. P. (1932). The birds of the Belgian Congo. *Bull. Amer. Mus. Nat. Hist.* **65**.

Child, G. S. (1965). Some notes on the mammals of Kilimanjaro. *Tanganyika Notes and Records* **64**.

Church, J. C. T. and Griffin, E. R. (1968). Myobaterium buruli lesions in the fruit bat web. *J. Pathol. Bacteriol.* **96**.

Courrier, R. (1927). Etude sur le déterminisme des caractères sexuels secondaires chez quelques mammifères à activité testiculaire périodique. *Archs Biol. Paris* **37**.

Crolla, D. (1968). Nectar feeding bats. *Animal Kingdom* **73**.

Cunningham van Someren, G. R. (1972). Some fruit bats eat leaves. *E. Afr. Nat. Hist. Soc. Bull.* (Feb.).

Dalquest, W. W. (1965). Mammals from the Save River, Mozambique with descriptions of two new bats. *J. Mammal.* **46**.

Davis, R. (1969). Wing loading in pallid bats. *J. Mammal.* **50**.

De Beaux, O. (1922a). Mammiferi abissini e somali. *Atti Soc. Ital. Sci. Nat. Milano* **61** (1).

De Beaux, O. (1922b). Collezioni zoologiche fatto nell'Uganda dal Dott. E. Bayon XVII: Mammiferi II: Chiropetra. *Ann. Mus. Civ. Stor. Nat. Genova* **3** (9).

De Beaux, O. (1923a). Di alcuni chirotteri africani del Museo Civico di Milano. *Atti. Soc. Ital. Sci. Nat. Milano* **62**.

De Beaux, O. (1923b). Mammiferi della Somalia Italiana. Raccolta del Maggiore Vittorio Tedesco Zammarano nel Museo. Civico di Milano. *Atti Soc. Ital. Sci. Nat. Milano* **62**.

De Beaux, O. (1924). Mammiferi della Somalia Italiana. *Atti Soc. Lig. Sci. Lett.* **3** (1).

Dekeyser, P. L. (1955). "Les Mammifères de l'Afrique Noire Française." Inst. Fr. Afr. Noire, Dakar.

Dekeyser, P. L. and Villiers, A. (1954). Essai sur le peuplement zoologique terrestre de l'ouest africain. *Bull. Inst. Fr. Afr. Noire* A**16** (3).

De Winton, W. E. (1897). On a collection of small mammals from Uganda. *Ann. Mag. Nat. Hist.* **6** (20).

De Winton, W. E. (1901). Notes on bats of the genus *Nyctinomus* found in Africa. *Ann. Mag. Nat. Hist.* **7** (7).

Dickson, J. M. and Green, D. G. (1969). "Keeping Bats as Laboratory Animals." Scientific Report of the Zoological Society, London, 1967—69.

Dinale, G. (1964). Studi sui chirotteri italiani. *Atti Soc. Ital. Sci. Nat. Milano.*

Dobson, G. E. (1875). A monograph of the genus *Taphozous*. *Proc. Zool. Soc. London.*

Dobson, G. E. (1878). Catalogue of the Chiroptera in the collection of the British Museum, London.

Dobson, G. E. (1881). On the structure of the pharynx, larynx and the hyoid bones in the Epomophori with remarks on its relation to the habits of these animals. *Proc. Zool. Soc. London.*

Dougras, A. M. (1967). Natural history of *Macroderma gigas*. *West Austr. Naturalist* **10**.

Dwyer, P. D. (1963a). The breeding biology of *Miniopterus schreibersi blepotis* (Temminck) Chiroptera in northeastern New South Wales. *Aust. J. Zool.* **2**.

Dwyer, P. D. (1963b). Seasonal changes in pelage of *Miniopterus schreibersi blepotis* in northeastern New South Wales. *Aust. J. Zool.* **2**.

Dwyer, P. D. (1964). Seasonal changes in activity and weight of *Miniopterus schreibersi blepotis* in northeastern New South Wales. *Aust. J. Zool.* **12**.

Dwyer, P. D. (1965). Flight patterns of some Australian bats. *Vic. Nat.* **82**.

Dwyer, P. D. (1966a). Mortality factors of the bent-wing bat. *Vic. Nat.* **83**.

Dwyer, P. D. (1966b). The population pattern of *Miniopterus schreibersi* in northeastern New South Wales. *Aust. J. Zool.*

Dwyer, P. D. (1968a). The biology, origin and adaptation of *Miniopterus australis* in New South Wales. *Aust. J. Zool.* **16**.

Dwyer, P. D. (1968b). The little bent-wing bat: evolution in progress. *Aust. Nat. Hist.* **16**.

Dwyer, P. D. and Hamilton-Smith, E. (1965). Breeding caves and maternity colonies of the bent-wing bat in southeastern Australia. *Helictite* **4**.

Eisentraut, M. (1936). Beitrag zur Mechanik der Fledermaus Fluges. *Z. Wiss. Zool.* **148**.

Eisentraut, M. (1938). Die Wärmerregulation tropischer Fledermäuse. *Sher. Gesch. Naturf. Fr. Berlin* **2**.

Eisentraut, M. (1940). Von Wärmehaushalt tropischer Chiropteren. *Biol. Zhl.* **60**.

Eisentraut, M. (1942). Beitrag zur Oekologie Kameruner Chiropteren. *Mitt. Zool. Mus. Berlin* **25**.

Eisentraut, M. (1945). Biologie der Flederhunde (Megachiroptera). *Biol. Gen.* **18**.

Eisentraut, M. (1950). Dressurversuche zur Festellung eines optischen Orientierungvermögens der Fledermäuse. *Jah. Ver. Vaterl. Naturk. Württenberg.*

Eisentraut, M. (1956a). Der Langzungen-Flughund *Magaloglossus woermanni* ein Blütenbesucher. *Z. Morph. Oek. Tiere* **45**.

Eisentraut, M. (1956b). Beitrag zur Chiropteran fauna von Kamerun. *Zool. Jb* **84**.

Eisentraut, M. (1957a). Aus dem Leben der Fledermäuse und Flughunde. Jena. *J. Jena Rev.*

Eisentraut, M. (1957b). Beitrag zur Säugetierfauna des Kamerungebirges und Verbreitung der Arten in den verschiedenen Höhenstufen. *Zool. Jb. (Syst.)* **85**.

Eisentraut, M. (1958). Beitrag zur Chiropeterenfauna Ostafrikas. *Veröffentl. Ueberseemus Bremen.* (A) **3** (1).

Eisentraut, M. (1959). Der Rassenkreis *Rousettus aegyptiacus* E. Geoff. *Bonn Zool. Beitr.* **10**.

Eisentraut, M. (1960). Berichte und Ergebnisse von Markierungsversuchen an Fledermäusen in Deutchland und Österreich. *Bonn. Zool. Beitr.*

Eisentraut, M. (1963). "Die Wirbeltiere des Kamerunsgebirge Verlag," Paul Parey, Hamburg and Berlin.

Eisentraut, M. (1964). La faune des Chiropetères de Fernando Po. *Mammalia* **28** (4).

Eisentraut, M. (1965). Der Rassenkreis *Rousettus angolensis* (Bocage). *Bonn. Zool. Beitr.* **16**.

Ellerman, J. R. and Morrison-Scott, T. C. S. (1951). Checklist of Palaearctic and Indian Mammals 1758 to 1946. British Museum (Nat. Hist.) London.

Ellerman, J. R., Morrison-Scott, T. C. S. and Hayman, R. W. (1953). Southern African Mammals 1758—1951: A Reclassification. British Museum (Nat. Hist.) London.

Erkert, S. (1970). Der Einflug des Lichtes auf die Activität von Flughunden.

Z. Vergl. Physiol. **67**.

Fain, A. (1953). Notes sur une collection de rongeurs, insectivores et chauves-souris capturés dans la région d'endemie pesteuse de Blukwa (Ituri Congo Belg). *Rev. Zool. Bot. Afr.* **48**.

Farney, J. and Fleharty, E. D. (1969). Aspect ratio loading, wing-span and membrane area of bats. *J. Mammal.* **50**.

Felten, H. (1956). Fledermäuse fressen Skorpione. *Natur. Volk.* **86**.

Felten, H. (1962). Bemerkungen zu Fledermäusen der Gattungen *Rhinopoma* und *Taphozous. Senck. Biol.* **43** (2).

Fenton, M. B. and Peterson, R. L. (1972). Further notes on *Tadarida aloysiisabaudiae* and *Tadarida russata. Can. J. Zool.* **50** (1).

Festa, E. (1909). Chirotteri ed insettivori. Spedizione al Ruwenzori di S.A.R. il Principe L. Amedeo di Savoia. Parte Sci. I. Milan.

Garg, B. L. (1968). On the systematic evolution of two genera of Megachiroptera with reference to the palato-dental index and their phylogenetic inter-relationship with an insectivore bat. *Ind. Sci. Congr. Ass. Proc. Sess.* **55** (3).

Garret, M. H. (1966). Sensory hairs in bats. *J. Tenn. Acad. Sci.* **41** (65).

Gaunt, W. (1967). Observations upon the developing dentition of *Hipposideros caffer. Acta Anat.* **68**.

Geoffroy, S. H. (1818). Description de l'Egypte. *Histoire naturelle, Paris* **2**.

George, J. C. (1966). The evolution of the bird and bat pectoral muscles. *Pavo* **3**.

Granvick, H. (1924). On mammals from the eastern slopes of Mt Elgon, Kenya Colony. *Lunds Univ. Arskr. N.F. Avd* **2**, 21 (3).

Grassé, P. P. (ed.) (1955). "Traité de Zoologie," Vol. 17. Mammifères, Systématique. Masson et Cie, Paris.

Gray, J. I. (1955). The Flight of Animals. Annual Report Smithsonian Institution, Washington.

Griffin, D. R. (1946). The mechanism by which bats produce supersonic sounds. *Anat. Rec.* **96**.

Griffin, D. R. (1950). The navigation of bats. *Scient. Amer.* **183** (2).

Griffin, D. R. (1951a). At what distance can a flying bat perceive small objects. *J. Mammal.* **32**.

Griffin, D. R. (1951b). Audible and ultrasonic sounds of bats. *Experientia*.

Griffin, D. R. (1952). Mechanisms in the bat larynx for production of ultrasonis sounds. *Proc. Fedn. Am. Socs. Exp. Biol.* **11** (59).

Griffin, D. R. (1953a). Acoustic orientation in tropical bats. *Science* **118**.

Griffin, D. R. (1953b). Bat sounds under natural conditions, with evidence for echolocation of insect prey. *J. Exp. Zool.* **123**.

Griffin, D. R. (1953c). Sensory physiology and the orientation of animals. *Scient. Amer.* **41**.

Griffin, D. R. (1958). "Listening in the Dark. The Acoustic Orientation of Bats and Men." New Haven.

Griffin, D. R. (1962). Comparative studies of the orientation of sounds of bats. *In* "Biological Acoustics," Symp. Zool. Soc. London, No. 7. (Haskell, P. T. and Fraser, F. C. eds), Academic Press, London.

Griffin, D. R. and Galambos, R. (1941). The sensory basis of obstacle avoidance by flying bats. *J. Exp. Zool.* **86**.

Griffin, D. R. and Galambos, R. (1942). Obstacle avoidance by flying bats;

the cries of bats. *J. Exp. Zool.* **89**.

Griffin, D. R. and Galambos, R. (1943). Flight in the dark; a study of bats. *Scient. Monthly* **56**.

Griffin, D. R., Webster, F. A. and Michael, C. R. (1960). The echolocation of flying insects by bats. *Anim. Behav.* **8**.

Griffin, D. R., Dunning, D. C., Cahlander, D. A. and Webster, F. A. (1962). Correlated orientation sounds and ear movements of horseshoe bats. *Nature, London* **196**.

Gyldenstolpe, N. (1928). Zoological results of the Swedish Expedition to central Africa, 1921. Vertebrata 5. Mammals from the Birunga volcanoes north of Lake Kivu. *Ark. Zool. Stockholm* A**20** (4).

Hahn, W. L. (1908). Some habits and sensory adaptations of cave-dwelling bats. *Biol. Bull.* **15**.

Harris, B. J. and Baker, H. G. (1959). Pollination of flowers by bats in Ghana. *Niger. Field.*

Harrison, D. L. (1949). The cranial vault in Chiroptera. *Bull. Brit. Ornith. Club. London* **69** (7).

Harrison, D. L. (1957a). A note on the occurrence of the woolly bat *Kerivoula cuprosa* Thomas in Kenya. *Rev. Zool. Bot. Afr.* **55**.

Harrison, D. L. (1957b). Notes on African bats, 1: Some systematic and anatomical notes on the African bats of the genus *Otomops* Thomas. *Durban Mus. Novit.* **5**.

Harrison, D. L. (1957c). Notes on African bats, 2: Some observations on the relationship between the African slit-faced bats, *Nycteris hispida* Schreber and *Nycteris aurita* K. Andersen. *Durban Mus. Novit.* **5** (2).

Harrison, D. L. (1958). A note on successive pregnancies in an African bat. *Mammalia* **22**.

Harrison, D. L. (1959). Report on the bats in the collection of the Natural Museum of Southern Rhodesia, Bulawayo. *Occ. Papers Natn. Mus. S. Rhod.* B**23**.

Harrison, D. L. (1960). Notes on some central and eastern African bats. *Durban Mus. Novit.* **6**.

Harrison, D. L. (1961a). A checklist of the bats of the Kenya Col. *J. E. Afr. Nat. Hist. Soc.* **23**.

Harrison, D. L. (1961b). Notes on southern and East African bats. *Durban Mus. Novit.* **6**.

Harrison, D. L. (1962). On bats collected on the Limpopo River with the description of a new race of the tomb bat, *Taphozous sudani*, Thomas 1915. *Occ. Papers Natn. Mus. S. Rhod.* B**26**.

Harrison, D. L. (1963a). On the occurrence of the leaf-nosed bat *Triaenops afer* Peters 1877, in Mozembique. *Durban Mus. Novit.* **7**.

Harrison, D. L. (1963b). Observations on the North African serotine bats, *Eptesicus serotinus isabellinus*, Temminck 1840. *Zool. Meded. Leiden* **38**.

Harrison, D. L. (1963c). A note on the occurrence of the forest bat *Kerivoula smithi* Thomas 1880, in Kenya. *Mammalia,* **27**.

Harrison, D. L. (1964). "The Mammals of Arabia," 1: "Insectivora, Chiroptera, Primates." Vol. XI. 1—192. Benn, London.

Harrison, D. L. and Fleetwood, J. D. L. (1960). A new race of the flat-headed bat, *Platymops barbatogularis* Harrison, from Kenya, with observations on the anatomy of the gular sac and genitalia. *Durban Mus. Novit.* **5**.

Hartman, C. G. (1933). On the survival of spermatozoa in the female genital tract of the bat. *Quart. Rev. Biol.* **8**.

Hartridge, H. (1920). The avoidance of obstacles by bats in their flight. *J. Physiol.* **54**.

Hayman, R. W. (1935). A note on *Hipposideros cyclops* Temminck, and its synonym *Hipposideros langi* Allen. *Ann. Mag. Nat. Hist.* **10** (15).

Hayman, R. W. (1937). Mammals collected by the Lake Rudolf Rift Valley Expedition 1934. *Postscript Ann. Mag. Nat. Hist.* **10** (19).

Hayman, R. W. (1938). A new crested bat, *Chaerephon. Ann. Mag. Nat. Hist.* **11** (I).

Hayman, R. W. (1951). A new African Molossid bat. *Rev. Zool. Bot. Afr.* **45**.

Hayman, R. W. (1954). Notes on some African bats, mainly from the Belgian Congo. *Rev. Zool. Bot. Afr.* **50**.

Hayman, R. W. (1957). Further notes on African bats. *Rev. Zool. Bot. Afr.* **56**.

Hayman, R. W. (1960). A note on the bat *Cloeotis percivali* Thomas. *Rev. Zool. Bot. Afr.* **61**.

Hayman, R. W. (1963). Mammals from Angola, mainly from the Lunda district. *Publções Cult. Co. Diam. Angola, Lisboa* **66**.

Hayman, R. W. (1967). Chiroptera. I "Preliminary Identification Manual for African Mammals." (Meester, J. ed.), Smithsonian Institution, Washington.

Hayman, R. W. and Harrison, D. L. (1966). A note on *Tadarida* (*Chaerephon*) *bivittata* Heuglin. *Z. Säugetierk.* **31**.

Hayman, R. W. and Hill, J. E. (1972). Order Chiroptera. *In* "The Mammals of Africa, an Identification Manual" (Meester, J., ed.). Smithsonian Institution, Washington.

Hayman, R. W. and Jones, T. S. (1950). A note on pattern variation in the Vespertilionid bat, *Glauconycteris poensis* (Gray). *Ann. Mag. Nat. Hist.* **3** (12).

Hayman, R. W., Misonne, X. and Verheyen, W. (in press). The bats of the Congo and of Ruanda and Burundi. *Ann. Mus. Roy. Afr. Centr.*

Heim de Balsac, H. (1965). Quelques enseignements d'ordre faunistique tirés de l'étude du régime alimentaire de *Tyto alba* dans l'ouest de l'Afrique. *Alauda* **33** (4).

Henderson, B. E., Tulcei, P. M., Lule, M. and Mutere, F. A. (1968). "Isolations from Bats!" East African Institute for Virus Research, Entebbe, Uganda Annual Report.

Henson, O. W. (1967). *In* "Animal Sonar Systems." (Busnel, R. G. ed.), France Lab. Physiol. Acoustique. Jouy en Josas.

Heran, I. (1965). Skull deformation in *Eidolon helvum. Vestnik Cesk.* (in German) *Spolecnosti Zool.* **19**.

Hesse, R. (1937). (Reference in Moreau, R. E. and Pakenham, R. H. W. (1940). *Proc. Zool. Soc. London* **A110**.)

Heuglin, Th. von (1869). "Reise in das Gebiet des Weissen Nil und seiner Westlichen Züflusse in den Jahren 1862—64." Leipzig. Heidelberg.

Hill, J. E. (1942). A new bat of the *Rhinolophus philippinensis* group from Mt Ruwenzori, Africa. *Amer. Mus. Novit.* **1180**.

Hill, J. E. (1963). A revision of the genus *Hipposideros. Bull. Brit. Mus. Nat.*

Hist. Zool. **11**.

Hill, J. E. and Carter, T. D. (1941). The mammals of Angola, Africa. *Bull. Amer. Mus. Nat. Hist.* **78** (1).

Hollister, N. (1918). East African Mammals in the United States Natural History Museum, I: Insectivora Chiroptera and Carnivora. *Bull. U.S. Natn. Mus.* **99** (1).

Hoogstraal, H. (1962). A brief review of the contemporary land mammals of Egypt (including Sinai), I: Insectivora and Chiroptera. *J. Egypt. Publ. Hist. Ass.* **37**.

Hooper, J. (1966). Ultrasonic voices of bats. *New Scientist* (Feb. 24).

Horst, R. and Wimsatt, W. A. (1964). Observations on the gular gland of *Molossus nigricans*. Paper to 44th Ann. Meeting of American Society of Mammalogists.

Huggel, H. (1958). Zum Studium der Biologie von *Eidolon helvum* Kerr: Aktivitat und Lebensrhythmus während eines ganzen Tages. *Verhl. Schweitz. naturf. Gesch.* **138**.

Huggel-Wolf, H. and Huggel-Wolf, M. L. (1965). La biologie *d'Eidolon helvum* Kerr (Megachiroptera). *Acta Tropica* **22** (1).

Jablonowski, J. (1899). Die löffelförmigen Haare der Molossi. *Ash. Zool. Anthrop. Ethn. Mus. Dresden* **7** (7).

Jaeger, P. (1945). Épanouissement et pollination de la fleur du Baobab. *Compt. Rend. Acad. Sci. Paris* **220**.

Jaeger, P. (1954). Les aspects actuels du problème de cheiroptèrogamie. *Bull. Inst. Fr. Afr. Noire* **A16**.

Jepsen, G. L. (1966). Early Eocene bat from Wyoming. *Science* **154**.

Jepsen, G. L. (1970). Bat origins and evolution. *In* "Biology of Bats" (Wimsatt, A. W., ed.). Academic Press, New York, London.

Jones, T. S. (1961). Notes on bat-eating snakes. *Niger. Field* **26**.

Jones, T. S. (1962). Twins in an African bat (*Scotophilus nigrita*). *Mammalia* **26**.

Kaisilia, J. (1966). The Egyptian fruit-bat, *Rousettus aegyptiacus* Geoffr. visiting flowers of *Bombax malabaricum*. *Ann. Zool. Fenn.* **3**.

Kay, L. and Pickvance, T. J. (1963). Ultrasonic emissions of the lesser horseshoe bat, *Rhinolophus hipposideros* (Bech). *Proc. Zool. Soc. London* **141**.

Knuth, P. (1897). Neue Beobachtung ueber Fledermäusblütige Pflanzen. *Bot. Zbl. Cassel.* **72**.

Knuth, P. (Transl. Ainsworth Davis, J. R.) (1906). "Handbook of Flower Pollination." Clarendon Press, Oxford.

Kock, D. (1969). "Die Fledermäus Fauna des Sudan." Waldemar Kramer, Frankfurt am Main.

Kolb, A. (1958). Food composition of *Myotis myotis*. *Z. Säugetierk.* **23**.

Koopman, K. F. (1965). Status of forms described or recorded by J. A. Allen in the American Museum Congo Expedition collection of bats. *Amer. Mus. Novit.* **2219**.

Koopman, K. F. (1966). Taxonomic and distributional notes on S. African bats. *Puku Occ. Papers (Dept. Game Fish., Zambia)* **4**.

Krzanowski, A. (1967). The magnitude of islands and size of bats. *Acta Zool. Cracoviensia* **11**.

Kuhn, T. S. (1962). "The Structure of Scientific Revolutions." University of Chicago Press.

Kulzer, E. (1957). Ueber die Orientierung der Fledermäuse. *Aus. Heimat.* **65** (7—8).

Kulzer, E. (1958). Untersuchungen ueber die Biologie von Flughunden der Gattung Rousettus (Gray) *Z. Morph. Oek. Tiere* **47**.

Kulzer, E. (1959). Fledermäuse aus OstAfrika. *Zool. Jb. Syst.* **87**.

Kulzer, E. (1962a). Fledermäuse aus Tanganyika. *Z. Säugetierk* **27** (3).

Kulzer, E. (1962b) Ostafrikanische Fledermäuse. *Natur. Mus.* **92** (4).

Kulzer, E. (1962c). Ueber die Jugendentwicklung der Angola Bulldog-fledermäus, *Tadarida* (*Mops*) *condylura. Säugetierk. Mitt.* **10** (3).

Kulzer, E. (1965a). Der Thermostat der Fledermäuse. *Natur. Mus.* **954** (8).

Kulzer, E. (1965b). Bulldog-Fledermäuse. *Die Natur.* **5**.

Kulzer, E. (1965c). Temperaturregulation bei Fledermäusen (Chiroptera) verschiedener Klimazonen. *Z. Vergl. Physiol.* **50**.

Kulzer, E. (1966a). Thermoregulation bei Fledermäusen. *Natur. Mus. Frankfurt* **96**.

Kulzer, E. (1966b). Die Geburt bei Flughunden der Gattung *Rousettus* (Gray). *Z. Säugetierk* **31** (3).

Kulzer, E. (1967). Die Herztätigkeit bei lethargischen und winterschlafenden Fledermäusen. *Z. Vergl. Physiol.* **56**.

Kulzer, E. (1968a). The bat heart in hibernation. *Naturw. Rundsch.* **21**.

Kulzer, E. (1968b). Der Flug des afrikanischen Flughundes, *Eidolon helvum. Natur. Mus.* **98**.

Kulzer, E. (1969a). The behaviour of *Eidolon helvum* in captivity. *Z. Säugetierk.* **34** (3).

Kulzer, E. (1969b). African fruit-eating cave bats. Part I *Afr. Wildlife* **23**.

Kulzer, E. (1969c). The hibernation of bats. *Z. Naturwiss.* **7**.

Kuzyakin, A. P. (1950). Letucie Myshi. Moscow.

Lang, H. and Chapin, J. P. (1917). The American Museum Congo Expedition Collection of bats. III: Field Notes. *Bull. Amer. Mus. Nat. Hist.* **37**.

Lanza, B. (1961). Alcune particolarità delle pliche palatine e dell'accrescimento post-embrionale del cranio di *Epomophorus. Novit. Zool. Ital.* **68**.

Lanza, B. and Harrison, D. L. (1963). A new description of the type specimen of *Nyctinomus aloysii-sabaudiae* Festa 1907. *Z. Säugetierk.* **28**.

Laurent, P. (1940). Le crâne de *Rhinolophus maclaudi* Pousargues. *Bull. Mus. Natn. Hist. Nat. Paris* **2** (12).

Lawrence, B. (1964). Notes on the horseshoe bats *Hipposideros caffer, ruber* and *beatus. Breviora* **207**.

Lawrence, B. and Novick, A. (1963). Behaviour as a taxonomic clue; relationships of *Lissonycteris* (Chiroptera). *Breviora* **184**.

Leedal, P. (1971). Notes on the Songwe limestone caves near Mbeya Tanzania. Natural History Group Mimeograph, Mbeya.

Leen, N. and Novick, A. (1969). "The World of Bats." Edita, Lausanne.

Leger, M. and Baury, A. (1932). De l'emploi de la chauve-souris comme animal réactif de la peste. *Bull. Soc. Pathol. Exor.* **16**.

Lombard, G. L. (1968). The cape fruit bat. *Fauna and Flora S. Afr.* **19**.

Long, C. A. and Kamensky, P. (1961). Osteometric variation and function of the high-speed wing of the free-tailed bat. *Amer. Midland Nat.* **77** (2).

Loveridge, A. (1922). Notes on East African Mammalia—other than horned

Ungulates—collected and kept in captivity 1915—19. Part II. *J. E. Afr. Uganda Nat. Hist. Soc.* **17**.

Loveridge, A. (1933). Field notes in Allen and Loveridge: Reports on the scientific results of an expedition to the southwestern highlands of Tanganyika Territory, II: Mammals. *Bull. Mus. Comp. Zool. Harvard*, **75**.

Loveridge, A. (1937). Field notes in Allen and Lawrence: Scientific results of an expedition to the rain forest region in East Africa. I: Introduction and Zoogeography. *Bull. Mus. Comp. Zool. Harvard* **79**.

Loveridge, A. (1942). Field notes in Allen and Loveridge: Scientific results of a 4th expedition to forested areas in East and central Africa, I: Mammals, *Bull. Mus. Comp. Zool. Harvard*, **89**.

Lumsden, W. H. R., Williams, M. C. and Mason, P. J. (1961). A virus from insectivorous bats in Uganda. *Ann. Trop. Med. Parasitol.* **55**.

Macalister, A. (1872). Myology of the Chiroptera. *Phil. Trans. Roy. Soc.* **162**.

Madkour, G. A. (1961). The structure of the facial area in the mouse-tailed bat, *Rhinopoma hardwickei cystops* Thomas. *Bull. Zool. Soc. Egypt.* **16**.

Malzy, P. and Jagord, H. (1959). A propos des rousettes. *Notes Afr.* **87**.

Marshall, F. H. A. (1937). On the changeover in oestrus cycle in animals, after the transference across the equator and further observations on the incidence of the breeding season and the factors controlling periodicity. *Proc. Roy. Soc. London* B122.

Marshall, I. and Corbet, J. (1959). The breeding biology of equatorial vertebrates. Reproduction of the bat, *Chaerophon hindei* (*Tadarida pumila*) at lat. 0° 26′ N. *Proc. Zool. Soc. London* **132**.

Matthews, L. H. (1941). Notes on the genitalia and reproduction of some African bats. *Proc. Zool. Soc. London* B111 (3—4).

Matschie, P. (1895). Säugetiere Deutsch Ostafrikas 3. Berlin.

Matschie, P. (1899a). Die Fledermäuse des Berliner Museums. *Naturkunde* I.

Matschie, P. (1899b). Beiträge zur Kenrtnis von *Hypsignathus monstrosus* Allen. *Sitz. Ber. Ges. Naturf. Freunde Berlin.*

Mayr, E. (1949). "Systematics and the Origin of Species." Columbia University Press, New York.

Mertens, R. (1938). Zoologische Eindrücke von einer Kamerunreise, 3: Der Hammerkopf-Flughund. *Natur. Volk.* **68**.

Metselaar, D., Williams, M. C., Simpson, D. I. H., West, R. and Mutere, F. A. (1969). Mt Elgon bat virus. A hitherto undescribed virus from *Rhinolophus hildebrandtii eloquens* K. Anderson. *Arch. Ges. Virusforsch.* **26**.

Miller, G. S. (1905). A new bat from German East Africa. *Proc. Biol. Soc. Wash.* **18**.

Miller, G. S. (1907). The families and genera of bats. *Bull. U.S. Natn. Mus.* **57**.

Mohres, F. P. (1952). Ueber eine neue Art von Ultraschallorientierung bei Fledermäusen, *Ver. Zool. Gesch. For.* 1951.

Mohres, F. P. (1953a). Ultraschallorientierung auch bei Flughunden (Megachiroptera) Pteropodidae. *Naturwiss.* **40**.

Mohres, F. P. (1953b). Ueber die Ultraschallorientierung der Hufeisennasen. *Z. Vergl. Physiol.* **34**.

Mohres, F. P. (1956a). Ueber die Orientierung der Flughunde (Pteropodidae). *Z. Vergl. Physiol.* **38**.

Mohres, F. P. (1956b). Untersuchungen ueber die Ultraschallorientierung von vier afrikanischen Fledermäusfamilien. *Ver. Zool. Ges.* **19**.

Mohres, F. P. (1960). Sonic orientation of bats and other animals. *In* "Sensory Specialization in Response to Environmental Demands," Symp. Zool. Soc. London, No. 3. (Lowenstein, O. ed.), Academic Press, London.

Mohres, F. P. and Kulzer, E. (1955a). Ein neuer kombinierter Typ der Ultraschallorientierung bei Fledermäusen. *Naturwiss.* **42**.

Mohres, F. P. and Kulzer, E. (1955b). Untersuchungen ueber die Ultraschallorientierung von vier afrikanischen Fledermäusfamilien. *Zool. Anz.* Suppl. 19.

Mohres, F. P. and Kulzer, E. (1957). *Megaderma.* Ein konvergenter Zwischentyp der Ultraschallpeilung bei Fledermäusen. *Naturwiss.* **44**.

Monard, A. (1939). Résultats de la mission scientifique du Dr Monard en Guinée Portugaise, 1937—38, III: Chiroptères. *Arq. Mus. Bocage* **10**.

Moreau, R. E. (1966). "The Bird Faunas of Africa and its Islands." Academic Press, London, New York.

Moreau, R. E. and Packenham, R. H. W. (1940). The land vertebrates of Pemba, Zanzibar and Mafia: a zoogeographic study. *Proc. Zool. Soc. London* A **110**.

Moreau, R. E., Hopkins, G. H. and Hayman, R. W. (1946). The type localities of some African mammals. *Proc. Zool. Soc. London* **115**.

Motta Manno, G. (1951). Esperienze per l'individuazione dell'organo produttore degli ultrasuoni nei Pipistrelli mediante lo studio del volo cieco. *Boll. Soc. Ital. Zool. Biol. Sperim.* **28**.

Mutere, F. A. (1965). The biology of the fruit-bat *Eidolon helvum*. Ph.D. Thesis. University of East Africa.

Mutere, F. A. (1966a). Breeding patterns of some bats in Uganda. Annual Report East African Institute for Virus Research, Entebbe, Uganda.

Mutere, F. A. (1966b). On tie bats of Uganda. *Uganda J.* **30**.

Mutere, F. A. (1967). Breeding cycles in tropical bats. *J. Appl. Ecol.* **5**.

Mutere, F. A. (1968). The breeding biology of *Rousettus aegyptiacus* living at 0° 22′ S. *Acta Tropica* **25**.

Mutere, F. A. (1969). Reproduction in two species of free-tailed bats (Molossidae). *Proc. E. Afr. Acad.*

Neuweiler, G. (1970). Neurophysiologische Untersuchungen zum Echoortungssystem der grossen Hufeisennase *Rhinolophus ferrumequinum* Schreber 1774. *Z. Vergl. Physiol.* **67**.

Nelson, J. E. (1964). Vocal communication in Australian flying foxes (Pteropodidae Megachiroptera). *Austr. J. Zool.* **13**.

Noack, T. (1893). Neue Beiträge zur Kenntnis der Säugetierfauna von OstAfrika. *Zool. J. Syst.* **7**.

Norberg, U. M. (1969). An arrangement giving a stiff leading edge to the hard wing in bats. *J. Mammal.* **50**.

Norberg, U. M. (1970a). Hovering flight of *Plecotus auritus* Linneus. *Proc. 2nd Int. Bat Res. Conf. Bijdragen Dierkunde* **40** (1).

Norberg, U. M. (1970b). Functional osteology and myology of the wing of *Plecotus auritus* Linneus. *Arkiv. Zool.* **22** (12).

Novick, A. (1955). Laryngeal muscles of the bat and production of ultrasonic sounds. *Amer. J. Physiol.* **183**.

Novick, A. (1958). Orientation in paleotropical bats, I: Microchiroptera. *J. Exp. Zool.* **138**.

Novick, A. (1969). *In* "World of Bats" (Leen and Novick). Edita, Lausanne.

Novick, A. and Griffin, D. R. (1961). Laryngeal mechanisms in bats for the production of orientation sounds. *J. Exp. Zool.* **148**.

Osmaston, H. A. (1953). Kalinzu forest fruit-bats. *J. E. Afr. Nat. Hist. Soc.* **22**.

Osmaston, H. A. (1965). Pollen and seed dispersal in *Chlorophora excelsa* and other Moraceae and in *Parkia filicoidea* Mimosaceae, with special reference to the role of the fruit-bat *Eidolon helvum*. *Commonwlth Forest. Rev.* **44** (2), (120).

Panouse, J. B. (1951). Les chauves-souris du Maroc. *Trav. Inst. Sci. Charif. Zool.*

Patterson, A. P. and Hardin, J. W. (1969). Flight speeds of five species of Vespertilionid bats. *J. Mammal.* **50**.

Pennycuick, C. J. (1972). "Animal Flight," E. Arnold, London.

Peterson, R. L. (1965). A review of the flatheaded bats of the family Molossidae from S. America and Africa. *Life Sci. Cont. Roy. Ontario Mus.* **64**.

Peterson, R. L. (1967). A new record for the African Molossid bat *Tadarida aloyssii-sabaudiae*. *Can. J. Zool.* **45**.

Peterson, R. L. (1971). The systematic status of the African Molossid bats, *Tadarida bemmeleni*, *Tadarida cistura*. *Can. J. Zool.* **49** (10).

Peterson, R. L. (1972). Systematic status of the African Molossid bats, *Tadarida congica*, *T. niagarae* and *T. trevori*. *Life Sci. Cont. Roy. Ontario Mus.* **85**.

Peterson, R. L. and Harrison, D. L. (1970). *Tadarida lobata, Life Sci. Roy. Ontario Mus. Occ. Papers* **16**.

Peterson, R. L. and Smith, D. (1973). A new species of *Glauconycteris, Life Sci. Roy. Ontario Mus. Occ. Paper* No. 22.

Peterson, Russel (1964). "Silently by Night." Longmans, New York.

Pierce, G. W. and Griffin, D. R. (1938). Experimental determination of supersonic notes emitted by bats. *J. Mammal.* **19**.

Piveteau, J. (1958). "Traité de Paléontologie." Masson and Cie, Paris.

Poole, E. (1936). Relative wing ratios of bats and birds. *J. Mammal.* **17**.

Porsch, O. (1932). Crescentia eine Fledermäusblume, *Oesterr. Bot. Zeitschr.* **80**.

Poulet, A. R. (1970). Les Rhinopomatidae de Mauretanie. *Mammalia* **34** (2).

Pye, J. D. (1960). A theory of echolocation by bats. *J. Laryng. Otol.* **74**.

Pye, J. D. (1967). Synthesising the waveforms of bats' pulses. "Animal Sonar Systems" (Busnel, R. G., ed.). France Lab. Physiol. Acoustique. Jouy en Josas.

Pye, J. D. (1968a). Animal sonar in air. *Ultrasonics* **6**.

Pye, J. D. (1968b). "Hearing in Bats." Ciba Symposium on hearing mechanisms in Vertebrates.

Pye, J. D. (1971). Bats and fog. *Nature, London* **229** (5286).

Pye, J. D. (1972). Bimodal distribution of constant frequencies in some hipposiderid bats (Mammalia: Hipposideridae) *J. Zool. London* **166**, 323—335.

Pye, J. D. and Flinn, M. (1964). Equipment for detecting animal ultrasound.

Ultrasonics **2**.

Pye, J. D. and Roberts, L. H. (1970). Ear movements in a hipposiderid bat. *Nature, London* **225**.

Pye, J. D., Flinn, M. and Pye, A. (1962). Correlated orientation sounds and ear movements of horseshoe bats. *Nature, London* **196**, 1186

Quay, W. B. (1969). Structure and evolutionary implications of the musculi arrectores pilorum in Chiroptera. *Anat. Rec.* **163**.

Rahm, U. (1965). Distribution et écologie de quelques mammifères de l'est du Congo. *Zool. Afr.* **1** (1).

Rahm, U. and Christiaensen, A. (1963). Les mammifères de la région occidentale du Lac Kivu. *Ann. Mus. Roy. Afr. Centr. Ann. Sér.* 8, *Sci. Zool.* **118**.

Ramaswami, L. S. and Anand Kumar, T. C. (1966). Effects of androgenous hormones on the reproduction structures of the female *Rhinopoma*. *Acta Anat.* **63**.

Reeder, W. C. and Cowles, R. B. (1951). Aspects of thermoregulation in bats. *J. Mammal.* **32**.

Rees, A. (1964). A checklist of the mammals and amphibia of Ulanga district. *Tanzania Notes and Records* **63**.

Reuben, R. (1963). Note on the breeding season of *Rhinopoma hardwickei* Gray. *J. Bombay Nat. Hist. Soc.* **60**.

Revilloid, P. (1916). À propos de l'adaptation au vol chez les microchiroptères. *Ver. Naturf. Ges. Basel.* **27**.

Revilloid, P. (1917—22). Contribution à l'étude des Chiroptères des terrains tertiares. *Mem. Soc. Pal. Suisse* **43** (44—45).

Roberts, A. (1951). "The Mammals of South Africa." The Mammals of South Africa Book Fund, Johannesburg.

Roberts, L. H. (1972). Variable resonance in constant frequency bats. *J. Zool. London* **166**.

Robin, M. H. A. (1881). Recherches anatomiques sur les mammifères de l'ordre des Chiroptères. *Ann. Sci. Nat.* **6**.

Roeder, K. (1965). Moths and ultrasound. *Scient. Amer.* **212** (94).

Roeder, K. (1967). Predator and prey. Moth and bat sonar. *Bull. Ent. Soc. Ann.* **13**.

Romankowa, A. (1963). Comparative study of the skeleton of the hyoid apparatus in some bat species. *Acta Theriol.* **7**.

Romer, A. S. (1945). "Vertebrate Palaeontology." Chicago University Press.

Rosenberg, A. (1840). (Reference in Eisentraut, M. (1945). *Biol. Gen.* **18**.)

Rosevear, D. R. (1953). Checklist and Atlas of Nigerian Mammals with a Foreword on Vegetation. Government Printer, Lagos.

Rosevear, D. R. (1962). A review of some African species of *Eptesicus* Rafinesque. *Mammalia* **26**.

Rosevear, D. R. (1965). "The Bats of West Africa." Trustees of the British Museum, London.

Rowlatt, U. (1967). Functional anatomy of the heart of the fruit-eating bat *Eidolon helvum* Kerr. *J. Morphol.* **123**.

Ruxton, A. E. (1926). On mammals collected by Capt. C. R. S. Pitman, Chief Warden, Entebbe, Uganda. *Ann. Mag. Nat. Hist.* **9** (18).

Ryan, R. M. (1966). A new and some imperfectly known Australian *Chali-*

nilobus, and the taxonomic status of African *Glauconycteris J. Mammal.* **47**.

Ryburg, O. (1947). Studies on bats' parasites. *Svensk. Natur. Stockholm.*

Sanborn, C. C. (1936). Descriptions and records of African bats. *Publ. Field Mus. Nat. Hist. Zool.* **20**.

Sanderson, I. T. (1940). The mammals of the North Cameroons forest area, being the results of the Percy Sladen Expedition to the Mamfe Division of the British Cameroons. *Trans. Zool. Soc. London* **24**.

Saville, D. B. O. (1957). Adaptive evolution in the avian wing. *Evolution* **2**.

Schneider, H. and Mohres, F. P. (1960). Die Ohrbewegungen der Hufeisen-fledermäuse Rhinolophidae und der Mechanismus des Bildorens. *Z. Vergl. Physiol.* **44** (1).

Schneider, R., Jurg Kugn, H. and Kelemen, G. (1967). Der Larynx der *Hypsignathus monstrosus* Allen 1861. Ein Unikum in der Morphologie des Kehlkopfes. *Z. Wiss. Zool.*

Schnitzler, H. (1967a). Doppler effect compensation in horseshoe bats. *Wissenschaft.* **54**.

Schnitzler, H. U. (1967b). "Discrimination of thin wires by flying horseshoe bats Rhinolophidae." NATO Adv. Study Inst. Symp. Animal sonar systems.

Schnitzler, H. U. (1968). Die Ultraschall Ortungslaute der Hufeisen Fleder-mäuse Rhinolophidae in verschiedenen Orientierungssituationen. *Z. Vergl. Physiol.* **57**.

Schnitzler, H. U. (1970). Echoortung bei der Fledermäus *Chilonycteris rubi-ginosa. Z. Vergl. Physiol.* **68**.

Schouteden, H. (1943). Notes sur deux chiroptères congolais, *Otomops* et *Plerotes. Rev. Zool. Bot. Afr.* **37**.

Schouteden, H. (1948). Faune du Congo Belge et du Ruanda-Urundi I. Mammifères. *Ann. Mus. Roy. Congo Belge Sér.* 8*, Sci. Zool.* **1**.

Shepherd, R. C. and Williams, M. C. (1964). Studies on viruses in East African bats: Haemagglutination inhibition and circulation of arbo-viruses. *Zoonoses Res.* **3**.

Shrivastava, R. K. (1962). Contribution à l'étude du muscle deltoide des chiroptères. *Mammalia* **26**.

Shortridge, G. C. (1934). "The Mammals of South-west Africa." 2 Vols. W. Heinemann, London.

Simonetta, B. (1960). Osservazioni sul cranio dei chirotteri e sulle loro affinita sistematiche. *Monit. Zool. Ital.* **68** (1).

Simpson, D. I. H., Williams, M. C., O'Sullivan, J. P., Cunningham, J. C. and Mutere, F. A. (1968). Studies on arboviruses and bats in East Africa. Isolation and haemagglutination inhibition studies on bats collected in Kenya and throughout Uganda. *Ann. Trop. Med. Parasitol.* **62**.

Simpson, G. G. (1945). The principles of classification and a classification of mammals. *Bull. Amer. Mus. Nat. Hist.* **85** (1).

Southern, H. N. (ed.) (1964). "The Handbook of British Mammals." Black-wells, Oxford.

Spallanzani, L. (1774—94). Lettere sopra il sospetto di un nuovo senso nei Pipistrelli. *In* "Le Opere di Lazzaro Spallanzani." Vol. 3. Milano.

Sprague, J. M. (1943). The hyoid region of the placental mammals with special reference to bats. *Amer. J. Anat.* **72**.

Sprague, J. M. (1949). A study of the hyoid region in the Chiroptera. University of Harvard Thesis.

Stark, J. (1943). Beitrag zur Kenrtnis der Morphologie und Entwicklungeschichte des Chiropeteres Craniums. *Z. Anat.* **112**.

Start, A. N. (1966). A note on the occurrence of *Tadarida africana* Dobson 1876 in Kenya. *Mammalia* **30**.

Stiles, C. W. and Nolan, M. O. (1931). Key catalogue of parasites reported for Chiroptera with their possible public health significance. *U.S. Publ. Health. Ser. Nat. Inst. Health. Bull.* **155**.

Storch, G. (1968). Funktionmorphologische Untersuchungen an der Kaumuskulatur und korrelierten Schädelstrukturen der Chiropteren. *Abh. Senck. Naturf. Ges.* **517**.

Swynnerton, G. H. (1958). Fauna of the Serengeti National Park. *Mammalia,* **22**.

Swynnerton, G. H. S. and Hayman, R. W. (1951). A checklist of the land mammals of the Tanganyika Territory and Zanzibar Protectorate. *J. E. Afr. Nat. Hist. Soc.* **20** (1).

Tamsitt, J. R. (1967). Niche and species diversity in neotropical bats. *Nature* **213**.

Tate, H. H. (1941). Results of the Archbold Expedition, 35. A review of the genus *Hipposideros* with special reference to Indo-Australian species. *Bull. Amer. Mus. Nat. Hist.* **78**.

Tate, H. H. (1942). Results of the Archbold Expedition, 47. Review of the Vespertilioninae with special attention to genera and species of the Archbold collection. *Bull. Amer. Mus. Nat. Hist.* **80**.

Taylor, R. M., Worth, T. H., Hurlbut, H. S. and Riek, F. (1956). A study of the ecology of West Nile. *Egypt-Amer. J. Trop. Med. Hyg.* **5**.

Tesh, R. B. *et al.* (1967). Bats as laboratory animals. *Health. Lab. Sci.* **4**.

Thomas, O. (1905). On a collection of mammals from southern Africa. *Ann. Mag. Nat. Hist.*

Thomas, O. (1915a). Notes on bats of the genus *Coleura. Ann. Mag. Nat. Hist.* **15** (8).

Thomas, O. (1915b). On three new bats obtained by Mr Willoughby Lowe. *Ann. Mag. Nat. Hist.* **15** (8).

Thomas, O. and Hinton, M. A. C. (1923). On the mammals obtained in Darfur by the Lynes Lowe Expedition. *Proc. Zool. Soc. London.*

Thomas, O. and Schwann, H. (1904). On mammals collected during the Uganda boundary commission by the late Mr W. G. Doggett and presented to the British Museum by Col. C. Delme Radcliffe. *Proc. Zool. Soc. London* **1**.

Thomas, O. and Wroughton, R. C. (1910). Ruwenzori expedition reports, 17. Mammals. *Trans. Zool. Soc. London* **19** (5).

Toschi, A. (1956). Missione del Prof. G. Scortecci in Somalia. *Att. Soc. Ital. Sci. Nat.* **95**.

Twente, J. W. Jr. (1955a). Aspects of a population study of cavern-dwelling bats. *J. Mammal.* **36**.

Twente, J. W. Jr. (1955b). Some aspects of habitat selection and other behaviour of cave-dwelling bats. *Ecology* **36**.

Usinger, R. L. (1966). Monograph on Cimicidae. American Entomology Society, Maryland.

Van der Pijl, L. (1936). Fledermäuse und Blumen. *Flora* **131**.

Van der Pijl, L. (1957). The dispersal of plants by bats (Cheiropterochory). *Acta Bot. Nederlandica* **6**.

Van Deusen, H. M. (1968). Carnivorous habits of *Hypsignathus monstrosus*. *J. Mammal.* **49** (2).

Vaughan, T. A. (1966). Aspects of the functional morphology of Molossid bats. Paper to 44th Ann. Meet. Amer. Soc. Mammalogists.

Verschuren, J. (1957). Ecologie, biologie et systématique des chiroptères. *Éxplorat. Parcs Natn. Congo Belge.* Inst. Parcs Natn. Congo Belge, Brussels.

Verschuren, J. (1967). Introduction à l'écologie et à la biologie des chiroptères. *Éxplorat. Parc Natn. Albert.* Inst. Parcs Natn. Congo Belge, Brussels.

Vesey-Fitzgerald, B. (1947). The senses of bats. *Endeavour* **6**.

Walker, E. P. (1964). "Mammals of the World." Baltimore.

Walker, A. (1969). True affinities of *Propotto leakeyi* Simpson 1967. *Nature* **223**, 5206.

Webster, F. A. (1965). The bat and ultrasonic principles. Acoustical control of airborne interceptors by bats. *Proc. Intern. Congr. Techn. Blindness.*

Webster, F. A. and Brazier, O. G. (1965). Experimental studies on target detection, evaluation and interception by echolocating bats. Aerospace Med. Res. Lab. Wright Patterson Airforce base, Ohio.

Webster, F. A. and Brazier, O. G. (1968). Experimental studies on echo-locating mechanisms in bats. *U.S. Airforce Tech. Doc. Rep. TDR.* 67.

Werner, H. J. (1966). Observations on the facial glands of *Tadarida brasiliensis. Proc. Louis. Acad. Sci.* **29**.

Werner, H. J., Dalquest, W. W. and Roberts, J. H. (1950). Histological aspects of the glands of the bat *Tadarida cynocephala* (Le Conte). *J. Mammal.* **31**.

Williams, J. (1967). "A Field Guide to the National Parks of East Africa." Collins, London.

Williams, M. C., Simpson, D. I. H. and Shepherd, R. C. (1964a). Studies on viruses in East African bats. Virus isolation. *Zoonoses Res.* **3**.

Williams, M. C., Simpson, D. I. H. and Shepherd, R. C. (1964b). Bats and arboviruses in East Africa. *Nature* **203**.

Wimsatt, W. A. (1970). "Biology of Bats." Academic Press, New York, London.

Wynne-Edwards, V. C. (1962). "Animal Dispersion in Relation to Social Behaviour." Oliver and Boyd, London.

Systematic Index

D

E

Subject Index

Axli